Signals and Communication Technology

This series is devoted to fundamentals and applications of modern methods of signal processing and cutting-edge communication technologies. The main topics are information and signal theory, acoustical signal processing, image processing and multimedia systems, mobile and wireless communications, and computer and communication networks. Volumes in the series address researchers in academia and industrial R&D departments. The series is application-oriented. The level of presentation of each individual volume, however, depends on the subject and can range from practical to scientific.

Indexing: All books in "Signals and Communication Technology" are indexed by Scopus and zbMATH

For general information about this book series, comments or suggestions, please contact Mary James at mary.james@springer.com or Ramesh Nath Premnath at ramesh.premnath@springer.com.

More information about this series at http://www.springer.com/series/4748

Frank Nielsen
Editor

Progress in Information Geometry

Theory and Applications

Springer

Editor
Frank Nielsen
Sony Computer Science Laboratories, Inc.
Tokyo, Japan

ISSN 1860-4862 ISSN 1860-4870 (electronic)
Signals and Communication Technology
ISBN 978-3-030-65458-0 ISBN 978-3-030-65459-7 (eBook)
https://doi.org/10.1007/978-3-030-65459-7

This Springer imprint is published by the registered company Springer Nature Switzerland AG
The registered company address is: Gewerbestrasse 11, 6330 Cham, Switzerland

To Audrey Rena,

To Julien Reo

Preface

This edited book project entitled *Progress in Information Geometry: Theory and Applications* (PIGTA) started in October 2019, right after the fourth biannual conference on the *Geometric Science of Information* (GSI) which took place at the end of August 2019 in Toulouse (France). The proceedings of the fourth GSI were published by Springer in the LNCS series (volume 11712, https://www.springer.com/gp/book/9783030269791). The international conference was a nice opportunity to interact with colleagues willing to contribute to this book. In the past, let us quickly mention that we carried out similar projects after conference/workshop events as follows:

- The book "*Matrix Information Geometry*" was published in 2013 after holding the Indo-French workshop in 2011 on this theme (https://www.springer.com/gp/book/9783642302312),
- The book "*Geometric Theory of Information*" was published in 2014 after the First International Conference on Geometric Science of Information (GSI) held at École des Mines (Paris, France) in August 2013.
- The book "*Computational Information Geometry: For Image and Signal Processing*" was published in 2017 (https://www.springer.com/gp/book/9783319470566) after the event held at the International Centre for Mathematical Sciences (ICMS, Edinburgh, UK) in September 2015.
- The book "*Geometric Structures of Information*" was published (https://www.springer.com/gp/book/9783030025199) in 2019 after the Third International Conference on Geometric Science of Information (GSI) held at École des Mines ParisTech (Paris, France) in November 2017.

This book project was originally called *Advances in Information Geometry* (AIG) but we decided to rename it later to avoid confusion since the book title coincided with another planned international event scheduled to be held in Japan in March 2020 (https://sites.google.com/view/aig2020/). However, the AIG conference event was unfortunately canceled due to the worldwide spread of the coronavirus (COVID-19).

Preparing this edited book during the pandemic was a different experience than in normal times: We had to allocate extra time for some contributors and reviewers, and to take into account personal situations to adapt the project. We are very grateful to the contributors and reviewers for their devoted times and efforts during this exceptional period, where most colleagues have to work remotely from home and had to quickly adapt to the evolving situation.

This edited book is organized into ten chapters, which are organized into two parts as follows: The first part deals with recent advances on the *fundamentals* of information geometry and the second part explores *connections* or *potential interactions* of information geometry with other domains (e.g., time-series analysis or optimal transport).

The first part on the fundamentals of information geometry is structured as follows:

- Professor Pistone further explores non-parametric information geometry in the chapter entitled "Information Geometry of Smooth Densities on the Gaussian Space: Poincaré Inequalities".
- Professors de Andrade, Vieira, and Cavalcante report the latest results in deformed exponential families in the chapter entitled "On Normalization Functions and φ-Families of Probability Distributions".
- Professors Zhang and Khan investigate affine connections with torsions in the chapter entitled "Affine Connections with Torsion in (Para-)complexified Structures".
- Professors Goto and Hino study the master equations and expectation variables of the moment dynamical system derived from the master equations using para-contact metric manifolds in the chapter entitled "Contact Hamiltonian Systems for Probability Distribution Functions and Expectation Variables: A Study Based on a Class of Master Equations".
- Professor Barbaresco considers homogeneous bounded domains and invariant Koszul form and its relationship with information geometry in the chapter entitled "Invariant Koszul Form of Homogeneous Bounded Domains and Information Geometry Structures".
- Professors Matsuzoe and Takatsu study gauge freedom in the chapter entitled "Gauge Freedom of Entropies on q-Gaussian Measures".
- Professor Nielsen investigates right-angles in (possibly mixed) geodesic triangles in dually flat spaces in the chapter entitled "On Geodesic Triangles with Right Angles in a Dually Flat Space".

The second part on connections or possible interactions of information geometry with other areas is structured as follows:

- Professors Nielsen and Sun introduce a novel metric distance based on optimal transport in the chapter entitled "Chain Rule Optimal Transport".
- Professors Cole and Shiu present some work on Topological Data Analysis (TDA) in theoretical physics in the chapter entitled "Towards the "Shape" of Cosmological Observables and the String Theory Landscape with Topological Data Analysis".

- Professors Marti, Nielsen, Bińkowski, and Donnat surveys the literature on time-series clustering in the chapter entitled "A Review of Two Decades of Correlations, Hierarchies, Networks and Clustering in Financial Markets".

We hope this book will be valuable to scholars working in the field of information geometry and geometric science of information.

Tokyo, Japan
October 2020

Frank Nielsen

Contents

Chapter 1
Information Geometry of Smooth Densities on the Gaussian Space: Poincaré Inequalities

Giovanni Pistone

Abstract We derive bounds for the Orlicz norm of the deviation of a random variable defined on $\mathbb{R}^n$ from its Gaussian mean value. The random variables are assumed to be smooth, and the bound itself depends on the Orlicz norm of the gradient. We shortly discuss possible applications to non-parametric Information Geometry.

Keywords Gaussian Poincaré-Wirtinger Inequality · Gaussian Space · Non-parametric Information Geometry · Orlicz Spaces

1.1 Introduction

In a series of papers [13, 24, 26, 27] we have explored a version of the non-parametric Information Geometry (IG) for smooth densities on $\mathbb{R}^n$. Especially, we have considered the IG associated to Orlicz spaces on the Gaussian space. The analysis of the Gaussian space is discussed, for example, in [16, 21]. This set-up provides a simple way to construct a statistical manifold modelled on Banach spaces of smooth densities. Other modelling options are in fact available, for example the global analysis methods of [10], but we prefer to work with assumptions that allow for the use of classical infinite dimensional differential geometry modelled on B-spaces as in [11].

The present note focuses on technical results about useful differential inequalities and does not consider in detail the applications. However, we have in mind two main examples of potential applications. The first one is the statistical estimation method based on Hyvärinen's divergence,

$$\mathrm{DH}\,(P|Q) = \frac{1}{2}\int |\nabla \log P(x) - \nabla \log Q(x)|^2 \; P(x)\,dx\;, \tag{1.1}$$

The author is supported by de Castro Statistics, Collegio Carlo Alberto, Turin, Italy. He is a member of GNAMPA-INDAM.

G. Pistone (✉)
de Castro Statistics, Collegio Carlo Alberto, Piazza Arbarello 8, 10122 Torino, Italy
e-mail: giovanni.pistone@carloalberto.org

F. Nielsen (ed.), *Progress in Information Geometry*,
Signals and Communication Technology,
https://doi.org/10.1007/978-3-030-65459-7_1

where $|\cdot|$ denotes the Euclidean norm of $\mathbb{R}^n$, P, Q are positive probability densities of the n-dimensional Lebesgue space, see in [8, 13]. The second one is the Otto's inner product [14, 22], which is defined by

$$\langle f, g \rangle_P = \int \nabla f(x) \cdot \nabla g(x) \, P(x) \, dx \, , \tag{1.2}$$

where P is a probability density and f, g are smooth random variables such that $\mathbb{E}_P[f] = \mathbb{E}_P[g] = 0$.

We focus on the exponential representation of positive densities $P = p \cdot \gamma = \mathrm{e}^{u - K(u)} \cdot \gamma$, where γ is the standard Gaussian density. The sufficient statistics u is assumed to belong to an exponential Orlicz space and $\int u(x) \gamma(x) \, dx = 0$. The set of all such couples (p, u) is called statistical bundle. There are other ways to represent positive densities, namely, those that use deformed exponential functions, $p \propto \exp_A$ [18]. This approach is intended to avoid the difficulty of the exponential growth and, for this reason, provides a somehow simpler treatment of smoothness, see [19, 20]. We do not further discuss here this interesting formalism.

This paper is organized as follows. In Sect. 1.2, we provide a recap of basic facts about non-parametric IG and introduce the Gaussian case. The results about Poincaré-Wirtinger inequalities are gathered in Sect. 1.3. This section contains the main contributions of the paper. A collection of simple examples of possible applications concludes the paper.

1.2 Statistical Bundle Modelled on Orlicz Spaces

First, we review below the theory of Orlicz spaces in order to fix convenient notation. The full theory is offered, for example, in [17, Chap. II] and [1, Chap. VII].

1.2.1 Orlicz Spaces

In this paper, we will need the following special type of Young function. Cf. [17, Sect. 7] for a more general case.

Assume $\phi \in C[0, +\infty[$ is such that: (1) $\phi(0) = 0$; (2) $\phi(u)$ is strictly increasing; (3) $\lim_{u \to +\infty} \phi(u) = +\infty$. The primitive function

$$\Phi(x) = \int_0^x \phi(u) \, du \, , \quad x \geq 0 \, ,$$

is strictly convex and will be called a *Young function*. Cf. [1, Sect. 8.2], where ϕ is assumed to be right-continous and non-decreasing.

The inverse function $\psi = \phi^{-1}$ has the same properties (1) to (3) as ϕ, so that its primitive

$$\Psi(y) = \int_0^y \psi(v)\, dv \ , \quad y \geq 0 \ ,$$

is again a Young function. The couple (Φ, Ψ), is a couple of *conjugate* Young functions. The relation is symmetric and we write both $\Psi = \Phi_*$ and $\Phi = \Psi_*$. The Young inequality holds true,

$$\Phi(x) + \Psi(y) \geq xy \ , \quad x, y \geq 0 \ ,$$

and the Legendre equality holds true ,

$$\Phi(x) + \Psi(\phi(x)) = x\phi(x) \ , \quad x \geq 0 \ .$$

Here are the specific cases we are going to use:

$$\Phi(x) = \frac{x^p}{p} \ , \quad \Psi(y) = \frac{y^q}{q} \ , \quad p, q > 1 \ , \quad \frac{1}{p} + \frac{1}{q} = 1 \ ; \tag{1.3}$$

$$\exp_2(x) = \mathrm{e}^x - 1 - x \ , \quad (\exp_2)_*(y) = (1+y)\log(1+y) - y \ ; \tag{1.4}$$

$$(\cosh - 1)(x) = \cosh x - 1 \ , \quad (\cosh - 1)_*(y) = \int_0^y \sinh^{-1}(v)\, dv \ ; \tag{1.5}$$

$$\mathrm{gauss}_2(x) = \exp\left(\frac{1}{2}x^2\right) - 1 \ . \tag{1.6}$$

Given a Young function Φ, and a probability measure μ, the Orlicz space $L_\Phi\left(\mu\right)$ is the Banach space whose closed unit ball is $\left\{f \in L^0(\mu) \,\middle|\, \int \Phi(|f|)\, d\mu \leq 1\right\}$. This defines the Luxemburg norm,

$$\|f\|_{L_\Phi(\mu)} \leq \alpha \quad \text{if, and only if,} \quad \int \Phi(\alpha^{-1}\,|f|)\, d\mu \leq 1 \ .$$

From the Young inequality, it holds

$$\int |uv|\, d\mu \leq \int \Phi(|u|)\, d\mu + \int \Phi_*(|v|)\, d\mu \ .$$

This provides a separating duality $\langle u, v\rangle_\mu = \int uv\, d\mu$ of $L_\Phi\left(\mu\right)$ and $L_{\Phi_*}\left(\mu\right)$ such that

$$\langle u, v\rangle_\mu \leq 2\, \|u\|_{L_\Phi(\mu)}\, \|v\|_{L_{\Phi_*}(\mu)} \ .$$

From the conjugation between Φ and Ψ, an equivalent norm can be defined, namely, the Orlicz norm

$$\|f\|_{L_\Phi(\mu)^*} = \sup\left\{\langle f, g\rangle_\mu \,\middle|\, \|f\|_{L_\Psi(\mu)} \leq 1\right\} \ .$$

Domination relation between Young functions imply continuous injection properties for the corresponding Orlicz spaces. We say that Φ_2 *eventually dominates* Φ_1, written $\Phi_1 \prec \Phi_2$, if there is a constant κ such that $\Phi_1(x) \leq \Phi_2(\kappa x)$ for all x larger than some $\bar{x}$. As, in our case, μ is a probability measure, the continuous embedding $L_{\Phi_2}(\mu) \to L_{\Phi_1}(\mu)$ holds if, and only if, $\Phi_1 \prec \Phi_2$. See a proof in [1, Theorem 8.2]. If $\Phi_1 \prec \Phi_2$, then $(\Phi_2)_* \prec (\Phi_1)_*$. With reference to our examples (1.4) and (1.5), we see that $\exp_2$ and $(\cosh -1)$ are equivalent. They both are eventually dominated by gauss_2 (1.6) and eventually dominate all powers (1.3).

A special case occurs when there exists a function C such that $\Phi(ax) \leq C(a)\Phi(x)$ for all $a \geq 0$. This is true, for example, for a power function and in the case of the functions $(\exp_2)_*$ and $(\cosh -1)_*$. In such a case, the conjugate space and the dual space are equal and bounded functions are a dense set.

The spaces corresponding to case (1.3) are ordinary Lebesgue spaces. The cases (1.4) and (1.5) provide isomorphic B-spaces $L_{(\cosh -1)}(\mu) \leftrightarrow L_{\exp_2}(\mu)$ which are of special interest for us as they provide the model spaces for our non-parametric version of IG, see Sect. 1.2.3.

A function f belongs to $L_{\cosh -1}(\mu)$ if, and only if, it is *sub-exponential*, that is, there exist constants $C_1, C_2 > 0$ such that

$$\mathbb{P}_\mu(|f| \geq t) \leq C_1 \exp(-C_2 t) \ , \quad t \geq 0 \ .$$

Sub-exponential random variable are of special interest in applications because they admit an explicit exponential bounds in the Law of Large Numbers. Random variables whose square is sub-exponential are called sub-gaussian. There is a large literature on this subject, see, for example, [4, 30–32].

We will be led to use a further notation. For each Young function Φ, the function $\overline{\Phi}(x) = \Phi(x^2)$ is again a Young function such that $\|f\|_{L_{\overline{\Phi}}(\mu)} \leq \lambda$ if, and only if, $\left\| |f|^2 \right\|_{L_\Phi(\mu)} \leq \lambda^2$. We denote the resulting space by $L^2_\Phi(\mu)$. For example, gauss_2 and $\overline{\cosh -1}$ are $\prec$-equivalent , hence the isomorphisn $L_{\mathrm{gauss}_2}(\mu) \leftrightarrow L^2_{(\cosh -1)}(\mu)$.

As an application of this notation, consider that for each increasing convex Φ it holds $\Phi(fg) \leq \Phi((f^2+g^2)/2) \leq (\Phi(f^2)+\Phi(g^2))/2$. It follows that when the $L^2_\Phi(\mu)$-norm of f and of g is bounded by one, the $L_\Phi(\mu)$-norm of f, g, and fg, are all bounded by one. The need to control the product of two random variables in $L_{(\cosh -1)}(\mu)$ appears, for example, in the study of the covariant derivatives of the statistical bundle, see [6, 7, 14, 28].

1.2.2 Calculus of the Gaussian Space

From now on, our base probability space is the Gaussian probability space $(\mathbb{R}^n, \gamma)$, $\gamma(z) = (2\pi)^{n/2} \exp\left(-|z|^2/2\right)$. We will use a few simple facts about the analysis of the Gaussian space, see [15, Chap. V].

Let us denote by $C^k_{\text{poly}}(\mathbb{R}^n)$, $k = 0, 1, \dots$, the vector space of functions which are differentiable up to order k and which are bounded, together with all derivatives, by a polynomial. This class of functions is dense in $L^2(\gamma)$. For each couple $f, g \in C^1_{\text{poly}}(\mathbb{R}^n)$, we have

$$\int f(x)\, \partial_i g(x)\, \gamma(x)\, dx = \int \delta_i f(x)\, g(x)\, \gamma(x)\, dx \ ,$$

where the divergence operator δ_i is defined by $\delta_i f(x) = x_i f(x) - \partial_i f(x)$. Multidimensional notations will be used, for example,

$$\int \nabla f(x) \cdot \nabla g(x)\, \gamma(x)\, dx = \int f(x)\, \delta \cdot \nabla g(x)\, \gamma(x)\, dx \ , \quad f, g \in C^2_{\text{poly}}(\mathbb{R}^n) \ ,$$

with $\delta \cdot \nabla g(x) = x \cdot \nabla g(x) - \Delta g(x)$.

For example, in this notation, the divergence of Eq. (1.1) with $P = p \cdot \gamma$, $Q = q \cdot \gamma$, and $p, q \in C^2_{\text{poly}}(\mathbb{R}^n)$, becomes

$$\frac{1}{2} \int \nabla \log \frac{p(x)}{q(x)} \cdot \nabla \log \frac{p(x)}{q(x)}\, p(x)\, \gamma(x)\, dx = \\ \frac{1}{2} \int \log \frac{p(x)}{q(x)}\, \delta \cdot \left(\nabla \log \frac{p(x)}{q(x)}\, p(x) \right) \gamma(x)\, dx \ .$$

The inner product Eq. (1.2) becomes, with $P = p \cdot \gamma$ and $f, g, p \in C^2_{\text{poly}}(\mathbb{R}^n)$,

$$\int \nabla f(x) \cdot \nabla g(x)\, p(x)\, \gamma(x)\, dx = \int f(x) \delta \cdot \nabla (g(x) p(x))\, \gamma(x)\, dx \ .$$

Hermite polynomials $H_\alpha = \delta^\alpha 1$ provide an orthogonal basis for $L^2(\gamma)$ such that $\partial_i H_\alpha = \alpha_i H_{\alpha - e_i}$, e_1 the i-th element of the standard basis of $\mathbb{R}^n$. In turn, this provides a way to prove that there is a closure of both operator ∂_i and δ_i on a domain which is an Hilbert subspace of $L^2(\gamma)$. Such a space is denoted by D^2 in [15]. Moreover, the closure of ∂_i is the infinitesimal generator of the translation operator, [3, 16]. The space D^2 is a Sobolev Space with Gaussian weight based on the L^2 norm, [1]. By replacing that norm with a $(\cosh - 1)$ Orlicz norm, one derives the applications to IG that are presented in [13, 27].

1.2.3 *Exponential Statistical Bundle*

We refer to [25, 27] for the definition of maximal exponential manifold $\mathcal{E}(\gamma)$, and of statistical bundle $S\mathcal{E}(\gamma)$. Below we report the results that are necessary in the context of the present paper.

A key result is the proof of the following statement of necessary and sufficient conditions, see [5] and [29, Theorem 4.7].

Proposition 1 *For all* $p, q \in \mathcal{E}(\gamma)$ *it holds* $q = \mathrm{e}^{u-K_p(u)} \cdot p$, *where* $u \in L_{(\cosh-1)}(\gamma)$, $\mathbb{E}_p[u] = 0$, *and* u *belongs to the interior of the proper domain of the convex function* K_p. *This property is equivalent to any of the following:*

1. p *and* q *are connected by an open exponential arc;*
2. $L_{(\cosh-1)}(p) = L_{(\cosh-1)}(q)$ *and the norms are equivalent;*
3. $p/q \in \cup_{a>1} L^a(q)$ *and* $q/p \in \cup_{a>1} L^a(p)$.

Item 2 ensures that all the fibers of the statistical bundle, namely $S_p\mathcal{E}(\gamma)$, $p \in \mathcal{E}(\gamma)$, are isomorphic. Item 3 gives a explicit description of the exponential manifold. For example, let p be a positive probability density with respect to γ, and take $q = 1$ and $a = 2$. Then sufficient conditions for $p \in \mathcal{E}(\gamma)$ are

$$\int p(x)^2\, \gamma(x)\, dx < \infty \quad \text{and} \quad \int \frac{1}{p(x)}\, \gamma(x)\, dx < \infty\,.$$

It is interesting to note that there is, so to say, a bound above and a bound below.

1.3 Bounding the Orlicz Norm with the Orlicz Norm of the Gradient

We discuss now inequalities related to the classical Gauss-Poincaré inequality,

$$\int \left(f(x) - \int f(y)\,\gamma(y)\,dy\right)^2 \gamma(x)\,dx \leq \int |\nabla f(x)|^2\, \gamma(x)\,dx\,, \tag{1.7}$$

where $f \in C^1_{\mathrm{poly}}(\mathbb{R}^n)$. A proof is given, for example, in [21, Sect. 1.4] and will follow as a particular case in an inequality to be proved below.

In terms of norms, the inequality above is equivalent to $\|f - \overline{f}\|_{L^2(\gamma)} \leq \||\nabla f|\|_{L^2(\gamma)}$, where $\overline{f} = \int f(y)\,\gamma(y)\,dy$. One can check whether the constant 1 is optimal, by taking $f(x) = \sum_i x_i$ and observing that the two sides both take the value $\sqrt{n}$.

This is an example of differential inequality of high interest. For example, if $p \in C^2_{\mathrm{poly}}$ is a probability density with respect to γ, then the χ^2-divergence of $P = p \cdot \gamma$ from γ is bounded as follows.

$$D_{\chi^2}(P|\gamma) = \int (p(x)-1)^2\,\gamma(x)\,dx \leq \int |\nabla p(x)|^2\,\gamma(x)\,dx = \int \delta \cdot \nabla p(x)\; p(x)\,\gamma(x)\,dx\,,$$

where $\delta \cdot \nabla p(x) = x \cdot \nabla p(x) - \Delta p(x)$. As $\int \delta \cdot \nabla p(x)\, \gamma(x)\, dx = 0$, the RHS is equal to

$$\int \delta \cdot \nabla p(x)(p(x)-1)\, \gamma(x)\, dx \leq \frac{1}{2}\int (\delta \cdot \nabla p(x))^2\, \gamma(x)\, dx + \frac{1}{2}\int (p(x)-1)^2\, \gamma(x)\, dx\ ,$$

so that, in conclusion,

$$\int (p(x)-1)^2\, \gamma(x)\, dx \leq \int (\delta \cdot \nabla p(x))^2\, \gamma(x)\, dx\ .$$

1.3.1 Ornstein-Uhlenbeck Semi-group

Generalisation of Eq. (1.7) can be derived from the Ornstein-Uhlenbeck semi-group which is defined on each $C^k_{\mathrm{poly}}(\mathbb{R}^n)$, $k = 0, 1, \ldots$, by the Mehler formula

$$P_t f(x) = \int f(\mathrm{e}^{-t}x + \sqrt{1-\mathrm{e}^{-2t}}y)\, \gamma(y)\, dy\ , \quad t \geq 0, \quad f \in C^k_{\mathrm{poly}}(\mathbb{R}^n)\ , \tag{1.8}$$

see [15, V-1.5] and [21, Sect. 1.3]. Notice that $P_0 f = f$ and $P_\infty f = \overline{f}$.

If X, Y are independent standard Gaussian random variables in $\mathbb{R}^n$, then

$$X_t = \mathrm{e}^{-t}X + \sqrt{1-\mathrm{e}^{-2t}}Y, \quad Y_t = \sqrt{1-\mathrm{e}^{-2t}}X - \mathrm{e}^{-t}Y \tag{1.9}$$

are independent standard Gaussian random variables for all $t \geq 0$. It is well known, and easily checked, that the infinitesimal generator of the Ornstein-Uhlembeck semi-group is $-\delta \cdot \nabla$, that is, for each $f \in C^2_{\mathrm{poly}}(\mathbb{R}^n)$, it holds

$$\frac{d}{dt}P_t f(x) = \int \nabla f(\mathrm{e}^{-t}x + \sqrt{1-\mathrm{e}^{-2t}}y) \cdot \left(-\mathrm{e}^{-t}x + \frac{\mathrm{e}^{-2t}}{\sqrt{1-\mathrm{e}^{-2t}}y}\right) \gamma(y)\, dy \tag{1.10}$$

$$= -(\delta \cdot \nabla) P_t f(x) \tag{1.11}$$

$$= -P_t(\delta \cdot \nabla) f(x)\ . \tag{1.12}$$

See [15, V.1.5].

These computations are well known in stochastic calculus, see, for example [9, Sect. 5.6]. In fact, because of Eq. (1.11), the function $p(x,t) = P_t p(x)$ is a solution of the equation

$$\frac{\partial}{\partial t}p(x,t) - \Delta p(x,t) + x \cdot \nabla p(x,t) = 0\ , \quad p(x,0) = p(x)\ ,$$

which is the Kolmogorov equation for the diffusion $dX_t = -X_t + \sqrt{2}dW_t$. Similarly, the function $u(x) = \int_0^\infty \mathrm{e}^{-t} P_t f(x)\, dt$ is a solution of the equation

$$\delta \cdot \nabla u(x) + u(x) = f(x)\ .$$

By the change of variable Eq. (1.9) and Jensen's inequality, it easily follows that for each convex function Φ it holds

$$\int \Phi(P_t f(x))\, \gamma(x)\, dx \leq \int \Phi(f(x))\, \gamma(x)\, dx\ . \tag{1.13}$$

That is, for all $t \geq 0$, the mapping $f \mapsto P_t f$ is non-expansive for the norm of each Orlicz space $L_\Phi(\gamma)$.

We will discuss now a first set of inequalities that involves convexity and differentiation as it is in Eq. (1.7). This set depends on the following proposition.

Proposition 2 *For all $\Phi : \mathbb{R}$ convex and all $f \in C^1_{poly}(\mathbb{R}^n)$, it holds*

$$\begin{aligned} &\int \Phi\left(f(x) - \int f(y)\,\gamma(y)\,dy\right)\gamma(x)\,dx \leq \\ &\qquad \iint \Phi\left(\frac{\pi}{2}\nabla f(x)\cdot y\right)\,\gamma(x)\gamma(y)\,dxdy = \\ &\frac{1}{\sqrt{2\pi}}\iint \Phi\left(\frac{\pi}{2}\,|\nabla f(x)|\,z\right)\,\mathrm{e}^{-z^2/2}\gamma(x)\,dzdx = \int \widetilde{\Phi}\,(|\nabla f(x)|)\,\gamma(x)\,dx\ , \end{aligned} \tag{1.14}$$

where $\widetilde{\Phi}$ is the convex function defined by

$$\widetilde{\Phi}(a) = \int \Phi\left(\frac{\pi}{2}az\right)\gamma(z)\,dz\ . \tag{1.15}$$

Proof It follows from Eqs. (1.8) and (1.10) that

$$\begin{aligned} f(x) - \int f(y)\,\gamma(y)\,dy &= P_0 f(x) - P_\infty f(x) = -\int_0^\infty \frac{d}{dt}P_t f(x)\,dt = \\ \frac{\pi}{2}\int_0^\infty p(t)\,dt \int &\left(\nabla f(\mathrm{e}^{-t}x + \sqrt{1-\mathrm{e}^{-2t}}y)\right)\cdot\left(\sqrt{1-\mathrm{e}^{-2t}}x - \mathrm{e}^{-t}y\right)\gamma(y)\,dy\ , \end{aligned}$$

where $p(t) = \frac{2}{\pi}\frac{\mathrm{e}^{-t}}{\sqrt{1-\mathrm{e}^{-2t}}}$ is a probability density on $t \geq 0$. After that, the application of Jensen inequality and the change of variable (1.9), gives Eq. (1.14). See more details in [27].

The arguments used here differs from those used, for example, in [21], which are based on the equation for the infinitesimal generator Eqs. (1.11) and (1.12). We will come to that point later. Notice that we can take $\Phi(s) = s^2$ and derive a Poincaré inequality with a non-optimal constant > 1.

We can prove now a set of inequalities of the Poincaré type. The first example is $\Phi(s) = \mathrm{e}^s$. In such a case, the equation for the moment generating function of the Gaussian distribution gives

$$\widetilde{\Phi}(a) = \int \exp\left(\frac{\pi}{2} az\right) \gamma(z)\, dz = \exp\left(\frac{\pi^2 a^2}{8}\right) ,$$

so that the inequality (1.14) becomes

$$\int \exp\left(f(x) - \overline{f}\right) \gamma(x)\, dx \leq \int \exp\left(\frac{\pi^2}{8} |\nabla f(x)|^2\right) \gamma(x)\, dx .$$

More clearly, we can change f to $\frac{2\kappa}{\pi} f$ and write

$$\begin{aligned} \int \exp\left(\frac{2\kappa}{\pi}\left(f(x) - \overline{f}\right)\right) \gamma(x)\, dx &\leq \\ \int \exp\left(\frac{\kappa^2}{2} |\nabla f(x)|^2\right) \gamma(x)\, dx &= \\ (2\pi)^{n/2} \int \exp\left(-\frac{1}{2}\left(|x|^2 - \kappa^2 |\nabla f(x)|^2\right)\right) dx &. \end{aligned} \tag{1.16}$$

The inequality above is non-trivial only if the RHS is bounded, that is

$$|\nabla f(x)| < \kappa^{-1} |x| , \quad x \in \mathbb{R}^n ,$$

that is, the function f is Lipschitz. We have found that $f \in C^1(\mathbb{R}^n)$ and globally Lipschitz implies that f is sub-exponential in the Gaussian space.

The first case of bound for Orlicz norms we consider is the Lebesgue norm, $\Phi(s) = s^{2p}$, $p > 1/2$. In such a case,

$$\widetilde{\Phi}(a) = \left(\frac{\pi}{2}\right)^{2p} m(2p)\, a^{2p} ,$$

where $m(2p)$ is the $2p$-moment of the standard Gaussian distribution. It follows that

$$\left\| f - \int f(y) \gamma(y)\, dy \right\|_{L^{2p}(\gamma)} \leq \frac{\pi}{2} (m(2p))^{1/2p} \left\| |\nabla f| \right\|_{L^{2p}(\gamma)} .$$

The cases $\Phi(a) = a^{2p}$ are special in that we can use the in the proof the multiplicative property $\Phi(ab) = \Phi(a)\Phi(b)$. The argument generalizes to the case where the convex function Φ is a Young function whose increase is controlled through a function C, $\Phi(uv) \leq C(u)\Phi(v)$, and, moreover, such that there exists a $\kappa > 0$ for which

$$\int C\left(\frac{\pi}{2}\kappa u\right)\gamma(u)\,du \leq 1\,,$$

then Eq. (1.15) becomes

$$\widetilde{\Phi}(\kappa a) = \int \Phi\left(\frac{\pi}{2}\kappa a z\right)\gamma(z)\,dz \leq \int C\left(\frac{\pi}{2}\kappa z\right)\gamma(z)\,dz\ \Phi(a) \leq \Phi(a)\,.$$

By using this bound in Eq. (1.14), we get

$$\int \Phi\left(\kappa\left(f(x) - \int f(y)\gamma(y)\,dy\right)\right)\gamma(x)\,dx \leq \int \Phi\left(|\nabla f(x)|\right)\gamma(x)\,dx\ .$$

Assume now that $\||\nabla f|\|_{L_\Phi(\gamma)} \leq 1$ so that the LHS does not exceed 1. Then $\kappa\left\|f - \overline{f}\right\|_{L_\Phi(\gamma)} \leq 1$, which, in turn, implies the inequality

$$\left\|f - \overline{f}\right\|_{L_\Phi(\gamma)} \leq \kappa^{-1}\,\||\nabla f|\|_{L_\Phi(\gamma)}\ .$$

For example, for $(\exp_2)_*(y) = (1+y)\log(1+y) - y$ we can take $C(u) = \max(|u|\,, |u|^2)$ and we want a $\kappa > 0$ such that

$$\int \max\left(\frac{\pi}{2}\kappa\,|u|\,, \left(\frac{\pi}{2}\kappa\,|u|\right)^2\right)\gamma(u)\,du \leq 1\,.$$

Such a κ exists because C is γ-integrable, continous, and $C(0) = 0$. For example, as $C(u) \leq u + u^2$, $u \geq 0$, we have

$$\int C\left(\frac{\pi}{2}\kappa u\right)\gamma(u)\,du = 2\int_0^\infty C\left(\frac{\pi}{2}\kappa u\right)\gamma(u)\,du \leq$$
$$\pi\kappa\int_0^\infty u\gamma(u)\,du + \frac{\pi^2}{2}\kappa^2\int_0^\infty u^2\gamma(u)\,du = \sqrt{\frac{\pi}{2}}\kappa + \frac{\pi^2}{4}\kappa^2$$

and we can take $k > 0$ satisfying $\sqrt{\frac{\pi}{2}}\kappa + \frac{\pi^2}{4}\kappa^2 = 1$.

For us, it is of particular interest the case of the Young function $\Phi = \cosh - 1$, for which there is no such bound. Instead, we use Eq. (1.16) with κ and $-\kappa$ to get

$$\int (\cosh - 1)\left(\frac{2\kappa}{\pi}\left(f(x) - \overline{f}\right)\right)\gamma(x)\,dx \leq$$
$$\int \operatorname{gauss}_2\left(\kappa\,|\nabla f(x)|\right)\gamma(x)\,dx\,. \qquad (1.17)$$

Now, if $\kappa = \||\nabla f|\|^{-1}_{L_{\mathrm{gauss}_2}(\gamma)}$, then the LHS is smaller or equal then 1, and hence $2\kappa/\pi \left\| f - \overline{f} \right\|_{L_{\cosh -1}(\gamma)} \leq 1$. It follows that

$$\left\| f - \overline{f} \right\|_{L_{\cosh -1}(\gamma)} \leq \frac{\pi}{2} \||\nabla f|\|_{L_{\mathrm{gauss}_2}(\gamma)} \ .$$

Our last case of this series is the Young function $\mathrm{gauss}_2(x) = \exp\left(\frac{1}{2}|x|^2\right) - 1$. Assume $f \in C^0_{\mathrm{poly}}(\mathbb{R}^n) \cap L_{\mathrm{gauss}_2}(\gamma)$, that is, there exists a constant $\lambda > 0$ such that

$$\int \mathrm{gauss}_2(\lambda^{-1} f(x))\, \gamma(x)\, dx = \\ (2\pi)^{-n/2} \int \exp\left(-\frac{1}{2}\left(|x|^2 - \lambda^{-2} f(x)^2\right)\right) dx - 1 < +\infty \, .$$

This holds if, and only if, $|x|^2 > \lambda^{-2} |f(x)|^2$, $x \in \mathbb{R}^n$, that is, f is bounded by a linear function with coefficient $\lambda > \sup_x |f(x)| / |x|$. The case does not seem to be of our interest.

In fact, if we compute $\widetilde{\mathrm{gauss}_2}(\kappa a)$ from Eq. (1.15), we find

$$\int \mathrm{gauss}_2\left(\frac{\pi}{2}\kappa a z\right) \gamma(z)\, dz = \\ \frac{1}{\sqrt{2\pi}} \int \exp\left(-\frac{1}{2}\left(1 - \left(\frac{\pi}{2}\right)^2 \kappa^2 a^2\right) z^2\right) dz - 1 = \\ \left(1 - \left(\frac{\pi}{2}\right)^2 \kappa^2 a^2\right)^{-1/2} - 1 \, .$$

if the argument of $(\cdot)^{-1/2}$ is positive, $+\infty$ otherwise. The inequality Eq. (1.14) becomes

$$\int \mathrm{gauss}_2\left(\kappa(f(x) - \overline{f})\right) \gamma(x)\, dx \leq \int \left(1 - \left(\frac{\pi}{2}\right)^2 \kappa^2 |\nabla f(x)|^2\right)^{-1/2} \gamma(x)\, dx - 1 \, .$$

The function in the RHS does not belong to the class of Young function we are considering here and would require a special study.

In the following proposition we give a summary of the inequalities proved so far.

Proposition 3 *There exists constants C_1, $C_2(p)$, C_3 such that for all $f \in C^1_{\mathrm{poly}}(\mathbb{R}^n)$ the following inequalities hold:*

$$\left\| f - \int f(y)\, \gamma(y)\, dy \right\|_{L_{(\exp_2)_*}(\gamma)} \leq C_1 \||\nabla f|\|_{L_{(\exp_2)_*}(\gamma)} \, . \tag{1.18}$$

$$\left\| f - \int f(y)\, \gamma(y)\, dy \right\|_{L^{2p}(\gamma)} \leq C_2(p) \||\nabla f|\|_{L^{2p}(\gamma)} \, , \quad p > 1/2 \, . \tag{1.19}$$

$$\left\| f - \int f(y)\,\gamma(y)\,dy \right\|_{L_{(\cosh -1)}(\gamma)} \leq C_3 \, \|| \nabla f \||_{L_{\text{gauss}_2}(\gamma)} \, . \tag{1.20}$$

Other equivalent norms could be used in the inequalities above. For example, $L_{(\exp_2)_*}(\gamma) \leftrightarrow L_{(\cosh -1)_*}(\gamma)$ and $L_{\text{gauss}_2}(\gamma) \leftrightarrow L^2_{\cosh -1}(\gamma)$. We do not care in the present paper to define explicitly the relevant Gauss-Sobolev spaces as in [27]. But notice the special relevance of the space based on the norm $f \mapsto \||\nabla f\||_{L^2_{\cosh -1}(\gamma)}$.

1.3.2 Generator of the Ornstein-Uhlenbeck Semi-group

We consider now a further set of inequalities which are based on the use of infinitesimal generator $-\delta \cdot \nabla$ of the Ornstein-Uhlenbeck semigroup, see Eqs. (1.11) and (1.12). Compare, for example, [21, Sect. 1.3.7].

We have, for all $f \in C^2_{\text{poly}}(\mathbb{R}^n)$, that

$$f(x) - \overline{f} = -\int_0^\infty \frac{d}{dt} P_t f(x)\,dt = \int_0^\infty \delta \cdot \nabla P_t f(x)\,dt \, . \tag{1.21}$$

Note that

$$\nabla P_t f(x) = \nabla \int f(\mathrm{e}^{-t}x + \sqrt{1-\mathrm{e}^{-2t}}y)\,\gamma(y)\,dy = \\ \mathrm{e}^{-t} \int \nabla f(\mathrm{e}^{-t}x + \sqrt{1-\mathrm{e}^{-2t}}y)\,\gamma(y)\,dy = \mathrm{e}^{-t} P_t \nabla f(x) \, ,$$

so that

$$P_t \delta \cdot \nabla f(x) = \delta \cdot \nabla P_t f(x) = \mathrm{e}^{-t} \delta \cdot P_t \nabla f(x) \, .$$

Now, Eq. (1.21) becomes

$$f(x) - \overline{f} = \int_0^\infty \mathrm{e}^{-t} \delta \cdot P_t \nabla f(x)\,dt \, . \tag{1.22}$$

As

$$\int \delta \cdot \nabla f(x)\,\gamma(x)\,dx = 0 \, ,$$

the covariance of $f, g \in C^0_{\text{poly}}(\mathbb{R}^n)$ is

$$\mathrm{Cov}_\gamma(f, g) = \\ \int (f(x) - \overline{f}) g(x)\,\gamma(x)\,dx = \int (f(x) - \overline{f})(g(x) - \overline{g})\,\gamma(x)\,dx \, .$$

It follows that for all $f, g \in C^2_{\text{poly}}(\mathbb{R}^n)$ we derive from Eq. (1.22)

$$\operatorname{Cov}_\gamma(f, g) = \int_0^\infty \mathrm{e}^{-t} \int P_t \nabla f(x) \cdot \nabla g(x)\, \gamma(x)\, dx\ dt\ . \qquad (1.23)$$

We use here a result of [27, Proposition 5]. Let $|\cdot|_1$ and $|\cdot|_2$ be two norms on $\mathbb{R}^n$, such that $|x \cdot y| \le |x|_1\, |y|_2$. For a Young function Φ, consider the norm of $L_\Phi(\gamma)$ and the conjugate space endowed with the dual norm,

$$\|f\|_{L_{\Psi,*}(\gamma)} = \sup\left\{\int fg\,\gamma \,\middle|\, \int \Phi(g)\,\gamma \le 1\right\}.$$

The following inequality that includes the standard Poincaré case when $\Phi(u) = u^2/2$.

Proposition 4 *Given a couple of conjugate Young function Φ, Ψ, and norms $|\cdot|_1$, $|\cdot|_2$ on $\mathbb{R}^n$ such that $x \cdot y \le |x|_1\, |y|_2$, $x, y \in \mathbb{R}^n$, for all $f, g \in C^1_{\text{poly}}(\mathbb{R}^n)$, it holds*

$$\left|Cov_\gamma(f, g)\right| \le \left\| |\nabla f|_1 \right\|_{L_\Phi(\gamma)} \left\| |\nabla g|_2 \right\|_{L_{\Psi,*}(\gamma)}.$$

The case of our interest here is $\Phi = \cosh - 1$, $\Psi = (\cosh - 1)_*$. As $(\cos - 1)_* \prec (\cosh - 1)$, it follows, in particular, that $\operatorname{Cov}_\gamma(f, f)$ is bouded by a constant times $\||\nabla f|\|^2_{L_{\cosh-1}(\gamma)}$.

1.4 Discussion and Conclusions

We have collected here a list of possible applications of the information geometry of the Gaussian space that has been introduced in [13, 27] and further developed in the present paper.

1.4.1 Sub-exponential Random Variables

Let $f \in C^2_{\text{poly}}(\mathbb{R}^n)$ be a random variable of the Gaussian space. Assume moreover that f is globally Lipschitz, that is,

$$|\nabla f(x)| \le \|f\|_{\text{Lip}(\mathbb{R}^n)}\, |x|$$

where $\|f\|_{\text{Lip}(\mathbb{R}^n)}$ is the Lipschitz semi-norm, that is, the best constant. It follows from Eq. (1.17) that $f \in L_{(\cosh-1)}(\gamma)$ and the norm admits a computable bound.

If p is any probability density of the maximal exponential model of γ, that is, it is connected to 1 by an open exponential arc, then Proposition 1 implies that

$f \in L_{(\cosh -1)}(p)$, that is, f is sub-exponential under the distribution $P = p \cdot \gamma$. If the sequence $(X_n)_{n=1}^{\infty}$ is independent and with distribution $p \cdot \gamma$, then the sequence of sample means will converge,

$$\lim_{n\to\infty} \frac{1}{n} \sum_{j=1}^{n} f(X_j) = \int f(x)\, p(x)\, \gamma(x)\, dx \ ,$$

with an exponential bound on the tail probability. See, for example, [31, Sect. 2.8].

1.4.2 Hyvärinen Divergence

Here we adapt [23] to the Gaussian case. Consider the Hyvärinen divergence of Eq. (1.1) in the Gaussian case, that is, $P = p \cdot \gamma$ and $Q = q \cdot \gamma$. As a function of q is of the form

$$H(q) = \frac{1}{2} \int |\nabla \log p(x)|^2\, p(x)\, \gamma(x)\, dx +$$
$$\frac{1}{2} \int |\nabla \log q(x)|^2\, p(x)\, \gamma(x)\, dx - \int \nabla \log p(x) \cdot \nabla \log q(x)\, p(x)\, \gamma(x)\, dx \ ,$$

where the first term does not depend on q and the second term is an expectation with respect to $p \cdot \gamma$. As $\nabla \log p = p^{-1} \nabla p$, the third term equals

$$-\int \delta \cdot \nabla \log q(x)\, p(x)\, \gamma(x)\, dx \ ,$$

which is again a p-expectation. To minimize the Hyvärinen divergence we must minimize the p-expected value of the local score

$$S(q, x) = \frac{1}{2} |\nabla \log q(x)|^2 - \delta \cdot \nabla \log q(x)$$

If p and q belong to the maximal exponential model of γ, then $q = \mathrm{e}^{u - K(u)}$ with $u \in L_{(\cosh -1)}(\gamma)$ and $\int u(x)\, \gamma(x)\, dx = 0$. The local score becomes $\frac{1}{2} |\nabla u|^2 - \delta \cdot \nabla u$. To compute the p-expected value of the score with an independent sample of $p \cdot \gamma$ we have interest to assume that the score is in $L_{(\cosh -1)}(\gamma)$, because this assumption implies the good convergence of the empirical means for all p, as it way explained in the section above.

Assume, for example, $\nabla u \in L^2_{(\cosh -1)}(\gamma)$. This implies directly $|\nabla u|^2 \in L_{(\cosh -1)}(\gamma)$. Moreover, we need to assume that the $L_{(\cosh -1)}(\gamma)$-norm of $\delta \cdot \nabla u$ is finite. Under such assumptions it seems reasonable to hope that the minimization on a suitable model of the sample expectation of the Hyvärinen score is consistent.

1.4.3 Otto's Metric

Let $P = p \cdot \gamma$ with p in the maximal exponential model of γ. Let f and g be in the p-fiber of the statistical manifold, that is, $f, g \in L_{(\cosh -1)}(p) = L_{(\cosh -1)}(\gamma)$ and $\int f(x) \gamma(x)\,dx = \int g(x) \gamma(x)\,dx = 0$. The Otto's inner product (1.2) becomes

$$\int \nabla f(x) \cdot \nabla g(x)\, p(x)\, \gamma(x)\, dx = \int f(x)\, \delta \cdot (p(x) \nabla g(x))\, \gamma(x)\, dx \; .$$

The LHS is well defined and regular if we assume $\nabla f, \nabla g \in L^2_{(\cosh -1)}(\gamma)$, because, in such a case, $|\nabla f|^2, |\nabla g|^2 \in L_{(\cosh -1)}(\gamma) = L_{(\cosh -1)}(p)$. The RHS provides the representation of the inner product in the inner product defined in $L_{(\cosh -1)}(\gamma)$. Note that the mapping $g \mapsto \delta \cdot (p \nabla g)$ is 1-to-1 if g is restricted by $\int g(x) p(x) \gamma(x)\,dx = 0$. The inverse of this mapping provides the natural gradient of the Otto's inner product in the sense of [2, 12].

1.4.4 Conclusion and Acknowledgments

In this paper we have derived bounds of the Orlicz norms of interest in IG based on the Orlicz norm of the gradient. The schematic examples above provide, in our opinion, a motivation for further study of this approach. There is a large literature on Sobolev spaces with weight that we have, regrettably, not used here. Its study would surely provide more precise and deep results than those presented here. I like to thank the Editor and the Referees for the very helpful and detailed review of this paper.

Disclaimer: Views and opinions expressed are those of the authors and do not necessarily represent official positions of their respective companies.

References

1. Adams, R.A., Fournier, J.J.F.: Sobolev spaces, Pure and Applied Mathematics (Amsterdam), vol. 140, 2nd edn. Elsevier/Academic Press, Amsterdam (2003)
2. Amari, S.I.: Natural gradient works efficiently in learning. Neural Comput. **10**(2), 251–276 (1998). https://doi.org/10.1162/089976698300017746
3. Bogachev, V.I.: Differentiable measures and the Malliavin calculus, Mathematical Surveys and Monographs, vol. 164. American Mathematical Society, Providence, RI (2010) https://doi.org/10.1090/surv/164
4. Buldygin, V.V., Kozachenko, Y.V.: Metric characterization of random variables and random processes, Translations of Mathematical Monographs, vol. 188. American Mathematical Society, Providence, RI (2000), translated from the 1998 Russian original by V. Zaiats

5. Cena, A., Pistone, G.: Exponential statistical manifold. Ann. Inst. Statist. Math. **59**(1), 27–56 (2007)
6. Chirco, G., Malagò, L., Pistone, G.: Lagrangian and Hamiltonian mechanics for probabilities on the statistical manifold, arXiv:2009.09431
7. Gibilisco, P., Pistone, G.: Connections on non-parametric statistical manifolds by Orlicz space geometry. IDAQP **1**(2), 325–347 (1998)
8. Hyvärinen, A.: Estimation of non-normalized statistical models by score matching. J. Mach. Learn. Res. **6**, 695–709 (2005)
9. Karatzas, I., Shreve, S.E.: Brownian motion and stochastic calculus, Graduate Texts in Mathematics, vol. 113. Springer-Verlag, New York, second edn. (1991). https://doi.org/10.1007/978-1-4612-0949-2
10. Kriegl, A., Michor, P.W.: The convenient setting of global analysis, Mathematical Surveys and Monographs, vol. 53. American Mathematical Society, Providence, RI (1997). https://doi.org/10.1090/surv/053
11. Lang, S.: Differential and Riemannian manifolds, Graduate Texts in Mathematics, vol. 160, 3rd edn. Springer-Verlag (1995)
12. Li, W., Montúfar, G.: Natural gradient via optimal transport. Inf. Geom. **1**(2), 181–214 (2018). https://doi.org/10.1007/s41884-018-0015-3
13. Lods, B., Pistone, G.: Information geometry formalism for the spatially homogeneous Boltzmann equation. Entropy **17**(6), 4323–4363 (2015)
14. Lott, J.: Some geometric calculations on Wasserstein space. Comm. Math. Phys. **277**(2), 423–437 (2008). https://doi.org/10.1007/s00220-007-0367-3
15. Malliavin, P.: Integration and probability, Graduate Texts in Mathematics, vol. 157. Springer-Verlag (1995), with the collaboration of Héléne Airault, Leslie Kay and Gérard Letac, Edited and translated from the French by Kay, With a foreword by Mark Pinsky
16. Malliavin, P.: Stochastic analysis, Grundlehren der Mathematischen Wissenschaften [Fundamental Principles of Mathematical Sciences], vol. 313. Springer-Verlag (1997)
17. Musielak, J.: Orlicz spaces and modular spaces. Lecture Notes in Mathematics, vol. 1034. Springer-Verlag (1983)
18. Naudts, J.: Generalised thermostatistics. Springer-Verlag London Ltd. (2011). https://doi.org/10.1007/978-0-85729-355-8
19. Newton, N.J.: A class of non-parametric statistical manifolds modelled on Sobolev space. Inf. Geom. **2**(2), 283–312 (2019). https://doi.org/10.1007/s41884-019-00024-z
20. Newton, N.J.: Sobolev statistical manifolds and exponential models. In: Geometric science of information, Lecture Notes in Comput. Sci., vol. 11712, pp. 443–452. Springer, Cham (2019)
21. Nourdin, I., Peccati, G.: Normal approximations with Malliavin calculus, Cambridge Tracts in Mathematics, vol. 192. Cambridge University Press, Cambridge (2012). https://doi.org/10.1017/CBO9781139084659, from Stein's method to universality
22. Otto, F.: The geometry of dissipative evolution equations: the porous medium equation. Comm. Partial Differ. Equa. **26**(1-2), 101–174 (2001)
23. Parry, M., Dawid, A.P., Lauritzen, S.: Proper local scoring rules. Ann. Statist. **40**(1), 561–592 (2012). https://doi.org/10.1214/12-AOS971
24. Pistone, G.: Examples of the application of nonparametric information geometry to statistical physics. Entropy **15**(10), 4042–4065 (2013). https://doi.org/10.3390/e15104042
25. Pistone, G.: Nonparametric information geometry. In: Nielsen, F., Barbaresco, F. (eds.) Geometric science of information, Lecture Notes in Comput. Sci., vol. 8085, pp. 5–36. Springer, Heidelberg (2013), first International Conference, GSI 2013 Paris, France, August 28-30, 2013 Proceedings
26. Pistone, G.: Translations in the exponential Orlicz space with Gaussian weight. In: Nielsen, F., Barbaresco, F. (eds.) Geometric Science of Information. pp. 569–576. No. 10589 in LNCS, Springer (2017), third International Conference, GSI 2017, Paris, France, November 7-9, 2017, Proceedings
27. Pistone, G.: Information geometry of the Gaussian space. In: Information geometry and its applications, Springer Proc. Math. Stat., vol. 252, pp. 119–155. Springer, Cham (2018)

28. Pistone, G.: Lagrangian function on the finite state space statistical bundle. Entropy **20**(2), 139 (2018). https://doi.org/10.3390/e20020139, http://www.mdpi.com/1099-4300/20/2/139
29. Santacroce, M., Siri, P., Trivellato, B.: New results on mixture and exponential models by Orlicz spaces. Bernoulli **22**(3), 1431–1447 (2016). https://doi.org/10.3150/15-BEJ698
30. Siri, P., Trivellato, B.: Robust concentration inequalities in maximal exponential models. Stat. Probab. Lett. **170**(0167–7152), 109001(2021). https://doi.org/10.1016/j.spl.2020.109001. http://www.sciencedirect.com/science/article/pii/S0167715220303047
31. Vershynin, R.: High-dimensional probability: an introduction with applications in data science, Cambridge Series in Statistical and Probabilistic Mathematics, vol. 47. Cambridge University Press, Cambridge (2018). https://doi.org/10.1017/9781108231596, with a foreword by Sara van de Geer
32. Wainwright, M.J.: High-dimensional statistics: a non-asymptotic viewpoint. Cambridge Series in Statistical and Probabilistic Mathematics, Cambridge University Press, Cambridge (2019). https://doi.org/10.1017/9781108627771

Chapter 2
On Normalization Functions and φ-Families of Probability Distributions

Luiza H. F. de Andrade, Francisca L. J. Vieira, and Charles C. Cavalcante

Abstract φ-families of probability distributions are a general representation to deformed exponential models which brings interesting insight on the geometrical aspects of the distribution family. In the φ-family, the analogue of the cumulant-generating function is a normalizing function. This function plays an important role in the statistical model and we investigate the behaviour of the function near the boundary of the domain of the parametrization in order to provide the precise conditions of validity of the φ- family model. We discuss the conditions for existence of the statistical model when considering φ-functions and the required conditions for the normalization function so we can provide a complete understanding about the investigated family of probability distribution.

2.1 Introduction

Deformed exponentials functions were initially studied in [13] and refer to the generalization of the classical exponential function by a given function ϕ of $[0, \infty)$, which is strictly positive on $[0, \infty)$. The function ϕ is used to define the $\ln_\phi$ (ϕ-logarithm) as:

$$\ln_\phi(u) = \int_1^u \frac{1}{\phi(x)} dx, \quad u > 0.$$

L. H. F. de Andrade
Department of Natural Sciences, Mathematics and Statistical,
Federal Rural University of Semi-Arid Region, Mossoró, RN, Brazil
e-mail: luizafelix@ufersa.edu.br

F. L. J. Vieira
Department of Mathematics, Regional University of Cariri, Juazeiro do Norte-CE, Brazil
e-mail: leidmar.vieira@urca.br

C. C. Cavalcante (✉)
Department of Teleinformatics Engineering, Federal University of Ceará, Fortaleza, CE, Brazil
e-mail: charles@ufc.br

F. Nielsen (ed.), *Progress in Information Geometry*,
Signals and Communication Technology,
https://doi.org/10.1007/978-3-030-65459-7_2

The function $\ln_\phi$ is concave, negative in $(0, 1)$, positive on $(1, +\infty)$ and its inverse is the function $\exp_\phi$, the so called deformed exponential [13].

The authors in [27] provided a generalization of exponential families of probability distributions $\mathcal{E}(p)$ [17, 18], based on a wider class of deformed exponential functions, calling it φ-families of probability distributions $\mathcal{F}_c^\varphi$ and the construction of these families is based on Musielak–Orlicz spaces [11].

Another generalization of exponential families of probability distributions, but in an infinite-dimensional setting, was studied in [29]. In [26], the authors studied the Δ_2-condition and its consequences on φ-families of probability distributions. More specifically, the behavior of the normalizing function near the boundary of its domain was analyzed, considering that the Musielak–Orlicz function Φ_c does not satisfy the Δ_2-condition.

To generalize the exponential families of probability $\mathcal{E}(p)$, the authors in [27] required the assumption that the deformed exponential function φ satisfies the condition that there exists a function $u_0 : T \to (0, \infty)$ such that $\int_T \varphi(c + \lambda u_0) d\mu < \infty$, for all $\lambda > 0$, for every function $c : T \to \mathbb{R}$, for which $\int_T \varphi(c) d\mu = 1$, considering that $\varphi(\cdot)$ is a probability density.

Going further on the understanding of the behavior of the φ-families, in [24] the authors investigated the importance of the existing condition imposed on the deformed exponential φ and its relevance to connect arcs in generalized statistical manifolds. Later, [1, 23] considered the impact of the same existing condition on the behavior of the normalizing function ψ near the boundary of the domain of parametrization of the φ-family of probability distributions. Still in this direction of the study of normalization functions, in [20] the authors found an example that has the same form of a deformed exponential function which does not satisfy the boundary condition. More recently, the authors in [8] also discussed some conditions for the normalization aspects on deformed exponentials. These aspects are of paramount importance on the understanding of a statistical family of distributions.

Our aim in this chapter is to analyze the behavior of the normalizing function near the boundary of its domain, considering deformed exponential functions which satisfy or do not satisfy the boundary condition. We study the cases in the purely atomic and in the non-atomic scenarios and discuss which are the constraints that need to be satisfied so the probability family may be properly defined.

This chapter is organized as follows. In Sect. 2.2 we revisit the deformed exponential functions and the construction of the φ-families of probability distributions and how the Δ_2-condition in Musielak–Orlicz functions influences the boundary of the domain of the parametrization. Section 2.3 is devoted to all the cases of the behavior of the normalizing function. Finally, in Sect. 2.4 we state our conclusions.

2.2 Revisiting Deformed Exponentials

In this section we revisit some fundamental concepts about the deformed exponential functions and probability distributions so we can proceed with the study about the characterization of them, in terms of the normalization function.

2.2.1 *Deformed Exponential Functions*

Definition 2.1 A *deformed exponential* is a convex function $\varphi\colon \mathbb{R} \to [0, \infty)$ such that $\lim_{u\to-\infty} \varphi(u) = 0$ and $\lim_{u\to\infty} \varphi(u) = \infty$.

It is easy to verify that the ordinary exponential and Tsallis q-exponential are deformed exponential functions [10]. The notion of deformed exponential $\exp_\phi$ was firstly introduced in [13] and studied after in [9, 14, 15].

Fixing an increasing function ϕ that is strictly positive in $(0, \infty)$ and $\frac{1}{\phi}$ is integrable we have the ϕ-logarithm defined as as:

$$\ln_\phi(u) = \int_1^u \frac{1}{\phi(x)} dx, \qquad u > 0.$$

The function $\ln_\phi$ is concave, negative in $(0, 1)$, positive on $(1, +\infty)$ and the inverse is the function $\exp_\phi$ [13]. Every deformed exponential $\exp_\phi(\cdot)$ satisfies the Definition 2.1 and the deformed exponential $\varphi(\cdot)$ proposed in [27] is a deformed exponential $\exp_\phi$ when $\varphi(0) = 1$.

Vigelis and Cavalcante in [27] used a σ-finite, non-atomic measure space (T, Σ, μ), and a deformed exponential φ to introduce a non-parametric generalization of the exponential statistical manifold proposed by [3, 18]. This generalization was subsequently investigated in [2, 20, 26] stressing the geometric aspects of the new family of distributions.

A very important aspect the authors realized in [27] is the need of an additional condition over the function to find the domain of the φ-families of probability distributions $\mathcal{F}_c^\varphi$, that is a generalization of the *exponential families of probabilites distributions* $\mathcal{E}(p)$ [3, 18, 19].

This condition is given as [27]:

Condition 2.1 *There exists a function* $u_0 : T \to (0, \infty)$ *such that*

$$\int_T \varphi(c + \lambda u_0) d\mu < \infty, \qquad \textit{for all } \lambda > 0, \tag{2.1}$$

for every function $c : T \to \mathbb{R}$, *for which* $\int_T \varphi(c) d\mu = 1$.

In Condition 2.1, the constraint $\int_T \varphi(c)d\mu = 1$ can be replaced by $\int_T \varphi(c)d\mu < \infty$. This fact was demonstrated in [20, Lemma 1] and, therefore, the condition in Eq. (2.1) can be rewritten as:

Condition 2.2 *There exists a function* $u_0 : T \to (0, \infty)$ *such that*

$$\int_T \varphi(c + \lambda u_0)d\mu < \infty, \qquad \textit{for all } \lambda > 0, \tag{2.2}$$

for every function $c : T \to \mathbb{R}$, *for which* $\int_T \varphi(c)d\mu < \infty$.

Conditions 2.1 and 2.2 are equivalent what give us several interesting directions to understand the geometry of the φ-families of distributions. Also, in [25, Proposition 8] the authors investigate and discuss the circumstances to u_0 function and why Conditions 2.1 and 2.2 are well-defined.

The "classical" exponential function $\exp(\cdot)$ satisfies the condition in Eq. (2.1) with $u_0 = 1$. The *Kaniadaki's* κ-exponential $\exp_\kappa : \mathbb{R} \to (0, \infty)$ for $\kappa \in (0, 1)$, defined as [5, 12]

$$\exp_\kappa(u) = \begin{cases} \left(\kappa u + \sqrt{1 + \kappa^2 u^2}\right)^{\frac{1}{\kappa}}, & \kappa \neq 0 \\ \exp(u), & \kappa = 0 \end{cases},$$

is a deformed exponential that satisfies (2.1) for any function u_0 such that $\int_T \exp_\kappa(u_0)d\mu < \infty$ [27, Example 1]. Additionally, the κ-exponential with $0 < \kappa < 1$ was used in the construction of a statistical manifold [16].

However, there exist deformed exponentials that do not satisfy Condition 2.1. An example was given in [20, Example 2], that is the following function,

$$\varphi(u) = \begin{cases} \exp\left((u + 1)^2/2\right), & u \geq 0, \\ \exp\left(u + 1/2\right), & u \leq 0. \end{cases} \tag{2.3}$$

If we multiply the function given in Eq. (2.3) by a constant, we obtain a function that does not satisfy the condition (2.1) and is still a deformed exponential as was defined by [13].

Let $\exp_\phi$ be a function $\exp_\phi : \mathbb{R} \to (0, \infty)$

$$\exp_\phi(u) = \begin{cases} \dfrac{\exp\left(\frac{(u+1)^2}{2}\right)}{\exp(1/2)} & u \geq 0 \\ \exp(u) & u \leq 0 \end{cases}, \tag{2.4}$$

where $\phi(x)$ is a strictly positive function such that $\frac{1}{\phi(x)}$ is integrable [15], defined as

$$\phi(x) = \begin{cases} (1 + 2\ln(x))^{1/2}\, x, & x \geq 1 \\ x, & 0 < x \leq 1 \end{cases}.$$

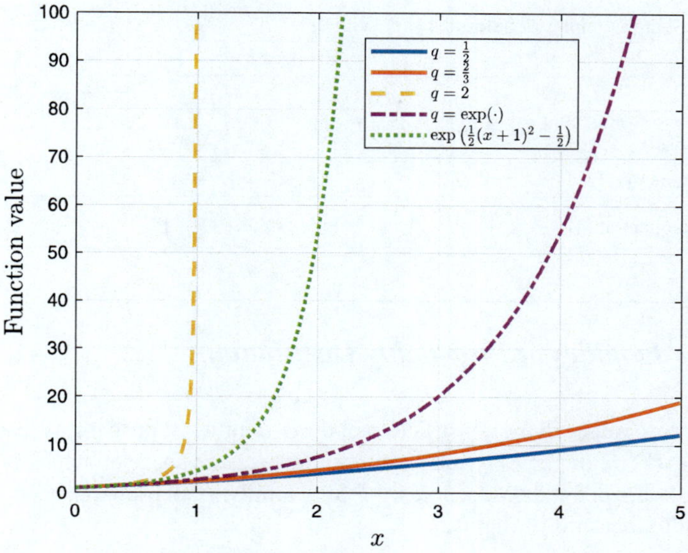

Fig. 2.1 Illustration of several deformed exponentials

Another example of a deformed exponential function is the q-exponential, $\exp_q : \mathbb{R} \to [0, \infty)$, for $q \in [0, \infty)$, is given by the following expression [22]

$$\exp_q(x) = [1 + (1-q)x]_+^{1/(1-q)},$$

that is, a $\exp_\phi(u)$ where $\phi(x) = x^q$, with $q > 0$ [13, Example 2]. This function, $\exp_q(x)$ satisfies Condition 2.1 for $q \in \mathbb{R}$, $0 < q < 1$, and since the argument of $\exp_q(\cdot)$ has to be positive, we have $\frac{-1}{1-q} \leq u$, for $0 < q < 1$.

Figure 2.1 shows the deformed exponential $\exp_q(x)$, for $q = \frac{1}{2}$, $q = \frac{2}{3}$, and $q \to 1$, that is, the "classical" exponential function $\exp(\cdot)$. We also have the deformed exponential given in Eq. (2.4), which does not satisfy the condition provided in (2.1) and the q-exponencial, for $q = 2$, in this case, the argument is $u < 1$, that also does not satisfy the Condition 2.1.

From Figure 2.1 one can note that, for $0 < q < 1$, the function $\exp_q(\cdot)$ satisfies the Condition 2.1 and for $q = 2$ the q-exponential does not satisfy (2.1). This is due to the fact that the q-exponential increases more rapidly to infinity than the function in the Eq. (2.4). Furthermore, once that q-exponential with $0 < q < 1$ satisfies Condition 2.1, in [7] this deformed exponential was used to construct non-parametric q-exponential statistical models. In Table 2.1 we show some examples of deformed exponential functions.

Finally, in [24] was shown that the condition expressed in Eq. (2.1) is a necessary and sufficient one to connect two probability densities by a φ-arc with a non-atomic measure.

Table 2.1 Some examples of deformed exponentials

Function	Expression
Kaniadaki's κ-exponential [5]	$\exp_\kappa(x) = \begin{cases} \left(\kappa x + \sqrt{1+\kappa^2 x^2}\right)^{\frac{1}{\kappa}}, & \kappa \neq 0 \\ \exp(x), & \kappa = 0 \end{cases}$
Tsallis exponential [21]	$\exp_q(x) = [1 + (1-q)x]_+^{1/(1-q)}, q \in [0, \infty)$
Newton exponential [4]	$\exp_\phi(x) = \left(\frac{1+\alpha x}{1-\alpha x}\right)^{\frac{1}{2\alpha}}, \quad -\frac{1}{\alpha} < x < \frac{1}{\alpha}$

2.2.2 φ-Families of Probability Distributions

In this section, we recall the construction of the φ-families of probability distributions in [27]. Let (T, Σ, μ) be a σ- finite, non-atomic measure space on which probability distributions are defined. We denote by $\mathcal{P}_\mu$ the family of all probability on T that are equivalent to measure μ, that is,

$$\mathcal{P}_\mu = \left\{ p \in L^0;\ p > 0 \text{ and } \int_T p \, d\mu = 1 \right\},$$

where L^0 is the linear space of all real-value. In this space, T may be seen of as the set of real numbers $\mathbb{R}$. These families are based on the replacement of the exponential function by a deformed exponential $\varphi(\cdot)$ that satisfies the condition provided in Eq. (2.1).

The φ-families of probability distributions were built based on the Musielak–Orlicz spaces [11]. Let φ be a deformed exponential. The Musielak–Orlicz function is defined in [27] by

$$\Phi_c(t, u) = \varphi(c(t) + u) - \varphi(c(t)), \tag{2.5}$$

where $c\colon T \to \mathbb{R}$ is a measurable function such that $\varphi(c)$ is μ-integrable. Then we have, the Musielak–Orlicz space L^{Φ_c}, the Musielak–Orlicz class $\tilde{L}^{\Phi_c}$ and the Morse-Transue space E^{Φ_c}, denoted, respectively, by L_c^φ, $\tilde{L}_c^\varphi$ and E_c^φ, which correspond to the following sets:

$$L_c^\varphi = \left\{ u \in L^0;\ \int_T \varphi(c(t) + \lambda u(t)) \, d\mu < \infty, \text{ for every } \lambda \in (-\epsilon, \epsilon) \right\},$$

$$\tilde{L}_c^\varphi = \left\{ u \in L^0;\ \int_T \varphi(c(t) + u(t)) \, d\mu < \infty \right\} \text{ and}$$

$$E_c^\varphi = \left\{ u \in L^0;\ \int_T \varphi(c(t) + \lambda u(t))\, d\mu < \infty \text{ for all } \lambda > 0 \right\}.$$

Clearly, $E_c^\varphi \subseteq \tilde{L}_c^\varphi \subseteq L_c^\varphi$.

Now, let $\mathcal{K}_c^\varphi$ be the convex set of all the functions $u \in L_c^\varphi$ such that $\varphi(c + \lambda u)$ is μ-integrable for every λ in a neighborhood of [0, 1], that is,

$$\mathcal{K}_c^\varphi = \left\{ u \in L_c^\varphi;\ \int_T \varphi(c + \lambda u) < \infty, \quad \text{for some } \lambda > 1 \right\}.$$

We know that $\mathcal{K}_c^\varphi$ is an open set in L_c^φ [27, Lemma 2] and, for $u \in \mathcal{K}_c^\varphi$, the function $\boldsymbol{\varphi}(c + u)$ is not necessarily in $\mathcal{P}_\mu$. Hence, the normalizing function $\psi\colon \mathcal{K}_c^\varphi \to \mathbb{R}$ is introduced in order to make the density $\boldsymbol{\varphi}(c + u - \psi(u)u_0)$ to be in $\mathcal{P}_\mu$ [27]. For any $u \in \mathcal{K}_c^\varphi$, $\psi(u) \in \mathbb{R}$ is the unique function which $\boldsymbol{\varphi}(c + u - \psi(u)u_0)$ is in $\mathcal{P}_\mu$ [27, Proposition 3].

Let

$$B_c^\varphi = \left\{ u \in L_c^\varphi :\ \int_T u\varphi'_+(t, c(t))d\mu = 0 \right\}$$

be a closed subspace of L_c^φ, thus for every $u \in \mathcal{B}_c^\varphi = B_c^\varphi \cap \mathcal{K}_c^\varphi$, by the convexity of φ, one has $\psi(u) \geq 0$ and $\boldsymbol{\varphi}(c + u - \psi(u)u_0) \in \mathcal{P}_\mu$.

For each measurable function $c\colon T \to \mathbb{R}$ such that $p = \varphi(c) \in \mathcal{P}_\mu$ is associated a parametrization $\boldsymbol{\varphi}_c\colon \mathcal{B}_c^\varphi \to \mathcal{F}_c^\varphi$, given by

$$\boldsymbol{\varphi}_c(u) = \boldsymbol{\varphi}(c + u - \psi(u)u_0), \tag{2.6}$$

where the operator $\boldsymbol{\varphi}$ acts on the set of real-value functions $u\colon T \to \mathbb{R}$ given by $\boldsymbol{\varphi}(u)(t) = \varphi(u(t))$ and the set $\mathcal{F}_c^\varphi = \varphi_c(\mathcal{B}_c^\varphi) \subseteq \mathcal{P}_\mu$ where $\mathcal{P}_\mu = \bigcup\{\mathcal{F}_c^\varphi : \boldsymbol{\varphi}(c) \in \mathcal{P}_\mu\}$ and the map $\boldsymbol{\varphi}_c$ is a bijection from $\mathcal{B}_c^\varphi$ to $\mathcal{F}_c^\varphi$.

In the following section, we will discuss the Δ_2-condition, which is a condition that the Musielak–Orlicz functions can satisfy or not. We will discuss this condition and its consequences.

2.2.3 The Δ_2- Condition and φ-Families of Probability Distributions

Condition 2.3 *(Δ_2 condition) A Musielak–Orlicz function Φ satisfies the Δ_2-condition ($\Phi \in \Delta_2$), if one can find a constant $K > 0$ and a non-negative function $f \in \tilde{L}_c^\varphi$ such that*

$$\Phi(t, 2u) \le K\Phi(t, u), \qquad \textit{for all } u \ge f(t) \ \textit{ and } \mu\textit{-a.e. } t \in T.$$

One can see that, if $\Phi \in \Delta_2$, then $\int_T \Phi(t, |u(t)|)\, d\mu < \infty$, for every $u \in L^\Phi$, in other words, it holds for the set of functions u which belong to the Musielak–Orlicz space. Then, L^Φ, $\tilde{L}^\Phi$ and E^Φ are equal as sets. Otherwise, if the Musielak–Orlicz function $\Phi_c(u) = \varphi(c(t) + u) - \varphi(c(t))$ does not satisfy the Δ_2-condition, then E_c^φ is a proper subspace of L_c^φ. It is easy to see this fact for Orlicz space [6, Theorem 10.1].

Moreover, we have the following result:

Lemma 2.1 [28, Remark 3.12] *Let Φ be a Musielak–Orlicz function not satisfying the Δ_2-condition and such that $\Phi(t, b_\Phi(t)) = \infty$ for μ-a.e. (almost everywhere) $t \in T$, where $b_\Phi(t) = \sup\{u \ge 0 : \Phi(t, u) < \infty\}$. Then we can find functions u_* and u^* in L^Φ such that*

$$\begin{cases} I_\Phi(\lambda u_*) < \infty, & \textit{for } 0 \le \lambda \le 1, \\ I_\Phi(\lambda u_*) = \infty, & \textit{for } 1 < \lambda, \end{cases} \tag{2.7}$$

and

$$\begin{cases} I_\Phi(\lambda u^*) < \infty, & \textit{for } 0 \le \lambda < 1, \\ I_\Phi(\lambda u^*) = \infty, & \textit{for } 1 \le \lambda. \end{cases}, \tag{2.8}$$

where $I_{\Phi_c}(u(t)) = \int_T \Phi_c(t, |u(t)|) d\mu$ for any $u \in L^0$.

Another important fact is that, given a deformed exponential $\varphi(\cdot)$ that satisfies the Condition 2.1, we can always find a Musielak–Orlicz function $\Phi_c(t, u) = \varphi(t, c(t) + u) - \varphi(t, c(t))$ that does not satisfy the Δ_2-condition. This is stated in the following Proposition.

Proposition 2.1 [26, Proposition 2] *Given any deformed exponential φ, we can find a measurable function $c \colon T \to \mathbb{R}$ with $\int_T \varphi(c) d\mu = 1$ such that the Musielak–Orlicz function $\Phi_c(t, u) = \varphi(t, c(t) + u) - \varphi(t, c(t))$ does not satisfy the Δ_2-condition.*

It is discussed in [26] that, given two Musielak–Orlicz functions $\Phi_c(t, u) = \varphi(c(t) + u) - \varphi(c(t))$ and $\Phi_b(t, u) = \varphi(b(t) + u) - \varphi(b(t))$, with $b, c : T \to \mathbb{R}$ functions such that $\int_T \varphi(b)\, d\mu = 1$ and $\int_T \varphi(c)\, d\mu = 1$, that satisfy the Δ_2-condition, then L_c^φ and L_b^φ are equal as sets and $\mathcal{F}_c^\varphi = \mathcal{F}_b^\varphi$ [26, Proposition 4].

In the next section, we will investigate the behavior of the normalizing function ψ near the boundary of the domain of the parametrization. This is important to keep the normalization function still valid for a given probability distribution.

2.3 The Behavior of ψ Near the Boundary of $\mathcal{B}_c^\varphi$

As we could see in the previous section, if the Musielak–Orlicz function $\Phi_c(t, u) = \varphi(c(t) + u) - \varphi(c(t))$ does not satisfy the Δ_2-condition, it implies that the Morse-Transue space E_c^φ is a proper subspace of L_c^φ. On its turn, this implies that the boundary of $\mathcal{B}_c^\varphi$, the domain of the parametrization (2.6), is not-empty.

Actually, by that fact that $\mathcal{B}_c^\varphi = B_c^\varphi \cap \mathcal{K}_c^\varphi$ is an open set in B_c^φ, we conclude that a function $u \in B_c^\varphi$ belongs to the boundary of $\mathcal{B}_c^\varphi$ if and only if $\int_T \varphi(c + \lambda u)\, d\mu < \infty$ for all $\lambda \in (0, 1)$, and $\int_T \varphi(c + \lambda u)\, d\mu = \infty$ for each $\lambda > 1$. It is important to know that if the function u belongs to the boundary of $\mathcal{B}_c^\varphi$, then u may or may not belong to the set Musielak–Orlicz class $\tilde{L}^{\Phi_c}$, that is, $\int_T \varphi(c + u)\, d\mu < \infty$, or $\int_T \varphi(c + u)\, d\mu = \infty$.

If the Musielak–Orlicz function $\Phi_c(t, u) = \varphi(c(t) + u) - \varphi(c(t))$ satisfies the Δ_2-condition, then $\int_T \varphi(c + u)\, d\mu < \infty$ for all $u \in L_c^\varphi$, so the set $\mathcal{B}_c^\varphi$ coincides with the closed set B_c^φ and its boundary is empty.

In this section, we will discuss the behavior of the normalizing function ψ in two cases. The first one supposes that the deformed exponential function $\varphi(\cdot)$ satisfies the condition stated in Eq. (2.2). On the other hand, the second case assumes that the deformed exponential function $\varphi(\cdot)$ does not satisfy the Condition 2.2. More specifically, given any function u in the boundary of $\mathcal{B}_c^\varphi$, denoted by $\partial\mathcal{B}_c^\varphi$, we want to know whether $\psi(\alpha u)$ converges to a finite value or not as $\alpha \uparrow 1$.

The results we are going to discuss in details in the next sections can be summarized in Table 2.2.

2.3.1 *Condition 2.2 is Satisfied*

Supposing that the Musielak–Orlicz function $\Phi_c(t, u) = \varphi(c(t) + u) - \varphi(c(t))$ does not satisfy the Δ_2-condition, then the boundary $\partial\mathcal{B}_c^\varphi$ is not-empty. We then study the behavior of ψ assuming, in this section, that the deformed exponential φ satisfies the Condition 2.2.

Table 2.2 Summary of results

Cond	Result
Condition 2 satisfied	We analyse the cases where a function $u \in \partial\mathcal{B}_c^\varphi$ belongs to the Musielak–Orlicz class. If so, we will show that the normalizing function converges to a finite value β near its boundary domain. On the other hand, we also show that if u does not belong to the Musielak–Orlicz class, the normalizing function will not converge (tends to infinity) near its boundary domain
Condition 2 is not satisfied	We may find a function $u \in \partial\mathcal{B}_c^\varphi$, such that it does not belong to the Musielak–Orlicz class but the normalizing function converges for a finite value β, near its boundary domain

We have that not all functions $u \in \partial\mathcal{B}_c^\varphi$ satisfy $\int_T \varphi(c+u)d\mu < \infty$ or $\int_T \varphi(c+u)d\mu = \infty$ as we can see in the next proposition, which is a consequence of the Lemma 2.1.

Proposition 2.2 [26, Proposition 5] *The boundary of $\mathcal{B}_c^\varphi$ is non-empty if and only if the Musielak–Orlicz function $\Phi_c = \varphi(t, c(t)+u) - \varphi(t, c(t))$ does not satisfy the Δ_2-condition. Moreover, in any of these cases, there exist functions w_* and w^* in $\partial\mathcal{B}_c^\varphi$ such that $\int_T \boldsymbol{\varphi}(c+w_*)d\mu < \infty$ and $\int_{\mathbb{T}} \boldsymbol{\varphi}(c+w^*)d\mu = \infty$.*

Proof Given non-negative functions u_* and u^* in L_c^φ satisfying (2.7) and (2.8) in 2.1, we consider the functions

$$w_* = u_* - \frac{\int_T u_*\boldsymbol{\varphi}'_+(c)d\mu}{\int_T u_0\boldsymbol{\varphi}'_+(c)d\mu}u_0, \qquad \text{and} \qquad w^* = u^* - \frac{\int_T u^*\boldsymbol{\varphi}'_+(c)d\mu}{\int_T u_0\boldsymbol{\varphi}'_+(c)d\mu}u_0,$$

which are in B_c^φ. Next, we show that w_* is in $\partial\mathcal{B}_c^\varphi$ and satisfies $\int_T \boldsymbol{\varphi}(c+w_*)d\mu < \infty$. For any $0 \le \lambda \le 1$, it is clear that

$$\int_T \boldsymbol{\varphi}(c+\lambda w_*)d\mu \le \int_T \boldsymbol{\varphi}(c+\lambda u_*)d\mu < \infty.$$

Now suppose that $\int_T \boldsymbol{\varphi}(c+\lambda_0 w_*)d\mu < \infty$ for some $\lambda_0 > 1$. In the view of $1 \le \int_T \boldsymbol{\varphi}(c+\lambda_0 w_*)\, d\mu < \infty$, we can find $\alpha_0 \ge 0$ such that $\int_T \boldsymbol{\varphi}(c+\lambda_0 w_* - \alpha_0 u_0)\, d\mu = 1$. By the definition of u_0, fixed any measurable function $\widetilde{c}$ such that $\int_T \boldsymbol{\varphi}(\widetilde{c})d\mu = 1$, we have that $\int_T \boldsymbol{\varphi}(\widetilde{c}+\alpha u_0)d\mu < \infty$ for all $\alpha \in \mathbb{R}$. Hence, considering $\widetilde{c} = c + \lambda_0 w_* - \alpha_0 u_0$ and

$$\alpha = \lambda_0 \frac{\int_T u_*\boldsymbol{\varphi}'_+(c)d\mu}{\int_T u_0\boldsymbol{\varphi}'_+(c)d\mu} + \alpha_0.$$

Hence, we obtain that $\int_T \boldsymbol{\varphi}(c+\lambda_0 u_*)d\mu = \int_T \boldsymbol{\varphi}(\widetilde{c}+\alpha u_0)d\mu < \infty$, which is a contradiction. Consequently, $\int_T \boldsymbol{\varphi}(c+\lambda w_*)d\mu = \infty$ for all $\lambda > 1$, and w_* belongs to $\partial\mathcal{B}_c^\varphi$ and satisfies $\int_T \boldsymbol{\varphi}(c+w_*)d\mu < \infty$.

Proceeding as above, we show that $\int_T \boldsymbol{\varphi}(c+\lambda w^*)d\mu < \infty$ for all $0 \le \lambda < 1$, and $\int_T \boldsymbol{\varphi}(c+\lambda w^*)d\mu = \infty$ for all $\lambda \ge 1$. This result implies that w^* belongs to $\partial\mathcal{B}_c^\varphi$ and is such that $\int_T \boldsymbol{\varphi}(c+w^*)d\mu = \infty$. □

The behavior of ψ depends on whether $\int_T \varphi(c+u)\, d\mu < \infty$ holds or not. We analyze such behavior in the next proposition.

Proposition 2.3 *Let u be a function in the boundary of $\mathcal{B}_c^\varphi$. For $\alpha \in [0,1)$, denote $\psi_u(\alpha) := \psi(\alpha u)$. If $\int_T \boldsymbol{\varphi}(c+u)\, d\mu < \infty$ then $\psi_u(\alpha) = \psi(\alpha u)$ converges to some $\beta \in (0,\infty)$ as $\lambda \uparrow 1$. On other side, if u is such that $\int_T \varphi(c+u)d\mu = \infty$, then $\psi(\alpha u) \to \infty$ as $\alpha \uparrow 1$.*

Proof Observing that the normalizing function ψ is convex with $\psi(0) = 0$, we may conclude that $\psi_u(\alpha) = \psi(\alpha u)$ is non-decreasing and continuous in $[0, 1)$. Moreover, $(\psi_u)'_+(\alpha)$ is non-decreasing in $[0, 1)$. Fix any function u in the boundary of $\mathcal{B}_c^\varphi$ such that $\int_T \varphi(c+u)\,d\mu < \infty$. Assume that $\psi(\alpha u)$ tends to ∞ as $\alpha \uparrow 1$. In this case, one can note that

$$\varphi(c + \alpha u - \psi(\alpha u)u_0) \leq \varphi(c + u\mathbf{1}_{\{u>0\}} - \psi(\alpha u)u_0) \to 0, \qquad \text{as } \alpha \uparrow 1.$$

Since $\varphi(c + \alpha u - \psi(\alpha u)u_0) \leq \varphi(c + u\mathbf{1}_{\{u>0\}})$, we can use the Dominated Convergence Theorem to write

$$\int_T \varphi(c + \alpha u - \psi(\alpha u)u_0)\,d\mu \to 0, \qquad \text{as } \alpha \uparrow 1,$$

which is a contradiction to $\int_T \varphi(c + \alpha u - \psi(\alpha u)u_0)d\mu = 1$. Thus $\psi(\alpha u)$ is bounded in $[0, 1)$, and $\psi(\alpha u)$ converges to some $\beta \in (0, \infty)$ as $\alpha \uparrow 1$.

Now, consider any function $u \in \partial\mathcal{B}_c^\varphi$ such that $\int_T \varphi(c+u)d\mu = \infty$, then suppose that, for some $\overline{\lambda} > 0$, the function u satisfies $\psi(\alpha u) \leq \overline{\lambda}$ for all $\alpha \in [0, 1)$. Denote $A = \{u \geq 0\}$. Observing that

$$\int_A \varphi(c + \alpha u - \overline{\lambda}u_0)d\mu \leq \int_T \varphi(c + \alpha u - \overline{\lambda}u_0)d\mu \leq \int_T \varphi(c + \alpha u - \psi(\alpha u)u_0)d\mu = 1,$$

we obtain that $\int_A \varphi(c + u - \overline{\lambda}u_0)d\mu < \infty$. In addition, it is clear that

$$\int_{T\setminus A} \varphi(c + u - \overline{\lambda}u_0)d\mu \leq \int_{T\setminus A} \varphi(c)d\mu \leq 1.$$

As a result, we have $\int_T \varphi(c + u - \overline{\lambda}u_0)d\mu < \infty$. From the condition (2.1), it follows that $\int_T \varphi(c+u)d\mu < \infty$, which is a contradiction. □

Thus, we conclude that for any $u \in \partial\mathcal{B}_c^\varphi$, if $\int_T \varphi(c+u)d\mu < \infty$, then $\psi(\alpha u) \to \beta$, and if $\int_T \varphi(c+u)d\mu = \infty$, then $\psi(\alpha u) \to \infty$, as $\alpha \uparrow 1$.

In the next section, we will discuss the behavior of the normalizing function ψ assuming that the deformed exponential function does not satisfy Condition 2.2.

2.3.2 *Condition 2.2 is Not Satisfied*

We know there are deformed exponential functions that do not satisfy Condition 2.2. Throughout this section it will be clear that this assumption is sufficient to ensure that the boundary of $\mathcal{B}_c^\varphi$ is not empty and, as a consequence, we can analyze the behavior of the normalizing function near the boundary of the parametrization domain.

In the following lemma, we find an equivalence for the inclusion of the Musielak–Orlicz classes and in the next proposition we use this lemma to find an equivalence for the occurrence of Condition 2.2.

Lemma 2.2 [11, Theorem 8.4] *Let Ψ and Φ be finite-value Musielak–Orlicz functions. Then, the inclusion $\tilde{L}^{\Phi} \subset \tilde{L}^{\Psi}$ is satisfied if and only if there exist $\alpha > 0$ and a non-negative function $f \in \tilde{L}^{\Phi}$ such that*

$$\alpha\Psi(t,u) \leq \Phi(t,u), \text{ for all } u > f(t).$$

Proposition 2.4 *A measurable function u_0 satisfies Condition 2.2 if and only if for some measurable function $c\colon T \to \mathbb{R}$ such that $\varphi(c)$ is μ-integrable, we can find constants $\overline{\lambda}, \alpha > 0$ and a non-negative function $f \in \tilde{L}^{\Phi_c}$ such that*

$$\alpha\Phi_c(t,u) \leq \Phi_{c-\overline{\lambda}u_0}(t,u), \qquad \text{for all } u > f(t). \tag{2.9}$$

Proof Suppose that u_0 satisfies Condition 2.2. Let $c\colon T \to \mathbb{R}$ be any measurable function such that $\int_T \varphi(c)d\mu < \infty$. As u is a measurable function with $\int_T \varphi(c - \overline{\lambda}u_0 + u)d\mu < \infty$ then

$$\int_T \varphi(c+u)d\mu = \int_T \varphi(c - \overline{\lambda}u_0 + u + \overline{\lambda}u_0)d\mu < \infty.$$

This result implies $\tilde{L}^{\Phi_{c-\overline{\lambda}u_0}} \subset \tilde{L}^{\Phi_c}$ and inequality (2.9) follows from Lemma 2.2.

Now suppose that inequality (2.9) is satisfied. By Lemma 2.2 we have $\tilde{L}^{\Phi_{c-\lambda u_0}} \subset \tilde{L}^{\Phi_c}$. Therefore, $u \in \tilde{L}^{\Phi_c}$ implies $u + \overline{\lambda}u_0 \in \tilde{L}^{\Phi_{c-\overline{\lambda}u_0}} \subset \tilde{L}^{\Phi_c}$. Or, equivalently, if u is a measurable function such that $\varphi(c+u)$ is μ-integrable, then $\varphi(c+u+\overline{\lambda}u_0)$ is μ-integrable. As a result, we conclude that $\int_T \varphi(c+u+\lambda u_0)d\mu < \infty$ for all $\lambda > 0$. Let $\widetilde{c}\colon T \to \mathbb{R}$ be any measurable function satisfying $\int_T \varphi(\widetilde{c})d\mu < \infty$. Denote $A = \{\widetilde{c} > c\}$. Thus, for each $\lambda > 0$, it follows that

$$\int_T \varphi(\widetilde{c} + \lambda u_0)d\mu = \int_T \varphi(c + (\widetilde{c} - c) + \lambda u_0)d\mu \leq \int_T \varphi(c + (\widetilde{c} - c)\chi_A + \lambda u_0)d\mu < \infty,$$

which shows that u_0 is stated under Condition 2.2. □

From Proposition 2.4 we have that Condition 2.2 is not satisfied if, and only if, there exists a measurable function $u\colon T \to \mathbb{R}$ such that $\int_T \varphi(c+u)d\mu = \infty$ but $\int_T \varphi(c+u-\overline{\lambda}u_0)d\mu < \infty$ for some $\overline{\lambda} > 0$. For our result we make use of the following lemmas.

Lemma 2.3 [11, Lemma 8.3] *Consider a non-atomic and σ-finite measure μ. If $\{u_n\}$ is a sequence of finite-value, non-negative, measurable functions, and $\{\alpha_n\}$ is a sequence of positive, real numbers, such that*

$$\int_T u_n d\mu \geq 2^n \alpha_n, \qquad \textit{for all } n \geq 1,$$

then an increasing sequence $\{n_i\}$ *of natural numbers and a sequence* $\{A_i\}$ *of pairwise disjoint, measurable sets can be found, such that*

$$\int_{A_i} u_{n_i} d\mu = \alpha_{n_i}, \qquad \textit{for all } i \geq 1.$$

For the next lemma we denote the functional $I_{\Phi_c} = \int_T \Phi_c(t, | u(t) |) d\mu$ *for any* $u \in L^0$.

Lemma 2.4 *Consider* $c\colon T \to [0, \infty)$ *a measurable function such that* $\int_T \varphi(c) d\mu < \infty$. *Suppose that, for each* $\lambda > 0$, *we cannot find* $\alpha > 0$ *and* $f \in \tilde{L}^{\Phi_c}$ *such that*

$$\alpha \Phi_c(t, u) \leq \Phi_{c-\lambda u_0}(t, u), \qquad \textit{for all } u > f(t). \tag{2.10}$$

Then, a strictly decreasing sequence $0 < \lambda_n \downarrow 0$, *and sequences* $\{u_n\}$ *and* $\{A_n\}$ *of finite-value, measurable functions, and pairwise disjoint, measurable sets, respectively, can be found such that*

$$I_{\Phi_c}(u_n \chi_{A_n}) = 1, \quad \textit{and} \quad I_{\Phi_{c-\lambda_n u_0}}(u_n \chi_{A_n}) \leq 2^{-n}, \qquad \textit{for all } n \geq 1. \tag{2.11}$$

Proof Let $\{\lambda_m\}$ be a strictly decreasing sequence such that $0 < \lambda_m \downarrow 0$. Define the non-negative functions

$$f_m(t) = \sup\{u > 0 : 2^{-m} \Phi_c(t, u) > \Phi_{c-\lambda_m u_0}(t, u)\}, \text{ for all } m \geq 1,$$

where we adopt the convention that $\sup \emptyset = 0$. Since (2.10) is not satisfied, we have that $I_{\Phi_c}(f_m) = \infty$ for each $m \geq 1$. For every rational number $r > 0$, define the measurable sets

$$A_{m,r} = \{t \in T : 2^{-m} \Phi_c(t, r) > \Phi_{c-\lambda_m u_0}(t, r)\},$$

and the simple functions $u_{m,r} = r \chi_{A_{m,r}}$.

For $r = 0$, we obtain the set $u_{m,r} = 0$. Let $\{r_i\}$ be an enumeration of the non-negative rational numbers with $r_1 = 0$. Define the non-negative, simple functions $v_{m,k} = \max_{1 \leq i \leq k} u_{m,r_i}$, for each $m, k \geq 1$. By the continuity of $\Phi_c(t, \cdot)$ and $\Phi_{c-\lambda_m u_0}(t, \cdot)$, it follows that $v_{m,k} \uparrow f_m$ as $k \to \infty$. From the Monotone Convergence Theorem for each $m \geq 1$, we can find some $k_m \geq 1$ such that the function $v_m = v_{m,k_m}$ satisfies $I_{\Phi_c}(v_m) \geq 2^m$.

Clearly, we have that $\Phi_c(t, v_m(t)) < \infty$ and $2^{-m}\Phi_c(t, v_m(t)) \geq \Phi_{c-\lambda_m u_0}(t, v_m(t))$. By Lemma 2.3, there exist an increasing sequence $\{m_n\}$ of indices and a sequence $\{A_n\}$ of pairwise disjoint, measurable sets such that $I_{\Phi_c}(v_{m_n}\chi_{A_n}) = 1$. Taking $\lambda_n = \lambda_m$, $u_n = v_{m_n}$ and A_n, we obtain (2.11). □

Next proposition ensures that, assuming Condition 2.2 is not satisfied, we have that the boundary of $\mathcal{B}_c^{\varphi}$ is not empty.

Proposition 2.5 *If the deformed exponential function φ does not satisfy Condition 2.2, then $\Phi_c \notin \Delta_2$.*

Proof Suppose that Condition 2.2 does not hold. Take $\lambda > 0$. Then, there exists a $n_0 \in \mathbb{N}$, such that $\lambda > \lambda_n$, for all $n \geq n_0$.

By Proposition 2.4 and Lemma 2.2, we can take $u = \sum_{n=n_0}^{\infty} u_n \chi_{A_n}$. Since $u \in L_c^{\varphi}$ and $I_{\Phi_c}(u) = \infty$, it follows the result. □

Next section analyzes the behavior of the normalizing function near the boundary of the parametrization domain in case the considered points are inside and outside the Musielak–Orlicz class.

2.3.3 Condition 2.2 is Not Satisfied: Behavior of the Normalizing Function

In this section we prove that if Condition 2.2 is not satisfied it is still possible to find $u \in \partial\mathcal{B}_c^{\varphi}$ such that $\int_T \varphi(c+u)d\mu = \infty$ and the normalizing function ψ converges to some finite value β. Moreover, we verify that for points u belonging to Musielak–Orlicz class $\widetilde{L}_c^{\Phi}$, regardless of occurrence of Condition 2.2, the normalizing function converges.

In the next proposition we study the behavior of the ψ in the case the boundary points are not in the Musielak–Orlicz class.

Proposition 2.6 *Assuming that Condition 2.2 is not satisfied in the definition of φ-function, then there exists $u \in \partial\mathcal{B}_c^{\varphi}$ such that $\int_T \varphi(c+u)d\mu = \infty$ but $\psi(\alpha u) \to \beta$, with $\beta \in (0, \infty)$, as $\alpha \uparrow 1$.*

Proof Let $\{\lambda_n\}$, $\{u_n\}$ and $\{A_n\}$ as in Lemma 2.4. Given any $\lambda > 0$, take $n_0 \geq 1$ such that $\lambda \geq \lambda_n$ for all $n \geq n_0$. Denote $B = T \setminus \bigcup_{n=n_0}^{\infty} A_n$, then we define $u = \sum_{n=n_0}^{\infty} u_n \chi_{A_n}$. From Eq. (2.11), it follows that

$$
\begin{aligned}
\int_T \varphi(c + u - \lambda u_0) d\mu &= \int_B \varphi(c - \lambda u_0) d\mu + \sum_{n=n_0}^{\infty} \int_{A_n} \varphi((c - \lambda u_0) + u_n) d\mu \\
&= \int_B \varphi(c - \lambda u_0) d\mu \\
&\quad + \sum_{n=n_0}^{\infty} \left\{ \int_{A_n} \varphi(c - \lambda u_0) d\mu + I_{\Phi_{c-\lambda u_0}}(u_n \chi_{A_n}) \right\} \\
&\le \int_T \varphi(c - \lambda u_0) d\mu + \sum_{n=n_0}^{\infty} 2^{-n} < \infty.
\end{aligned}
$$

Consequently, for $\alpha \in (0, 1)$, we can write

$$
\begin{aligned}
\int_T \varphi(c + \alpha u) d\mu &= \int_T \varphi\Big(c + \alpha(u - \lambda u_0) + (1 - \alpha)\frac{\alpha\lambda}{1-\alpha} u_0\Big) d\mu \\
&\le \alpha \int_T \varphi(c + u - \lambda u_0) d\mu + (1 - \alpha) \int_T \varphi\Big(c + \frac{\alpha\lambda}{1-\alpha} u_0\Big) d\mu \\
&< \infty.
\end{aligned}
$$

On the other hand, for $\alpha \ge 1$, it follows that

$$
\begin{aligned}
\int_T \varphi(c + \alpha u) d\mu &\ge \int_B \varphi(c) d\mu + \sum_{n=n_0}^{\infty} \int_{A_n} \varphi(c + u_n) d\mu \\
&\ge \int_B \varphi(c) d\mu + \sum_{n=n_0}^{\infty} \left\{ \int_{A_n} \varphi(c) d\mu + I_{\Phi_c}(u_n \chi_{A_n}) \right\} \\
&= \int_T \varphi(c) d\mu + \sum_{n=n_0}^{\infty} 1 = \infty.
\end{aligned}
$$

We can choose $\lambda' < 0$ such that

$$
w = \lambda' u_0 \chi_B + \sum_{n=n_0}^{\infty} u_n \chi_{A_n}
$$

satisfies $\int_T w \varphi'_+(c) d\mu = 0$. Clearly, $\int_T \varphi(c + w) d\mu = \infty$, $\int_T \varphi(c + \alpha w) d\mu < \infty$ for $\alpha \in (0, 1)$ and $\int_T \varphi(c + \alpha w) d\mu = \infty$ for $\alpha > 1$, that is, $w \in \partial \mathcal{B}_c^{\varphi}$ and $\int_T \varphi(c + w - \lambda u_0) d\mu < \infty$ for some fixed $\lambda > 0$.

Suppose that $\psi(\alpha w) \uparrow \infty$, then for all $K > 0$, there exists $\delta > 0$ such that $0 < |\alpha - 1| < \delta$ implies that $\psi(\alpha w) > K$. Let $\lambda'' > \lambda$ be such that $\int_T \varphi(c + w - \lambda'' u_0) d\mu < 1$, taking $K = \lambda''$ we have $\varphi(c + \alpha w - \psi(w)u_0) < \varphi(c + \alpha w_{\{w>0\}} - \lambda'' u_0) < \varphi(c + w_{\{w>0\}} - \lambda'' u_0)$, that is a μ-integrable function. Therefore by the Dominated Convergence Theorem we have

$$\lim_{\alpha \uparrow 1} \int_T \varphi(c + \alpha w - \lambda'' u_0) d\mu = \int_T \varphi(c + w - \lambda'' u_0) d\mu,$$

then

$$\begin{aligned} 1 = \lim_{\alpha \uparrow 1} & \int_T \varphi(c + \alpha w - \psi(\alpha w)u_0) d\mu \\ & \leq \lim_{\alpha \uparrow 1} \int_T \varphi(c + \alpha w - \lambda'' u_0) d\mu = \int_T \varphi(c + w - \lambda'' u_0) d\mu < 1, \end{aligned}$$

which is a contradiction.

If the Condition 2.2 is not satisfied, we elucidated the case of $u \in \partial \mathcal{B}_c^{\varphi}$ such that $\int_T \varphi(c + u) d\mu = \infty$.

We can then make the following question: What about the case of $u \in \partial \mathcal{B}_c^{\varphi}$ such that $\int_T \varphi(c + u) d\mu < \infty$, how is the behavior of the normalizing function ψ?

This behavior is elucidated in the next proposition.

Proposition 2.7 *Consider the deformed exponential function φ. Given $u \in \partial \mathcal{B}_c^{\varphi}$ such that $\int_T \varphi(c + u) d\mu < \infty$, we have that $\psi(\alpha u) \to \beta$, with $\beta \in (0, \infty)$ as $\alpha \uparrow 1$.*

Proof In fact, since $\int_T \varphi(c + u) d\mu < \infty$, we have $\int_T \varphi(c + u - \lambda u_0) d\mu < \infty$ for all $\lambda > 0$. Suppose that $\psi(\alpha u) \uparrow \infty$, as $\alpha \uparrow 1$. Then, for all $A > 0$, there exists $\delta > 0$, such that $0 < 1 - \alpha < \delta \Rightarrow \psi(\alpha u) > A$.

Since $\int_T \varphi(c + u - \lambda u_0) d\mu < \infty$ for all $\lambda > 0$, we have that there exists $\gamma > \lambda$, such that $\int_T \varphi(c + u - \gamma u_0) d\mu < 1$. In particular, take $A = \gamma$. Then, from the Dominated Convergence Theorem it follows that

$$\begin{aligned} 1 = & \lim_{\alpha \uparrow 1} \int_T \varphi(c + \alpha u - \psi(\alpha u)u_0) d\mu \\ \leq & \lim_{\alpha \uparrow 1} \int_T \varphi(c + \alpha u - \gamma u_0) d\mu \\ = & \int_T \varphi(c + u - \gamma u_0) d\mu \\ < & 1, \end{aligned}$$

which is an absurd. Therefore, we obtain the desired result. □

2.4 Conclusions

In this chapter, we investigated the behavior of the normalizing function near the boundary of the domain of the parametrization. When Condition 2.2 occurs it was proven that given any u belonging to the Musielak–Orlicz class, we have that $\psi(\alpha u)$ converges when α tends to 1 and when u does not belong to the Musielak–Orlicz class, it follows that $\psi(\alpha u)$ diverges when alpha tends to 1.

Assuming that Condition 2.2 does not occur, we proved that there exists a function u not belonging the Musielak–Orlicz class such that $\psi(\alpha u)$ converges when α tends to 1. Moreover, we observe that regardless the occurence of Condition 2.2, whenever u is in the Musielak–Orlicz class, $\psi(\alpha u)$ converges since α tends to 1.

These results allow a more general class of normalizing functions and describe the behaviour of it in the boundary domain and therefore make possible the characterization of a wider class of probability distributions for several conditions associated to the convergence of the normalizing function.

Acknowledgements This work was supported by Coordenação de Aperfeiçoamento de Pessoal de Nível Superior - Brasil (CAPES) - Finance Code 001 and Conselho Nacional de Desenvolvimento Científico e Tecnológico (CNPq) (Procs. 309472/2017-2 and 408609/2016-8).

Disclaimer: Views and opinions expressed are those of the authors and do not necessarily represent official positions of their respective companies.

References

1. Andrade, L.H.F.d., Vigelis, R.F., Vieira, F.L.J., Cavalcante, C.C.: Normalization and ϕ-function: Definition and Consequences, pp. 231–238. Springer International Publishing, Cham (2017)
2. Andrade, L.H., Vieira, F.L., F.Vigelis, R., Cavalcante, C.C.: Mixture and exponential arcs on generalized statistical manifold. Entropy **20**(3), 147 (2018)
3. Cena, A., Pistone, G.: Exponential statistical manifold. Ann. Inst. Statist. Math. **59**(1), 27–56 (2007). https://doi.org/10.1007/s10463-006-0096-y
4. J. Newton, N.: An infinite-dimensional statistical manifold modelled on Hilbert space. J. Funct. Anal. **263**(6), 1661–1681 (2012). https://doi.org/10.1016/j.jfa.2012.06.007
5. Kaniadakis, G.: Non-linear kinetics underlying generalized statistics. Physica A: Stat. Mech. Appl. **296**(3), 405–425 (2001). https://doi.org/10.1016/S0378-4371(01)00184-4
6. Krasnoseli'skii, M.A., Rutickii, J.B.: Convex functions and Orlicz spaces. Translated from the first Russian edition by Leo F. Boron, P. Noordhoff Ltd., Groningen (1961)
7. Loaiza, G., Quiceno, H.R.: A q-exponential statistical Banach manifold. J. Math. Anal. Appl. **398**(2), 466–476 (2013). https://doi.org/10.1016/j.jmaa.2012.08.046
8. Matsuzoe, H., Scarfone, A.M., Wada, T.: Normalization problems for deformed exponential families. In: Nielsen, F., Barbaresco, F. (eds.) Geometric Science of Information, pp. 279–287. Springer International Publishing, Cham (2019)
9. Matsuzoe, H., Wada, T.: Deformed algebras and generalizations of independence on deformed exponential families. Entropy **17**(8), 5729–5751 (2015). https://doi.org/10.3390/e17085729
10. Montrucchio, L., Pistone, G.: A Class of Non-parametric Deformed Exponential Statistical Models, pp. 15–35. Springer International Publishing, Cham (2019)

11. Musielak, J.: Orlicz spaces and modular spaces. Lecture Notes in Mathematics, vol. 1034. Springer, Berlin (1983)
12. Naudts, J.: Deformed exponentials and logarithms in generalized thermostatistics. Physica A: Stat. Mech. Appl. **316**(1), 323–334 (2002). https://doi.org/10.1016/S0378-4371(02)01018-X
13. Naudts, J.: Estimators, escort probabilities, and ϕ-exponential families in statistical physics. JIPAM. J. Inequal. Pure Appl. Math. **5**(4), Article 102, 15 (electronic) (2004)
14. Naudts, J.: Generalised Thermostatistics. Springer, London Ltd, London (2011). https://doi.org/10.1007/978-0-85729-355-8
15. Naudts, J., Zhang, J.: Rho-tau embedding and gauge freedom in information geometry. Inf. Geometry **1**(1), 79–115 (2018). https://doi.org/10.1007/s41884-018-0004-6
16. Pistone, G.: κ-exponential models from the geometrical viewpoint. Eur. Phys. J. B **70**(1), 29–37 (2009). https://doi.org/10.1140/epjb/e2009-00154-y
17. Pistone, G., Rogantin, M.P., et al.: The exponential statistical manifold: mean parameters, orthogonality and space transformations. Bernoulli **5**(4), 721–760 (1999)
18. Pistone, G., Sempi, C.: An infinite-dimensional geometric structure on the space of all the probability measures equivalent to a given one. Ann. Statist. **23**(5), 1543–1561 (1995). https://doi.org/10.1214/aos/1176324311
19. Santacroce, M., Siri, P., Trivellato, B.: New results on mixture and exponential models by Orlicz spaces. Bernoulli **22**(3), 1431–1447 (2016). https://doi.org/10.3150/15-BEJ698
20. de Souza, D.C., Vigelis, R.F., Cavalcante, C.C.: Geometry induced by a generalization of rényi divergence. Entropy **18**(11), 407 (2016). https://doi.org/10.3390/e18110407
21. Tsallis, C.: Introduction to Nonextensive Statistical Mechanics: Approaching a Complex World, 1st edn. Springer, New York (2009)
22. Umarov, S., Tsallis, C., Steinberg, S.: On a q-central limit theorem consistent with nonextensive statistical mechanics. Milan J. Math. **76**, 307–328 (2008). https://doi.org/10.1007/s00032-008-0087-y
23. Vieira, F.L.J., Vigelis, R.F., de Andrade, L.H.F., Cavalcante, C.C.: Deformed exponential and the behavior of the normalizing function. In: Nielsen, F., Barbaresco, F. (eds.) Geometric Science of Information, pp. 271–278. Springer International Publishing, Cham (2019)
24. Vigelis, R.F., de Andrade, L.H.F., Cavalcante, C.C.: On the Existence of Paths Connecting Probability Distributions, pp. 801–808. Springer International Publishing, Cham (2017)
25. Vigelis, R.F., de Andrade, L.H.F., C.Cavalcante, C.: Conditions for the existence of a generalization of Rényi divergence. Physica A: Stat. Mech. Appl. **558**, 124953 (2020). https://doi.org/10.1016/j.physa.2020.124953
26. Vigelis, R.F., Cavalcante, C.C.: The Δ_2-condition and ϕ-families of probability distributions. In: Geometric Science of Information, Lecture Notes in Computer Science, vol. 8085, pp. 729–736. Springer, Heidelberg (2013)
27. Vigelis, R.F., Cavalcante, C.C.: On ϕ-families of probability distributions. J. Theoret. Probab. **26**(3), 870–884 (2013). https://doi.org/10.1007/s10959-011-0400-5
28. Vigelis, R.F., Cavalcante, C.C.: Smoothness of the Orlicz norm in Musielak-Orlicz function spaces. Math. Nachr. **287**(8–9), 1025–1041 (2014)
29. Zhang, J., Hästö, P.: Statistical manifold as an affine space: a functional equation approach. J. Math. Psych. **50**(1), 60–65 (2006). https://doi.org/10.1016/j.jmp.2005.08.003

Chapter 3
Affine Connections with Torsion in (Para-)complexified Structures

Jun Zhang and Gabriel Khan

Abstract We investigate integrability conditions for an almost (para-)complex structure L, $L^2 = \pm\text{id}$, on manifolds that admit affine connections carrying torsion in general. The affine connections ∇ under consideration are not assumed to be (para-)complex $\nabla L \neq 0$. Two kinds of torsion matching conditions leading to integrability of L are analyzed, which are generalizations of the first and second canonical connections along the Gauduchon line. We also discuss (para-)holomorphicity of ∇ encoded by the second condition.

3.1 Introduction

On a differentiable manifold $\mathcal{M}$ of even dimension, one can separately consider three entities all taking inputs from sections $\Xi(T\mathcal{M})$ of the tangent bundle $T\mathcal{M}$—an affine connection $\nabla : \Xi(T\mathcal{M}) \times \Xi(T\mathcal{M}) \to \Xi(T\mathcal{M})$, a pseudo-Riemannian metric $g : \Xi(T\mathcal{M}) \times \Xi(T\mathcal{M}) \to C\mathcal{M}$, and a tangent bundle endomorphism-induced operator $L : \Xi(T\mathcal{M}) \to \Xi(T\mathcal{M})$. Here and below, we use L to denote either an *almost complex* operator J, $J^2 = -\text{id}$, or an *almost para-complex* operator K, $K^2 = \text{id}$ assuming the multiplicity of the $\pm$ 1 eigenvalues are equal. We use the term "(para-) complex" to mean "either complex or para-complex", corresponding to when $L = J$ or when $L = K$.

Enforcing compatibility of L with g

$$g(LX, Y) + g(X, LY) = 0, \tag{3.1}$$

J. Zhang (✉) · G. Khan
University of Michigan, Ann Arbor, USA
e-mail: junz@umich.edu

G. Khan
e-mail: gkhan@iastate.edu

G. Khan
Iowa State University, Ames, USA

F. Nielsen (ed.), *Progress in Information Geometry*,
Signals and Communication Technology,
https://doi.org/10.1007/978-3-030-65459-7_3

for any vector fields X, Y on $\mathcal{M}$, that is, $X, Y \in \Xi(T\mathcal{M})$, leads to an almost (para-) Hermitian structure of $(\mathcal{M}, g, L)$. Here, g is positive-definite and hence a Riemannian metric when $L = J$, and is of split signature and hence a Norden metric when $L = K$. When L is integrable, $(\mathcal{M}, g, L)$ becomes a *(para-)Hermitian manifold*. Compatibility of g and ∇ has been investigated in the context of affine differential geometry and information geometry, through the notion of "Codazzi-coupling" (definition to follow). A manifold $\mathcal{M}$ equipped with a metric g and a torsion-free connection ∇ is called a *statistical manifold* if (g, ∇) is Codazzi-coupled [1]. Finally, compatibility of L with (a torsion-free) ∇ has also been studied, in terms of Codazzi-coupling between them (e.g. [2, 3]). In [4], it was further shown that when a torsion-free connection ∇ is Codazzi-coupled both to g and to L, then $(\mathcal{M}, g, L)$ in fact turns into a (para-)Kähler manifold. Such a (para-)Kähler manifold is called *Codazzi-(para-)Kähler* manifold [4] because of the additional structure imposed by ∇. A special case is the class of "special Kähler manifold" of [2], with ∇ assumed to be curvature-free in addition to being torsion-free.

Given (g, L) on $\mathcal{M}$, one can define g-conjugate transformation (denoted ∇^*) and L-gauge transformation (denoted ∇^L), respectively, given any connection ∇ on $\mathcal{M}$. These transformations are involutive $(\nabla^*)^* = \nabla = (\nabla^L)^L$. Details were in [4] and will be given below. When Eq. (3.1) is satisfied, then these two transformations of ∇ are commutative $(\nabla^*)^L = (\nabla^L)^* \equiv \nabla^\dagger$, so that $\{\text{id}, *, L, \dagger\}$ form a 4-element Klein group of transformation of affine connections on $\mathcal{M}$ (Theorem 2.13 of [4]). Under this scenario, $\nabla^\dagger$ then becomes the conjugate connection of ∇ with respect to the fundamental form ω defined by $\omega(X, Y) = g(LX, Y)$; here (g, L, ω) is known as the "compatible triple."

How this quadruple of connections $(\nabla, \nabla^*, \nabla^L, \nabla^\dagger)$ interact with g and L on an almost (para-)Hermitian manifold $(\mathcal{M}, g, L)$ deserves further investigation. Since Codazzi-coupling of ∇ with either g or L ensures that ∇^* or ∇^L will have same torsion as that of ∇, enforcing Codazzi couplings of a torsion-free ∇ *both* with g *and* with L will lead to vanishing torsion for the entire quadruple of connections $(\nabla, \nabla^*, \nabla^L, \nabla^\dagger)$. In this case (see [4]) $\nabla^\dagger = \nabla$, so Codazzi-(para-)Kähler manifolds admit pairs of torsion-free connections. These manifolds are both *(para-)Kähler* manifolds (with integrable L and d-closed ω) and *statistical* manifolds (with a family of α-connections [5]). Manifolds where $\nabla^\dagger = \nabla$, or equivalently $\nabla\omega = 0$, are called *holomorphic statistical manifold* [6] (a notion originally due to Takashi Kurose) when torsion of ∇ is zero; in this case, ∇ is a symplectic connection. So Codazzi-(para-)Kähler manifolds are important examples of holomorphic statistical manifolds.

In a recent paper, Grigorian and Zhang [7] relaxed the restriction of torsion-freeness of ∇, and investigated integrable structures on an almost (para-)Hermitian manifold $(\mathcal{M}, g, L)$ that nevertheless admits affine connections with torsion. Connections with torsion on Hermitian manifolds have always been a topic of interest, e.g. [8, 9]. The work of [4, 7] investigated connections that are not necessarily parallel to L, $\nabla L \neq 0$. That is, neither ∇ nor ∇^L is (para-)complex, though $\frac{1}{2}(\nabla + \nabla^L)$ is always parallel to L:

$$\frac{1}{2}(\nabla + \nabla^L)L = 0.$$

Along with $\nabla^L := L^{-1}\nabla L$, the L-conjugate transformation of ∇, we can define *torsion-coupling* of (∇, L) as the following relation (Definition 1) between the torsions T^∇ and T^{∇^L} of ∇ and ∇^L:

$$T^\nabla(LX, Y) = L(T^{\nabla^L}(X, Y)). \tag{3.2}$$

Torsion-coupling is equivalent to Codazzi-coupling when torsion of ∇ vanishes. So torsion-coupling of ∇ with L provides an appropriate generalization to impose Codazzi-like coupling requirement for connections with non-vanishing torsion. It is shown ([7], Theorem 6) that ∇ is torsion-coupled to L if any only L is integrable *and* ∇ is (para-)holomorphic. Moreover, given a (para-)Hermitian manifold $(\mathcal{M}, g, L)$ with four connections $\nabla, \nabla^L, \nabla^*, \nabla^\dagger \equiv \nabla^{*L} = \nabla^{L*}$, all possibly carrying torsion, if any of the four is torsion-coupled to L and any of the four is Codazzi-coupled to g, then *all* four must be torsion-coupled to L *and* Codazzi-coupled to g ([7], Theorem 12). This leads to the definition of a *Codazzi-(para-)Hermitian* structure–which is a (para-)Hermitian manifold with all four of the aforementioned connections being (para-)holomorphic and Codazzi-coupled to g. In this case, their torsions satisfy $T = T^*$, $T^L = T^\dagger$, but $d\omega \neq 0$ in general, unless the torsions of the quadruple of the connections are all zero.

The torsion-coupling condition Eq. (3.2) on an almost (para-)complex manifold $(\mathcal{M}, L)$, as first investigated in [7], is interesting because the same equation (called the *2nd Matching Condition* or MC2 below) encodes two things simulataneously, both a constraint on L, i.e., L should be integrable, and a constraint on ∇, i.e., ∇ should be (para-)holomorphic. In this paper, we investigate this torsion-coupling condition further—by decomposing this condition into two parts, one which implies integrability of L, and another which implies (para-)holomorphicity of ∇. We investigate another coupling of L with ∇ (called the *1st Matching Condition* or MC1 below) that would lead to integrability condition of L and certain properties of ∇ in reference to the Gauduchon line [8]. Our goal in this paper is to understand how torsion of a connection interacts with a (para-)complex structure L on a manifold when the connection is (in general) not required to be parallel with respect to L.

3.2 Torsion of ∇ and Integrability of L

3.2.1 L Conjugation of ∇

Given a real manifold $\mathcal{M}$ of even dimensions $\dim \mathcal{M} = 2n$, we study affine connection ∇ (in general carrying torsion and curvature) on $\mathcal{M}$, i.e., ∇ operates on sections $\Xi(T\mathcal{M})$ of its tangent bundle $T\mathcal{M}$, i.e., vector fields with real dimension $2n$. Let $L : \Xi(T\mathcal{M}) \to \Xi(T\mathcal{M})$ be an a map induced by tangent bundle endomorphism, with L^{-1} denoting the inverse map. Then, the L-gauge transformed connection is defined as $\nabla^L := L^{-1} \circ \nabla \circ L$, or

$$\nabla^L_X Y = L^{-1}(\nabla_X(LY)) \tag{3.3}$$

for any vector fields X and Y on $\mathcal{M}$, i.e., X and Y are elements of $\Xi(T\mathcal{M})$. That ∇^L is indeed an affine connection needs verification, which we omit here. The L-gauge transformation $\nabla \to \nabla^L$ of any connection ∇ satisfies

$$(\nabla^{L_1})^{L_2} = \nabla^{L_1 \circ L_2},$$

with $\circ$ denoting the composition of L_1 and L_2 as operators. Identifying $\circ$ as the group multiplication operation and the identity transformation $L = \text{id}$ as the group identity element, L-gauge transformation of affine connections form a transformation group acting on the space of linear connections on $\mathcal{M}$.

It can be seen that $\nabla = \nabla^L$ if and only if $\nabla L = 0$. A connection ∇ is called *(para-)complex* if $\nabla L = 0$. In this paper, we do not generally make the assumption of (para-)complexity of ∇.

We define a vector-valued skew-symmetric bilinear form S, a $(1, 2)$-tensor, via the expression

$$S(X, Y) = (\nabla_X L)Y - (\nabla_Y L)X, \tag{3.4}$$

where

$$(\nabla_X L)Y = \nabla_X(LY) - L(\nabla_X Y).$$

The pair (∇, L) is said to be *Codazzi-coupled* if $S = 0$.

Recall that the torsion T^∇ of ∇ is defined as:

$$T^\nabla(X, Y) = \nabla_X Y - \nabla_Y X - [X, Y]. \tag{3.5}$$

We easily derive the identify

$$S(X, Y) = L(T^{\nabla^L}(X, Y) - T^\nabla(X, Y)). \tag{3.6}$$

This leads to the following well-known results.

Proposition 1 *(e.g., [10]) Let ∇ be an affine connection, and let L be an arbitrary tangent bundle isomorphism. Then the following statements are equivalent:*

(i) (∇, L) *is Codazzi-coupled.*
(ii) $T^\nabla(X, Y) = T^{\nabla^L}(X, Y)$.
(iii) (∇^L, L^{-1}) *is Codazzi-coupled.*

Corollary 1 *For the special case of (para-)complex operators* $L^2 = \pm\text{id}$,

(a) $\nabla^L = \nabla^{L^{-1}}$, *i.e., L-conjugate transformation is involutive,* $(\nabla^L)^L = \nabla$.
(b) (∇, L) *is Codazzi-coupled if and only if* (∇^L, L) *is Codazzi-coupled.*

Codazzi-coupling of (∇, L) on an almost-(para-)complex manifold $(\mathcal{M}, L)$ mirrors Codazzi-coupling of (∇, g) on a Riemannian manifold $(\mathcal{M}, g)$.

3.2.2 Nijenhuis Tensor N_L and Integrability

On an (even-dimensional) manifold $\mathcal{M}$ with an almost (para-)complex structure L, there is an associated Nijenhuis tensor

$$N_L(X, Y) = -L^2[X, Y] + L[X, LY] + L[LX, Y] - [LX, LY]. \tag{3.7}$$

When $N_L = 0$, the operator L is said to be integrable. Via the celebrated Newlander-Nirenberg theorem, in the complex case $L = J$, this is the obstruction to turning an almost complex structure J into a complex structure. In other words, vanishing of N_J is equivalent to $\mathcal{M}$ admitting local holomorphic charts, with J as the complex structure.

The following identity for the Nijenhuis tensor can be easily verified.

Lemma 1 *The Nijenhuis tensor N_L satisfies*

$$N(LX, Y) = N(X, LY) = -LN(X, Y).$$

Proof Because $L^2 = \pm\text{id}$, it can be taken out of the bracket,

$$\begin{aligned} N(LX, Y) &= -L^2[LX, Y] + L[LX, LY] + L[L^2X, Y] - [L^2X, LY] \\ &= -L^2[LX, Y] + L[LX, LY] + L^3[X, Y] - L^2[X, LY] \\ &= -L(L[LX, Y] - [LX, LY] - L^2[X, Y] + L[X, LY]) \\ &= -LN(X, Y). \end{aligned}$$

Likewise, $N(X, LY) = -LN(X, Y)$.

Proposition 2 *The Nijenhuis tensor N_L can be expressed in the follow forms:*

$$N_L(X, Y) = \left(T^{\nabla^L}(L(LX), Y) - LT^{\nabla}(LX, Y)\right) + \left(T^{\nabla^L}(LX, LY) - LT^{\nabla}(X, LY)\right) \tag{3.8}$$

and

$$N_L(X, Y) = \left(T^{\nabla^L}(L(LX), Y) + T^{\nabla^L}(LX, LY)\right) - L\left(T^{\nabla}(LX, Y) + T^{\nabla}(X, LY)\right). \tag{3.9}$$

Proof The relationships were derived in Lemma 3 [7]. We prove them below (to correct certain typos there).

Starting from Eqs. (3.4), (3.5), and (3.6), we have

$$\begin{aligned} \nabla_X(LY) - \nabla_Y(LX) &= (\nabla_X L)Y + L(\nabla_X Y) - ((\nabla_Y L)X + L(\nabla_Y X)) \\ &= S(X, Y) + L(\nabla_X Y - \nabla_Y X) \\ &= S(X, Y) + L([X, Y] + T^{\nabla}(X, Y)) \\ &= LT^{\nabla^L}(X, Y) + L[X, Y]. \end{aligned} \tag{3.10}$$

Substituting $X \to LX, Y \to LY$ in Eq. (3.10) yields

$$\nabla_{LX}(L^2Y) - \nabla_{LY}(L^2X) = LT^{\nabla^L}(LX, LY) + L[LX, LY],$$

or, after cancelling an L on both sides

$$L(\nabla_{LX}Y - \nabla_{LY}X) = T^{\nabla^L}(LX, LY) + [LX, LY].$$

Applying L operator on both sides of Eq. (3.10) and adding to the above, we have

$$\begin{aligned} &L(\nabla_X(LY) - \nabla_Y(LX) + \nabla_{LX}Y - \nabla_{LY}X) \\ &= L^2 T^{\nabla^L}(X, Y) + L^2[X, Y] + T^{\nabla^L}(LX, LY) + [LX, LY], \end{aligned}$$

or

$$\begin{aligned} &L([LX, Y] + T^{\nabla}(LX, Y) + [X, LY] + T^{\nabla}(X, LY)) \\ &= L^2 T^{\nabla^L}(X, Y) + T^{\nabla^L}(LX, LY) + L^2[X, Y] + [LX, LY]. \end{aligned}$$

Rearranging the terms yields

$$\begin{aligned} &-L^2[X, Y] + L[LX, Y] + L[X, LY] - [LX, LY] \\ &= L^2 T^{\nabla^L}(X, Y) + T^{\nabla^L}(LX, LY) - LT^{\nabla}(LX, Y) - LT^{\nabla}(X, LY). \end{aligned}$$

The lefthand side is nothing but $N_L(X, Y)$. So we obtain Eqs. (3.8) and (3.9).

3.2.3 MC1 Versus MC2

A careful examination of expressions (3.8) and (3.9) reveals two different scenarios under which N_L vanishes—these scenarios will be referred to as *1st Matching Condition* (MC1) and *2nd Matching Condition* (MC2) of (∇, L).

Definition 1 A pair (∇, L) is said to satisfy

(i) *1st Matching Condition*, or MC1, if the following holds:

$$T^{\nabla}(LX, Y) + T^{\nabla}(X, LY) = 0; \tag{3.11}$$

(ii) *2nd Matching Condition*, or MC2, if the following holds:

$$T^{\nabla}(LX, Y) = LT^{\nabla}(X, Y) + (\nabla_X L)Y - (\nabla_Y L)X, \tag{3.12}$$

or equivalently,

$$T^{\nabla}(LX, Y) = L(T^{\nabla^L}(X, Y)). \tag{3.13}$$

Remark. *1st Matching Condition* of torsion of ∇ with L, or MC1, resembles the compatibility condition between a symplectic form ω and L, namely, $\omega(LX, Y) = \omega(X, LY)$. For this reason, we also refer to MC1 as "torsion-compatibility." *2nd Matching Condition* of torsion of ∇ with L, or MC2, was referred to as "torsion-coupling" in [7], where a comprehensive study was conducted on how such coupling interacts with Codazzi coupling of (∇, g) in almost (para-)Hermtian manifolds $(\mathcal{M}, g, L)$.

Proposition 3 *The Nijenhuis tensor N_L vanishes if*

(a) MC1 (torsion-compatibility) holds for both ∇ and ∇^L; or
(b) MC2 (torsion-coupling) holds for either ∇ or ∇^L.

Proof With respect to statement (a), substituting the following torsion-compatibility conditions

$$T^{\nabla}(LX, Y) + T^{\nabla}(X, LY) = 0$$
$$L^2 T^{\nabla^L}(X, Y) + T^{\nabla^L}(LX, LY) = 0$$

into Eq. (3.9) leads to $N_L = 0$. With respect to statement (b), substituting Eq. (3.13), the torsion-coupling condition, into Eq. (3.8) leads to $N_L = 0$.

Proposition 4 *With respect to the following statements*

(i) $N_L = 0$;
(ii) (∇, L) satisfies MC1, i.e., Eq. (3.11);
(iii) (∇^L, L) satisfies MC1;
(iv) (∇, L) satisfies MC2, i.e., Eq. (3.12) or Eq. (3.13);
(v) (∇^L, L) satisfies MC2;

we have

(a) (iv) and (v) are equivalent;
(b) (iv) or (v) implies (i);
(c) any two of (i), (ii), (iii) implies the third;
(d) either (iv) or (v) together with (ii) or (iii) imply that

$$T^{\nabla} = T^{\nabla^L} = 0.$$

Proof That (∇, L) satisfies MC2 means

$$T^{\nabla^L}(LX, Y) = L^{-1}T^{\nabla}(L(LX), Y) = L^{-1}(T^{\nabla}(L^2X, Y)) = L(T^{\nabla}(X, Y)).$$

So, that (∇, L) satisfies MC2 is equivalent to that (∇^L, L) satisfies MC2. Hence (iv) and (v) are equivalent statements. And each implies $N_L = 0$.

That (ii) plus (iii) imply (i) is what statement (a) of Proposition 3 indicates. To show (i) and (ii) leads to (iii), we only need to mention that having assumed (ii), the formula for the Nijenhuis tensor reduces to

$$N_L(X,Y) = L^2 T^{\nabla^L}(X,Y) + T^{\nabla^L}(LX, LY).$$

Further assuming (i) leads to (iii).

Finally, we show that (iv) will lead to $T^\nabla(LX,Y) = T^\nabla(X, LY)$:

$$T^\nabla(LX,Y) = LT^{\nabla^L}(X,Y) = -LT^{\nabla^L}(Y,X) = -T^\nabla(LY,X) = T^\nabla(X,LY).$$

And likewise, $T^{\nabla^L}(LX,Y) = T^{\nabla^L}(X,LY)$. Therefore, further imposing (ii) or (iii) forces $T^\nabla = T^{\nabla^L} = 0$.

Remark. In the special case $\nabla = \nabla^L$, then ∇ is a (para-)complex connection $\nabla L = 0$. In this case, the last two terms of the righthand side of Eq. (3.12) vanish. So $T^\nabla(LX,Y) = LT^\nabla(X,Y)$. A complex connection that is metric and satisfies Eq. (3.12) (torsion-coupling condition) is known as a *second canonical connection* for Hermitian manifolds. It is also called the *Chern connection.* On the other hand, a complex connection that is metric and satisfies Eq. (3.11) (torsion-compatibility condition), $T^\nabla(LX,Y) + T^\nabla(X,LY) = 0$, is known as a *first canonical connection* for Hermitian manifolds. So our definitions of MC1 (torsion-compatibility) and MC2 (torsion-coupling) generalize the appropriate definitions of first and second canonical connections with torsion, to non-(para-)complex connections with torsion. Part (d) of Proposition 4 further claims that MC1 and MC2 are genuinely different types of coupling, if ∇ and ∇^L must carry non-zero torsions.

3.3 Torsion of ∇ Under (Para-)complexification

3.3.1 Splitting of $T\mathcal{M} \otimes \mathbb{L}$ by L

For a smooth manifold $\mathcal{M}$, an endomorphism of the tangent bundle $T\mathcal{M}$ induces a smooth map L of the tangent bundle. By definition, L is called an *almost complex structure* if $L^2 = -\text{id}$, or an *almost para-complex structure* if $L^2 = \text{id}$ and the multiplicities of the eigenvalues ± 1 are equal. We have used J and K to denote almost complex structures and almost para-complex structures, respectively, and use L when these two structures can be treated in a unified way. It is clear from our definition that such structures exist only when $\mathcal{M}$ is of even dimension.

Denote eigenvalues of L as $\pm\alpha$, where $\alpha = 1$ for $L = K$ and $\alpha = i$ for $L = J$, depending on the nature of L. We have

$$L^4 = \text{id}, \quad \alpha^4 = 1, \quad \alpha^2 L^2 = \text{id}.$$

Following the standard procedure, we (para-)complexify $T\mathcal{M}$ by tensoring with complex field $\mathbb{C}$ or the para-complex (a.k.a. split-complex) field $\mathbb{D}$, and use $T^{\mathbb{L}}\mathcal{M}$ to denote the resulting $T\mathcal{M} \otimes \mathbb{C} = \bigcup_{p\in\mathcal{M}} T_p\mathcal{M} \otimes \mathbb{C}$ or $T\mathcal{M} \otimes \mathbb{D} = \bigcup_{p\in\mathcal{M}} T_p\mathcal{M} \otimes \mathbb{D}$, depending on the type of L. In analogy with standard notation in the complex case, let $T_p^+\mathcal{M}$ and $T_p^-\mathcal{M}$ be the eigenbundles of L corresponding to the eigenvalues $\pm\alpha$, i.e., at each point $p \in \mathcal{M}$, the fiber is defined by

$$T_p^+\mathcal{M} := \{Z \in T_p^{\mathbb{L}}\mathcal{M} : L_p(Z) = \alpha Z\},$$
$$T_p^-\mathcal{M} := \{Z \in T_p^{\mathbb{L}}\mathcal{M} : L_p(Z) = -\alpha Z\}.$$

with

$$T^{\mathbb{L}}\mathcal{M} \equiv T\mathcal{M} \otimes \mathbb{L} = \bigcup_{p\in\mathcal{M}} T_p^+\mathcal{M} \oplus T_p^-\mathcal{M} = T^+\mathcal{M} \oplus T^-\mathcal{M}.$$

We will refer to vectors to be of "type +" or $(1,0)$, or "type −" or $(0,1)$, if they take values in $T_p^+\mathcal{M}$ or $T_p^-\mathcal{M}$, respectively. Moreover, define π^+ and π^- to be the projections of a (para-)complexified vector field to $T^+\mathcal{M} = \bigcup_p T_p^+\mathcal{M}$ and $T^-\mathcal{M} = \bigcup_p T_p^-\mathcal{M}$, respectively:

$$\pi^+(Z) = \frac{1}{2}(Z + \alpha^{-1}LZ), \qquad \pi^-(Z) = \frac{1}{2}(Z - \alpha^{-1}LZ).$$

As subbundles of the (para-)complexified tangent bundle $T^{\mathbb{L}}\mathcal{M}$, $T^+\mathcal{M}$ and $T^-\mathcal{M}$ are distributions. A distribution is called a foliation if it is closed under the Lie bracket $[\cdot,\cdot]$.

Let $T_p^{(n,m)}\mathcal{M}$ denotes the tensor product of n-copies of $T_p^+\mathcal{M} \equiv T_p^{(1,0)}\mathcal{M}$ and m-copies of $T_p^-\mathcal{M} \equiv T_p^{(0,1)}\mathcal{M}$:

$$T_p^{(n,m)}\mathcal{M} = \underbrace{(T_p^+\mathcal{M} \times \cdots \times T_p^+\mathcal{M})}_{n \text{ times}} \times \underbrace{(T_p^-\mathcal{M} \times \cdots \times T_p^-\mathcal{M})}_{m \text{ times}}.$$

In particular, $T_p^{(2,0)}\mathcal{M}$ will supply two vectors fields both drawn from $T_p^+\mathcal{M}$, $T_p^{(0,2)}\mathcal{M}$ will supply two vectors fields both drawn from $T_p^-\mathcal{M}$, and $T_p^{(1,1)}\mathcal{M}$ will supply two vectors fields drawn respectively from $T_p^+\mathcal{M}$ and $T_p^-\mathcal{M}$. That is, the duplet vector fields (X, Y) take in value, respectively, from $(+,+)$, from $(+,-)$ or $(-,+)$, or from $(-,-)$, which are shorthand notations for $T_p^{(2,0)}\mathcal{M}$, $T_p^{(1,1)}\mathcal{M}$, or $T_p^{(0,2)}\mathcal{M}$.

It is important to understand what (para-)complexification accomplishes. Take $L = J$, for example. Because $J^2 = -\text{id}$, the J operator on a $dim(\mathcal{M}) = 2n$ dimensional vector space (i.e., J maps a real-valued $2n$-vector to a real-valued $2n$-vector) does not have any eigenvector over the real field $\mathbb{R}$: that is, there is no $X \in T_p\mathcal{M}$ such that $JX = \alpha X$ where $\alpha = \pm i$ are the two eigenvalues of the $2n \times 2n$ real-valued operator J.

Given a vector $X \in T_p\mathcal{M}$ on the real even-dimensional manifold, (para-)complexification amounts to supplying it with an "imaginary" component, to yield an element $Z \in T_p^{\mathbb{L}}\mathcal{M} \equiv T_p\mathcal{M} \otimes \mathbb{L}$. Even better, we can turn X into either X^+ or X^-, where

$$X^+ = X + \alpha^{-1}LX \in T_p^+\mathcal{M}, \quad X^- = X - \alpha^{-1}LX \in T_p^-\mathcal{M}.$$

The (1,0) or "plus"-eigenspace $T_p^+\mathcal{M}$ and the (0,1) or "minus"-eigenspace $T_p^-\mathcal{M}$ are formed, respectively, by such X^+ and X^- having been supplemented with the imaginary parts $\alpha^{-1}LX$ or $-\alpha^{-1}LX$:

$$T_p^+\mathcal{M} := \{X^+ = X + \alpha^{-1}LX : X \in T_p\mathcal{M}\},$$
$$T_p^-\mathcal{M} := \{X^- = X - \alpha^{-1}LX : X \in T_p\mathcal{M}\}.$$

Note that for any (para-)complexified vector field $Z \in T^{\mathbb{L}}\mathcal{M}$, the vector field $Z^+ = Z + \alpha^{-1}LZ$ is always in $T_p^+\mathcal{M}$ and the vector field $Z^- = Z - \alpha^{-1}LZ$ is always in $T_p^-\mathcal{M}$. This can be easily shown:

$$LZ^+ = \alpha Z^+, \quad LZ^- = -\alpha Z^-$$

because

$$L(Z + \alpha^{-1}LZ) \equiv \alpha(Z + \alpha^{-1}LZ), \qquad L(Z - \alpha^{-1}LZ) \equiv -\alpha(Z - \alpha^{-1}LZ).$$

Thus, any vector $Z \in T_p^{\mathbb{L}}\mathcal{M}$ can be readily decomposed into $T_p^+\mathcal{M}$ and $T_p^-\mathcal{M}$. Without loss of generality, $T_p^+\mathcal{M}$ is formed by vectors of the form $X^+ = X + \alpha^{-1}LX$, and $T_p^-\mathcal{M}$ of vectors of the form $X^- = X - \alpha^{-1}LX$, when X exhausts the real vector field as sections of $T_p\mathcal{M}$.

3.3.2 *(Para-)holomorphicity of* ∇

The splitting of $T\mathcal{M} \otimes \mathbb{L}$ by L into direct sum of $T_p^+\mathcal{M}$ and $T_p^-\mathcal{M}$ subbundles gives rise to questions of whether/how ∇ respects such splitting.

3.3.2.1 (Para)-Dolbeault Operator $\bar{\partial}$

The (para-)Dolbeault operator $\bar{\partial}$ on $\mathcal{M}$ for a given L is defined as [8]

$$\bar{\partial}_X Y = \frac{1}{4}\left([X, Y] - L^{-1}[LX, Y] + L^{-1}[X, LY] - L^2[LX, LY]\right) \tag{3.14}$$

for any vector fields X and Y in $\Xi(T\mathcal{M})$. It can be checked that $\bar{\partial}_X Y$ is tensorial in X, such that $\bar{\partial}_{fX} Y = f\left(\bar{\partial}_X Y\right)$, and is a derivation in Y, such that the Leibniz rule is satisfied. It can be easily verified that

$$\bar{\partial}_{LX} Y = -L(\bar{\partial}_X Y), \qquad \bar{\partial}_X(LY) = L(\bar{\partial}_X Y).$$

Using definition Eq. (3.7) for N_L, we can rewrite the definition (3.14) of $\bar{\partial}$ as

$$\bar{\partial}_X Y = \frac{1}{2}\left([X, Y] - L^{-1}[LX, Y] + \frac{1}{2} L^2 N_L(X, Y) \right). \tag{3.15}$$

When we extend $\bar{\partial}$ to (para-)complexified tangent bundles, then we will find that

$$\begin{aligned} \bar{\partial}_{X^+} Y^+ &= \bar{\partial}_{X^-} Y^- = 0; \\ \bar{\partial}_{X^-} Y^+ &= \pi^+[X^-, Y^+]; \\ \bar{\partial}_{X^+} Y^- &= \pi^-[X^+, Y^-]. \end{aligned}$$

Stated otherwise, $\bar{\partial} Y^+$ is a vector-valued 1-form, of type $+$ as a vector and type $-$ as a 1-form, and conversely in types for $\bar{\partial} Y^-$.

It is important to note that this operator $\bar{\partial}$ is conventionally deployed only when the almost complex structure J is integrable (i.e., becomes a complex structure), whence $T^+\mathcal{M}$ becomes a holomorphic tangent bundle (with holomorphic coordinate z on the base manifold) and $T^-\mathcal{M}$ an anti-holomorphic tangent bundle (with anti-holomorphic coordinates $\bar{z}$ on the base manifold). It is also known as the *intrinsic Cauchy-Riemann operator* of J, since it only depends on the almost-complex structure J. $\bar{\partial}$ usually operates on $T^+\mathcal{M}$ only, and induces the differentiation of tangent vector fields in $T^+\mathcal{M}$ with respect to the anti-holomorphic coordinates $\frac{\partial}{\partial \bar{z}}$. The operation of $\bar{\partial}$ on $T^-\mathcal{M}$ with respects to $\frac{\partial}{\partial \bar{z}}$ will be identically zero.

In our current usage of $\bar{\partial}$, we do not assume integrability of L, so it makes sense to consider $\bar{\partial}$ on the whole (para-)complexified tangent bundle $T^{\mathbb{L}}\mathcal{M}$. In this more general usage, $\bar{\partial}$ is said to define a *pseudo-(para-)holomorphic* structure independent of whether L is integrable or not. For convenience and brevity, we drop the prefix "pseudo."

3.3.2.2 (Para-)holomorphic Connections

Given a connection ∇ operating on $T^{\mathbb{L}}\mathcal{M}$, we can ask the question whether ∇ as a covariant derivative is compatible with $\bar{\partial}$. To understand this we may define an alternative operator $\bar{\partial}^\nabla$, see [8]

$$\bar{\partial}^\nabla_X Y = \frac{1}{2}\left(\nabla_X Y - \nabla_{LX}(L^{-1} Y) \right) \tag{3.16}$$

and extend the vector fields X and Y to (para-)complexified ones in $T^{\mathbb{L}}\mathcal{M}$. It satisfies

$$\bar{\partial}^\nabla_X (LY) = -\bar{\partial}^\nabla_{LX}(Y).$$

Clearly,

$$\bar{\partial}^{\nabla}_{X^+} Y^+ = \bar{\partial}^{\nabla}_{X^-} Y^- = 0;$$
$$\bar{\partial}^{\nabla}_{X^-} Y^+ = \nabla_{X^-} Y^+;$$
$$\bar{\partial}^{\nabla}_{X^+} Y^- = \nabla_{X^+} Y^-.$$

In other words, operating on $Y^+ \in T^+\mathcal{M}$, $\bar{\partial}^{\nabla}$ takes the (0,1)-part of the vector-valued 1-form ∇Y^+ (and conversely the (1,0)-part when operating on $Y^- \in T^-\mathcal{M}$).

Again, we do not assumed integrability of L, so the same caveat applies to $\bar{\partial}^{\nabla}$ as for the operator $\bar{\partial}$ defined above—$\bar{\partial}^{\nabla}$ is defined on the full (para-)complexified tangent bundle $T^{\mathbb{L}}\mathcal{M}$, and therefore a restriction of it to $T^-\mathcal{M}$ would actually be denoted by ∂^{∇} if acting on $T^-\mathcal{M}$ as a (para-)holomorphic bundle.

On a (para-)holomorphic vector bundle, a connection is conventionally called *(para-)holomorphic* if the two Dolbeault operators $\bar{\partial}$ and $\bar{\partial}^{\nabla}$ coincide. We extend this notion to arbitrary connections on $T^{\mathbb{L}}\mathcal{M} = T^{(1,0)}\mathcal{M} \oplus T^{(0,1)}\mathcal{M}$ where ∇ does not necessarily preserve $T^{(1,0)}\mathcal{M}$ or $T^{(0,1)}\mathcal{M}$.

Definition 2 A connection ∇ is called (para-)holomorphic if $\bar{\partial}^{\nabla}_X Y = \bar{\partial}_X Y$ for any vector fields X and Y.

Using (3.14) and (3.16), we can also derive a necessary and sufficient condition for (para-)holomorphicity in terms of N_L, T^{∇}, and T^{∇^L}. In fact the following were proven in [7]:

Lemma 2 *Given an arbitrary pair* (∇, L)*, the following statements are equivalent*

(i) ∇ *is (para-)holomorphic;*
(ii) ∇^L *is (para-)holomorphic;*
(iii) the following holds

$$\frac{1}{2} N_L(X, Y) = L^2\, T^{\nabla^L}(X, Y) - L T^{\nabla}(LX, Y). \qquad (3.17)$$

Two connections ∇^1 and ∇^2 are said to be $\bar{\partial}$-balanced when $\bar{\partial}^{\nabla^1}_X Y = \bar{\partial}^{\nabla^2}_X Y$ holds. This Lemma implies that a sufficient condition for (∇, ∇^L) to be $\bar{\partial}$-balanced is that either of them is (para-)holomorphic. (Para-)holomorphicity of ∇ (and ∇^L) is stronger than (∇, ∇^L) being $\bar{\partial}$-balanced. These notions were investigated in [7].

3.3.3 *(Para-)complexifying N_L and T^{∇}*

3.3.3.1 Decomposition of N_L

Both $T^{(1,0)}\mathcal{M}$ and $T^{(0,1)}\mathcal{M}$ are foliations if and only if L is integrable, i.e., the integrability condition $N_L = 0$ is satisfied.

We recall that in the (para-)complex case, the $(1, 1)$-component of the Nijenhuis tensor vanishes identically. Heuristically, this follows from the fact that the Nijenhuis tensor is (apart from a multiplicative constant) the (0, 1)-part of the Lie bracket of two (1, 0)-vector fields $N(Z_1, Z_2) = \pi^-[\pi^+ Z_1, \pi^+ Z_2]$. Therefore, we do not expect there to be any (1, 1)-component. To see this directly, let X and Y be real vector fields, from which the (1,0)-vector X^+ and (0,1)-vector Y^- can be constructed. We have

$$\begin{aligned} \mathcal{N}^{(1,1)} &= N(X^+, Y^-) \\ &= -L^2[X^+, Y^-] + L[LX^+, Y^-] + L[X^+, LY^-] - [LX^+, LY^-] \\ &= -L^2[X^+, Y^-] + \alpha L[X^+, Y^-] - \alpha L[X^+, Y^-] + \alpha^2[X^+, Y^-] \\ &= 0 \end{aligned}$$

where the last step invoked $\alpha^2 L^2 = \text{id}$.

The (2, 0)-part of the Nijenhuis tensor, namely taking inputs from $T_p^{(2,0)}\mathcal{M} = T_p^+\mathcal{M} \times T_p^+\mathcal{M}$ or $(+,+)$, can be calculated as the following

$$\begin{aligned} \mathcal{N}^{(2,0)} &= N(X^+, Y^+) \\ &= -L^2[X^+, Y^+] + L[LX^+, Y^+] + L[X^+, LY^+] - [LX^+, LY^+] \\ &= -L^2[X^+, Y^+] + \alpha L[X^+, Y^+] + \alpha L[X^+, Y^+] - \alpha^2[X^+, Y^+] \\ &= -2L^2\left([X^+, Y^+] - \alpha^{-1}L[X^+, Y^+]\right) \\ &= -4\, L^2\, \pi^-([X^+, Y^+]) \\ &= -4\alpha^2\, \pi^-([X^+, Y^+]). \end{aligned}$$

And similarly

$$N^{(0,2)} = N(X^-, Y^-) = -4\alpha^2\, \pi^+([X^-, Y^-]).$$

On the other hand,

$$\begin{aligned} N^{(2,0)} &= N(X^+, Y^+) = N(X + \alpha^{-1}LX, Y + \alpha^{-1}LY) \\ &= N(X, Y) + \alpha^{-1}N(LX, Y) + \alpha^{-1}N(X, LY) + \alpha^{-2}N(LX, LY) \\ &= N(X, Y) - \alpha^{-1}LN(X, Y) - \alpha^{-1}LN(X, Y) + \alpha^{-2}L^2\, N(X, Y) \\ &= 2\, N(X, Y) - 2\alpha^{-1}LN(X, Y) = 4\pi^-(N(X, Y)), \end{aligned}$$

and similarly

$$N^{(0,2)} = 4\pi^+(N(X, Y)).$$

The above derivation shows that N_L, when defined over (para-)complexified vector fields, decompose into three parts:

(i) $\mathcal{N}^{(1,1)}$, which is identically zero;
(ii) $\mathcal{N}^{(2,0)}$, which may not be zero;
(iii) $\mathcal{N}^{(0,2)} \equiv \overline{N^{(2,0)}}$.

Therefore, to show that the (para-)complexified Nijenhuis tensor vanishes, it is sufficient to show that its $(2,0)$-part does. The nature of $N^{(2,0)}$ will be investigated in Sect. 3.3.5.1.

3.3.3.2 (Para-)complexifying MC1 and MC2

The torsion tensor T^{∇} of a connection ∇ on $\mathcal{M}$ is a 1,2-tensor, that is, it takes in two (real-valued) vector fields X, Y and outputs one (real-valued) vector-field denoted by $T^{\nabla}(X, Y)$. When the vector fields are (para-)complexified, we can treat T^{∇} as mapping two (para-)complexified vector fields to one (para-)complexified vector field as an output. Naturally, we can investigate its various components, taking inputs in, $T_p^+\mathcal{M} \times T_p^+\mathcal{M}$ or $(+,+)$ called "(2,0)-part", $T_p^+\mathcal{M} \times T_p^-\mathcal{M}$ or $(+,-)$ called "(1,1)-part", $T_p^-\mathcal{M} \times T_p^-\mathcal{M}$ or $(-,-)$ called "(0,2)-part".

Because torsion can be decomposed into (2,0)-, (1,1)-, or (0,2)-part, we can also investigate the 1st Matching Condition (MC1) and 2nd Matching Condition (MC2) between T^{∇} and L when torsion T^{∇} breaks down in terms of these separate components. We refer to this as "(para-)complexifying" torsion-coupling and torsion-compatibility conditions. In doing so, we hope to obtain a deeper appreciation of whether enforcing these conditions separately on each (para-)holomorphic/anti-(para-)holomorphic components affects integrability of L and/or (para-)holomorphicity of ∇.

With respect to the "torsion-coupling" of (∇, L) or MC2, we will show that the $(2,0)+(0,2)$-part is equivalent to the vanishing of the Nijenhuis tensor, while the $(1,1)$-part is equivalent to the holomorphicity of ∇. Since torsion-coupling of ∇ implies torsion-coupling of ∇^L, enforcing $(1,1)$-part of torsion-coupling also implies that ∇^L is holomorphic (just as ∇ is). Furthermore, the $(1,1)$-part of torsion-coupling implies that the $(1,1)$-component of the (para-)complexified torsion is 0, both for ∇ and ∇^L.

With respect to the "torsion-compatibility" of (∇, L) or MC1, we will show that the (para-)complexified version of this condition Eq. (3.11) implies that the $(2,0)+(0,2)$-part of the torsion must vanish; the torsion carried by ∇, if any, is entirely of $(1,1)$ type. Because the (2,0)- and (0,2)-part are equivalent, which is meant by (2,0) + (0,2)-part, we will use the phrase (2,0)-part or (2,0)-component from now on.

The next two subsections give the details of these assertions.

3.3.4 Torsion-Compatibility (MC1)

3.3.4.1 (1,1)-Part of Torsion-Compatibility

It is easy to verify that

$$T^\nabla(LX^+, Y^-) + T^\nabla(X^+, LY^-) = \alpha T^\nabla(X^+, Y^-) - \alpha T^\nabla(X^+, Y^-) \equiv 0$$

for any real vector fields X, Y. This is to say, the (1,1)-part of the torsion-compatibility condition is *always* satisfied.

3.3.4.2 (2,0)-Part of Torsion-Compatibility

Let us now require torsion-compatibility to hold for the (2,0)-part:

$$T^\nabla(LX^+, Y^+) + T^\nabla(X^+, LY^+) = 0.$$

Writing this out explicitly,

$$\begin{aligned} 0 &= 2\alpha T^\nabla(X^+, Y^+) = 2\alpha(T^\nabla(X + \alpha^{-1}LX, Y + \alpha^{-1}LY)) \\ &= 2\alpha(T^\nabla(X, Y) + \alpha^{-2}T^\nabla(LX, LY) + \alpha^{-1}T^\nabla(LX, Y) + \alpha^{-1}T^\nabla(X, LY)) \\ &= 2\,T^\nabla(LX, Y) + 2\,T^\nabla(X, LY) + \alpha^{-1}\left(2\,L^2\,T^\nabla(X, Y) + 2\,T^\nabla(LX, LY)\right). \end{aligned}$$

Therefore, we obtain

$$\begin{aligned} T^\nabla(LX, Y) + T^\nabla(X, LY) &= 0, \\ L^2 T^\nabla(X, Y) + T^\nabla(LX, LY) &= 0. \end{aligned}$$

The first equation is the torsion-compatibility we started with. The second equation is identical to the first one, after setting LX for X in the first equation. So no new conditions are obtained after (para-)complexifying torsion-compatibility. Summarizing the situation for both (1,1)- and (2,0)-components of the torsion tensor, (para-)complexifying torsion-compatibility (MC1) does not yield any new information beyond saying that the MC1 condition amounts to the (2,0)-component of torsion being zero.

3.3.4.3 Para-Complexifying Codazzi Coupling

Codazzi coupling of ∇ and L leads to, according to Proposition 1,

$$T^\nabla(X, Y) = T^{\nabla^L}(X, Y).$$

(Para-)complexifying the above, whether taking (2,0)- or (1,1)-component, leads to

$$\begin{aligned} T^\nabla(LX, Y) + T^\nabla(X, LY) &= T^{\nabla^L}(LX, Y) + T^{\nabla^L}(X, LY), \\ L^2\,T^\nabla(X, Y) + T^\nabla(LX, LY) &= L^2 T^{\nabla^L}(X, Y) + T^{\nabla^L}(LX, LY). \end{aligned}$$

Substituting $X \to LX$ in the first equation leads to the second equation. So (para-)complexifying Codazzi coupling leads to a relaxation of the relation $T^{\nabla}(X, Y) = T^{\nabla^L}(X, Y)$.

Codazzi coupling, along with MC1 applied to either ∇ or ∇^L, will lead to MC1 satisfied for both ∇ and ∇^L and hence, the integrability of L.

3.3.5 *Torsion-Coupling (MC2)*

In contrast to MC1, the situation is different when (para-)complexifying torsion-coupling (MC2)—after complexifying, the (2,0)- and (1,1)-part reveal independent information, as we will show now.

3.3.5.1 (2,0)-Part of Torsion-Coupling: Integrability

From Sect. 3.3.3.1, we see that the only non-vanishing part of the (para-)complexfied N_L is $N^{(2,0)}$ (and $N^{(0,2)}$). In this subsection, we relate $N^{(0,2)}$ to MC2 or torsion-coupling.

Proposition 5 *The (2,0)-part $N^{(2,0)}$ of N_L vanishes if and only if the (2,0)-part of torsion-coupling is enforced. That is,*

$$N^{(2,0)} := N_L(X^+, Y^+) = 0$$

if and only if

$$LT^{\nabla^L}(X^+, Y^+) = T^{\nabla}(LX^+, Y^+).$$

Proof Instead of enforcing the full torsion-coupling condition, we only assume that the (2, 0)-part of the torsion is coupled to L, and show that it is equivalent to the vanishing of the Nijenhuis tensor.

$$\begin{aligned}
&\frac{1}{2}N_L(X^+, Y^+)\\
&= \frac{1}{2}\left(T^{\nabla^L}(L^2X^+, Y^+) + T^{\nabla^L}(LX^+, LY^+) - LT^{\nabla}(LX^+, Y^+) - LT^{\nabla}(X^+, LY^+)\right)\\
&= \frac{1}{2}\left(L^2\,T^{\nabla^L}(X^+, Y^+) + \alpha^2\,T^{\nabla^L}(X^+, Y^+) - 2\alpha L\,T^{\nabla}(X^+, Y^+)\right)\\
&= L^2\,T^{\nabla^L}(X^+, Y^+) - \alpha LT^{\nabla}(X^+, Y^+)\\
&= L\left(LT^{\nabla^L}(X^+, Y^+) - T^{\nabla}(LX^+, Y^+)\right).
\end{aligned}$$

The expression in the parenthesis is the expression of torsion-coupling applied to (2,0)-part.

Let us now writing out explicitly the (2,0)-part of torsion-coupling

$$L\left(T^{\nabla^L}(X+\alpha^{-1}LX, Y+\alpha^{-1}LY)\right) = \alpha T^{\nabla}(X+\alpha^{-1}LX, Y+\alpha^{-1}LY)$$

or explicitly,

$$\begin{aligned}&L\left(T^{\nabla^L}(X,Y)+\alpha^{-2}T^{\nabla^L}(LX,LY)+\alpha^{-1}(T^{\nabla^L}(LX,Y)+T^{\nabla^L}(X,LY))\right)\\ &=\alpha\left(T^{\nabla}(X,Y)+\alpha^{-2}T^{\nabla}(LX,LY)\right)+T^{\nabla}(LX,Y)+T^{\nabla}(X,LY),\end{aligned}$$

from which we obtain two equations (separating real and imaginary components):

$$\begin{aligned}L^2T^{\nabla^L}(X,Y)+T^{\nabla^L}(LX,LY) &= L\left(T^{\nabla}(X,LY)+T^{\nabla}(LX,Y)\right),\\ L\left(T^{\nabla^L}(LX,Y)+T^{\nabla^L}(X,LY)\right) &= L^2T^{\nabla}(X,Y)+T^{\nabla}(LX,LY).\end{aligned}$$

The above two equations are equivalent to each other, since substituting LX for X in the first yields the second. Furthermore they exactly (!) express the condition $N_L = 0$. This shows that the (2,0)-part of the torsion-coupling condition is exactly $N_L = 0$. Since $N^{(1,1)} \equiv 0$, $N_L = 0$ is equivalent to $N^{(2,0)} = 0$.

3.3.5.2 (1, 1)-Part of the Torsion-Coupling: (Para-)holomorphicity

Let us now only impose torsion-coupling condition to the (1,1)-part, that is, when inputs are taken from $T_p^+\mathcal{M} \times T_p^-\mathcal{M}$ or $(+,-)$:

$$LT^{\nabla}(X+\alpha^{-1}LX, Y-\alpha^{-1}LY) = \alpha T^{\nabla^L}(X+\alpha^{-1}LX, Y-\alpha^{-1}LY).$$

Writing out explicitly,

$$\begin{aligned}&L\left(T^{\nabla}(X,Y)-\alpha^{-2}T^{\nabla}(LX,LY)+\alpha^{-1}T^{\nabla}(LX,Y)-\alpha^{-1}T^{\nabla}(X,LY)\right)\\ &=\alpha\left(T^{\nabla^L}(X,Y)-\alpha^{-2}T^{\nabla^L}(LX,LY)+\alpha^{-1}T^{\nabla^L}(LX,Y)-\alpha^{-1}T^{\nabla^L}(X,LY)\right)\end{aligned}$$

from which we obtain two equations (separating real and imaginary components):

$$\begin{aligned}L\left(T^{\nabla}(X,Y)-L^2T^{\nabla}(LX,LY)\right) &= T^{\nabla^L}(LX,Y)-T^{\nabla^L}(X,LY),\\ L\left(T^{\nabla}(LX,Y)-T^{\nabla}(X,LY)\right) &= L^2T^{\nabla^L}(X,Y)-T^{\nabla^L}(LX,LY).\end{aligned}$$

The above two equations are equivalent to each other—we can obtain the second from subsituting $X \to LX$ in the first. Rearranging the second:

$$T^{\nabla^L}(LX,LY)-LT^{\nabla}(X,LY) = L^2T^{\nabla^L}(X,Y)-LT^{\nabla}(LX,Y). \tag{3.18}$$

We call the above Eq. (3.18) "Condition H". It is the $(1, 1)$-part of the (para)complexified torsion-coupling condition, which is a slight relaxation to the standard torsion-coupling condition. If the lefthand side (and the righthand side) equals 0, that is the standard torsion-coupling condition Eq. (3.13).

Proposition 6 *Condition H or Eq. (3.18), which expresses the (1,1)-part of torsion-coupling, is satisfied if and only if* ∇ *is (para-)holomorphic.*

Proof Recall from Proposition 2

$$N_L(X,Y) = \underbrace{\left(T^{\nabla^L}(L^2X,Y) - LT^{\nabla}(LX,Y)\right)}_{A} + \underbrace{\left(T^{\nabla^L}(LX,LY) - LT^{\nabla}(X,LY)\right)}_{B}.$$

Eq. (3.18) amounts to requiring the underbracketed "A" and "B" are equal. Hence, it is equivalent to

$$\frac{1}{2}N_L(X,Y) = L^2T^{\nabla^L}(X,Y) - LT^{\nabla}(LX,Y).$$

By Lemma 2, this is equivalent to ∇ being (para-)holomorphic.

3.4 Summary and Discussions

Affine connections on almost Hermitian manifold are investigated with a typical assumption that these connections ∇ are complex: $\nabla J = 0$. When torsions are to be considered, it is widely known that two types of torsion exist that nevertheless can lead to integrable complex manifold (vanishing of N_J):

$$T(JX, JY) = T(X,Y), \qquad \text{torsion is (1,1)-type,} \tag{3.19}$$

$$T(JX, Y) = JT(X,Y), \qquad \text{torsion is (2,0)-type.} \tag{3.20}$$

If furthermore ∇ is a metric connection, $\nabla g = 0$, then the (1,1)-type is called the *first canonical connection* and the (2.0)-type the *second canonical connection*.

The second canonical connection is also known as Chern connection. It is a holomorphic connection, and is uniquely determined. Reference [11] classified 16 classes of almost Hermitian structures based on its Levi-Civita connection. In general, Levi-Civita connection and Chern connection are not the same on a Hermitian manifold; they are one and the same if the Hermitian manifold is Kähler, where the torsion of the Chern connection vanishes.

Our paper extends these considerations to arbitrary connections that are in general not parallel with respect to L, $\nabla L \neq 0$. The approach is through dualistic approach, by considering ∇^L, the conjugate connection with respect to L. The unique L-conjugate connection ∇^L has the property that $\frac{1}{2}(\nabla + \nabla^L)L = 0$. Either of these connections may carry torsion. So our paper characterizes the torsion tensors T^{∇} and T^{∇^L} that will lead to integrable (para-)complex structure. The case of $\nabla L = 0$ is recovered as the special case where $\nabla = \nabla^L$.

A Hermitian connection ∇ is one that preserves the Hermitian form h, $\nabla h = 0$, defined by $h(X, Y) = g(X, Y) + \sqrt{-1}\omega(X, Y)$. Equivalently, a Hermitian connection is one which is metric ($\nabla g = 0$) and complex ($\nabla J = 0$). In general, Hermitian connections carry non-zero torsion. In fact, there is a line of Hermitian connections called the *Gauduchon connections*, all with differerent torsions. This line passes through the first canonical connection and the second canonical connection which satisfies, respectively, Eq. (3.19) and Eq. (3.20). But there are infinitely many other Hermitian connections that do not fall on the Gauduchon line.

Our investigations generalize the first and second canonical connections by removing the assumption of their being (para-)complex connections $\nabla L \neq 0$. Specifically, MC2 generalizes Eq. (3.20) to involve both ∇, ∇^L; while MC1 is to be enforced separately on ∇ and ∇^L. The set of torsion-compatible connections (satisfying MC1) intersects the line of Gauduchon connections at the first canonical connection. This intersection is unique (and transversal) because for all other Gauduchon connections, the torsion is not $(1, 1)$. The set of torsion-coupled connections (satisfying MC2) intersects the line of Gauduchon connections at the Chern connection (the second canonical connection). This intersection is also unique and transversal because the Chern connection is the unique holomorphic Hermitian (complex and metric) connection.

MC1 condition expressed a constraint about the torsion of ∇, namely, the (2,0)- and (0,2)-component of T^{∇} must vanish. However, this constraint can be *separately* enforced upon ∇ and ∇^L. Only when both T^{∇} and T^{∇^*} satisfy MC1 would an integrable L result. In the presence of Codazzi coupling, which stipulates $T^{\nabla} = T^{\nabla^L}$, (∇, L) satisfies MC1 if and only if (∇^L, L) satisfies MC1. Thus MC1 articulates a generalization of the first canonical connection.

MC2 condition, on the other hand, expresses a "balance" of T^{∇} and T^{∇^L}, such that it involves the torsions of ∇ and ∇^L in an intertwined fashion. (∇, L) satisfies MC2 if and only if (∇^L, L) satisfies MC2. It is in this scenario that ∇ and ∇^L both become (para-)holomorphic connections. MC2 articulates a generalization of the second canonical connection.

$$\begin{cases} (2,0)\text{-part of MC2} \Leftrightarrow N^{(2,0)} = 0 \Leftrightarrow N_L(X, Y) = 0, \quad \text{since } N^{(1,1)} \equiv 0 \\ (1,1)\text{-part of MC2} \Leftrightarrow \ \text{Cond. H} \ \Leftrightarrow \quad \text{(para-)holomorphicity of } \nabla \end{cases}$$

MC1 and MC2 conditions are independent of each other. When torsion-coupling and torsion-compatibility are both imposed, then the connections must be torsion-free $T^{\nabla} = T^{\nabla^L} = 0$. These are torsion-free holomorphic connections yet generally not (para-)complex.

Acknowledgements The project is supported by DARPA/ARO Grant W911NF-16-1-0383 ("Information Geometry: Geometrization of Science of Information") and a sub-contract from AFOSR Grant FA9550-19-1-0213 ("Brain-Inspired Networks for Multifunctional Intelligent Systems and

Aerial Vehicles"), awarded to: Jun Zhang. Work completed when G. Khan is with the University of Michigan.

Disclaimer: Views and opinions expressed are those of the authors and do not necessarily represent official positions of their respective companies.

References

1. Lauritzen, S.L.: Statistical manifolds. In: Differential Geometry in Statistical Inference, volume 10 of IMS Lecture Notes Monograph Series, pp. 163–216. Institute of Mathematical Statistics (1987)
2. Hitchin, H.L.: The moduli space of complex Lagrangian submanifolds. Asian J. Math. **3**(1), 77–92 (1999)
3. Simon, U.: Affine differential geometry. In: Handbook of Differential Geometry, vol. 1, pp. 905–961. North-Holland (2000)
4. Fei, T., Zhang, J.: Interaction of Codazzi couplings with (para-)Kähler geometry. Res. Math. **72**, 2037–2056 (2017)
5. Zhang, J. Teng, F.: Information geometry with (para-)Kähler structures. Information Geometry and its Applications IV, pp. 297-321. Springer Proceedings in Mathematics and Statistics (PROMS) series, vol. 252 (2018)
6. Furuhata, H.: Hypersurfaces in statistical manifolds. Diff. Geom. Appli. **27**(3), 420–429 (2009)
7. Grigorian, S., Zhang, J.: (Para-)holomorphic and conjugate connections on (para-)Hermitian and (para-)Kähler manifolds. Res. Math. **74**(4), 150 (2019)
8. Gauduchon, P.: Hermitian connections and Dirac operators Boll. Un. Mat. Ital. B **7**(11), 257–288 (1997)
9. Yang, B., Zheng, F.: On compact Hermitian manifolds with flat Gauduchon connections. Acta Mathematica Sinica, English Series **34**, 1259–1268 (2018)
10. Schwenk-Schellschmidt, A., Simon, U.: Codazzi-equivalent affine connections. Res. Math. **56**(1–4), 211–229 (2009)
11. Gray, A., Hervella, L.M.: The sixteen classes of almost Hermitian manifolds and their linear invariants. Annali di Matematica pura ed applicata **123**(1), 35–58 (1980)

Chapter 4
Contact Hamiltonian Systems for Probability Distribution Functions and Expectation Variables: A Study Based on a Class of Master Equations

Shin-itiro Goto and Hideitsu Hino

Abstract Based on information and para-contact metric geometries, in this paper two classes of dynamical systems are formulated. One is for describing the time-development of probability distribution functions following continuous-time master equations, and the other one for describing expectation variables with respect to the distribution functions. These two classes of systems are described by contact Hamiltonian systems on contact metric manifolds. Here the systems for expectation variables are exactly derived from the master equations describing nonequilibrium processes. Their relaxation processes are characterized by the lengths of integral curves of contact Hamiltonian vector fields and the values of the Ricci tensor fields. At equilibrium states realized as the time-asymptotic limit of the contact Hamiltonian systems, possible information geometric structures are given.

Keywords Master equations · Para-contact metric manifolds · nonequilibrium statistical mechanics · Information geometry

4.1 Introduction

Information geometry is a geometrization of mathematical statistics, and its differential geometric aspects and applications to statistics have been investigated [1, 2]. Examples of such applications include statistical inference, quantum information, thermodynamics. In addition, links between equilibrium thermodynamics and information geometry have been studied [3]. Moreover, dynamical systems for probability distribution functions have been investigated from viewpoints of information geometry [4–6].

S. Goto (✉) · H. Hino
The Institute of Statistical Mathematics, 10-3 Midori-cho, Tachikawa, Tokyo 190-8562, Japan
e-mail: sgoto@ism.ac.jp

H. Hino
e-mail: hino@ism.ac.jp

F. Nielsen (ed.), *Progress in Information Geometry*,
Signals and Communication Technology,
https://doi.org/10.1007/978-3-030-65459-7_4

Contact geometry has been known as an odd-dimensional counterpart of symplectic geometry [7–9], and this geometry is applicable to various mathematical sciences. They include thermodynamics [10–12], fluid mechanics [13], optimization problems [14], and dissipative mechanical systems [15, 16]. In developing contact geometric thermodynamics [17, 18], links between contact geometry and information geometry have been discussed [3, 19]. In this context, when a metric tensor field is introduced on a contact manifold, it was found that para-Sasakian geometry is suitable for describing thermodynamics [20, 21], where para-Sasakian manifolds are para-contact metric manifolds satisfying some additional condition. It is then of interest to explore how para-contact metric manifolds describe dynamical systems for probability distribution functions inducing thermodynamics.

Master equations are used for modeling Markov processes [22] in the form of continuous-time first order differential equations and of discrete-time maps, and are used for describing probability distribution functions by modeling so-called transition matrices or rates. Let Γ be a set of finite states, and $p : \Gamma \times \mathbb{R} \to \mathbb{R}_{\geq 0}$ a time-dependent probability distribution function, where $t \in \mathbb{R}$ denotes time. Continuous-time master equations are then written as

$$\frac{\partial}{\partial t} p(j,t) = \sum_{j' \in \Gamma \setminus \{j\}} \left[w(j|j')\, p(j',t) - w(j'|j)\, p(j,t) \right], \tag{4.1}$$

where $w : \Gamma \times \Gamma \to I$, $(I := [0,1] \subset \mathbb{R})$ is such that $w(j|j')$ denotes a probability that a state jumps from j' to j, and is referred to as a transition matrix. These systems are used in various scientific disciplines including nonequilibrium statistical mechanics [22, 23]. Aside from interests in the natural sciences, its resolution has wide implications in mathematical engineering. One of them is to apply such processes to Monte-Carlo simulations, called Markov-Chain Monte-Carlo methods [24]. Further progress in building mathematical foundation of these methods and their applications continues to attract attention [25, 26].

Given master equations, one is interested in time-development of the expectation variables. Such variables are defined as follows. Given a set of prescribed functions $\mathcal{O}_a : \Gamma \to \mathbb{R}$, $(a = 1, \dots, n)$ the expectation variables $\langle \mathcal{O}_a \rangle (t)$ are defined as

$$\langle \mathcal{O}_a \rangle (t) = \sum_{j \in \Gamma} \mathcal{O}_a(j)\, p(j,t), \qquad a = 1, \dots, n,$$

where $\{p(j,t)\}$ obey (4.1), and n is independent of $|\Gamma|$ in general. These expectation variables are expected to play roles of time-dependent thermodynamic variables, and thus how information and contact geometric theories are applied to these variables should be clarified.

Summary of this Contribution

In this paper a class of nonequilibrium thermodynamic processes activated by continuous-time master equations are formulated on a para-contact metric manifold. Here master equations describe the time-development of probability distribution functions, and the time-development of expectation variables are called moment dynamical systems in this paper, where moment dynamical systems are obtained from master equations by integrating some functions over states with the probability distribution functions. Thus, roughly speaking, master equations represent microscopic scale systems, and it is assumed in this paper that moment dynamical systems can be treated as thermodynamic systems. To show explicit calculations, simple transition matrices of master equations are considered as a toy model. This model is referred to as a solvable model (and solvable master equations) in this paper, and is of the form

$$\frac{\partial}{\partial t} p(j,t) = p_\theta^{\mathrm{eq}}(j) - p(j,t), \qquad j \in \Gamma, \tag{4.2}$$

where p_θ^{eq} is a prescribed equilibrium distribution function. Although there is no interaction term in (4.2), it turns out that there is a link between (4.2) and a model that has interaction terms. This link is found by introducing the probability generating function (see Appendix A). Discussions on expectation variables in this paper have been in [27], and those involving an almost para-contact structure for expectation variables have been in [28]. In addition to the above discussions, more details, a contact Hamiltonian system for a class of the master equations, and how the Ricci tensor fields behave are discussed as additional contributions. Then the following is shown in this paper:

> *Claim* (Theorem 1): The solvable master equations and moment dynamical system derived from the solvable master equations are described on para-contact metric manifolds, and its convergence to the equilibrium states are characterized by the Mrugala metric fields and the Ricci tensor fields associated with the Levi-Civita connections.

Some of details are as follows.

The relations among master equations (equations for probability distribution functions), moment dynamical systems (equations for expectation variables), and their geometric descriptions above are summarized as the following diagram:

$$\begin{array}{ccc}
\text{Contact Ham Sys} & & \text{Contact Ham Sys} \\
(*)\wr\!\!\downarrow & & \wr\!\!\downarrow [27] \\
\text{Master Eq} & \overset{\text{integration}}{\underset{[27]}{\rightsquigarrow}} & \text{Moment Dyn Sys} \\
\| & & \| \\
\text{(Eq for Prob Dist)} & & \text{(Eq for Expect Var)}
\end{array}$$

where $(*)$ denotes one of the contributions developed in this paper, $A \rightsquigarrow B$ means that A induces B, Ham is the abbreviation for Hamiltonian, Sys for system, Prob

Table 4.1 Contact manifolds and their basic objects

System	Contact Mfd	Dimension	Coordinates	Contact Hamiltonian
General	$(\mathcal{C}, \lambda)$	$2m+1$	(x, y, z)	$h_{\varpi} = \varpi - z$
Master equations	$(\mathcal{C}_f^{\Gamma}, \lambda_f^{\Gamma})$	$2\|\Gamma\|+1$	$(p_{\mathrm{eq}}^f, \psi, I_f)$	$h_{I_f^{\mathrm{eq}}}^{\Gamma} = I_f^{\mathrm{eq}} - I_f$
Moment Dyn. Sys.	$(\mathcal{C}^{\mathcal{O}}, \lambda^{\mathcal{O}})$	$2n+1$	$(\theta, \langle \mathcal{O} \rangle_{\theta}, \Psi)$	$h_{\Psi^{\mathrm{eq}}}^{\mathcal{O}} = \Psi^{\mathrm{eq}} - \Psi$

for probability, Dist for distribution, and Var for variable. In both of the toy master equations and moment dynamical systems, relaxation processes are shown to exist, where relaxation process is a process towards an equilibrium state from nonequilibrium states. In addition, it is shown that such a relaxation process can be described by a contact Hamiltonian system, that a distance between nonequilibrium states and the equilibrium state is calculated with an introduced metric tensor field, and that the the Ricci tensor field is an indicator of how far a state is close to the equilibrium state. A part of these descriptions of time-development together with phase spaces are summarized as the following:

$$
\begin{array}{ccc}
\text{Nonequilibrium state} & \overset{(*),[20,28]}{=\!=\!=\!=} & \text{(Para) contact (metric) manifold} \\
\text{Relaxation} \downarrow \text{process} & & \text{Flow of contact Ham Sys} \downarrow (*),[28] \\
\text{Equilibrium state} & \underset{(*)[12,3]}{=\!=\!=\!=} & \text{Legendre submanifold / Dually flat space}
\end{array}
$$

where Legendre submanifod is a submanifold in contact manifold, and dually flat space is a manifold invented in information geometry.

Since various manifolds will be discussed in this paper, some of manifolds used in this paper and their objects are summarized in Table 4.1.

This paper is organized as follows. In Sect. 4.2, some definitions and facts used in the later sections are summarized. In Sect. 4.3, master equations with particular kernels (transition matrix elements) are introduced and their basic properties are summarized. In Sect. 4.4, a class of dynamical systems describing expectation variables is derived from the master equations. In Sect. 4.5, a contact geometric description of the master equations is described. In Sect. 4.6, a contact geometric description of the class of derived dynamical systems for the expectation variables is given. In Sect. 4.7, a possibility of extending the toy model studied in this paper is discussed. Finally, Sect. 4.8 summarizes this paper and discusses some future studies.

4.2 Underlying Geometry

In this section mathematical definitions and facts commonly used in the later sections are summarized. In this paper manifolds are assumed smooth and connected. In addition tensor fields are assumed smooth and real. The set of vector fields on a manifold $\mathcal{M}$ is denoted by $\mathcal{S}T\mathcal{M}$, that of p-forms denoted by $\mathcal{S}\Lambda^p\mathcal{M}$, the interior product with $X \in \mathcal{S}T\mathcal{M}$ is denoted by $\imath_X$, and the Lie derivative along $X \in \mathcal{S}T\mathcal{M}$ denoted by $\mathcal{L}_X$.

4.2.1 *Para-Contact Metric Manifolds*

In this subsection definitions and some existing statements are summarized, and the notations used in this paper are fixed (see [20, 29]). Note that there are several ways to define (para-) contact metric manifold.

Let $\mathcal{M}$ be a $(2m+1)$-dimensional manifold ($m \geq 1$). An *almost para-contact structure* on $\mathcal{M}$ is a triplet (ϕ, ξ, λ), where ξ is a vector field, λ a one-form, $\phi : \mathcal{S}T\mathcal{M} \to \mathcal{S}T\mathcal{M}$ a $(1,1)$-tensor field such that

$$\text{(i)}: \ \phi^2 = \mathrm{Id} - \lambda \otimes \xi, \ \ \text{(ii)}: \ \lambda(\xi) = 1, \ \text{ and (iii)}: \ \ker(\lambda) = \mathrm{Im}(\phi) = \mathcal{D}^+ + \mathcal{D}^-,$$

where $\ker(\lambda) := \{X \in \mathcal{S}T\mathcal{M} \,|\, \lambda(X) = 0\}$, $\mathrm{Im}(\phi) := \{\phi(X, -) \in \mathcal{S}T\mathcal{M} | X \in \mathcal{S}T\mathcal{M}\}$, $\mathcal{D}^\pm$ are eigen-spaces of $\mathcal{D} := \ker(\lambda)$ whose eigenvalues are ± 1, and Id is the identity operator. A pseudo Riemannian metric tensor field g satisfying

$$g(\phi X, \phi Y) = -\,g(X, Y) + \lambda(X)\,\lambda(Y), \qquad \forall X, Y \in \mathcal{S}T\mathcal{M}$$

is referred to as a *metric tensor compatible with an almost para-contact structure*. It is verified for non-compact manifolds that any almost para-contact structure admits a metric tensor field compatible with an almost para-contact structure. Then $(\mathcal{M}, \phi, \xi, \lambda, g)$ is referred to as an *almost para-contact metric manifold*.

On almost para-contact metric manifolds, one can show that

$$\lambda\phi = 0, \ \phi\xi = 0, \ \lambda(X) = g(X, \xi), \ g(\xi, \xi) = 1, \ \text{and } g(\phi X, Y) + g(X, \phi Y) = 0,$$

for $\forall X, Y \in \mathcal{S}T\mathcal{M}$. If g of an almost para-contact metric manifold satisfies

$$g(X, \phi Y) = \frac{1}{2}\mathrm{d}\lambda(X, Y), \qquad \forall X, Y \in \mathcal{S}T\mathcal{M} \tag{4.3}$$

then, $(\mathcal{M}, \phi, \xi, \lambda, g)$ is referred to as a *para-contact metric manifold*, where the convention of the numerical factor $\mathrm{d}\lambda(X, Y) = X\lambda(Y) - Y\lambda(X) - \lambda([X, Y])$, $([X, Y] := XY - YX)$ has been adapted. This convention corresponds to

$$\lambda' \wedge \lambda'' = \lambda' \otimes \lambda'' - \lambda'' \otimes \lambda',$$

for all one-forms λ' and λ''. When the other convention of the numerical factor $2\,\mathrm{d}\lambda(X, Y) = X\lambda(Y) - Y\lambda(X) - \lambda([X, Y])$ is adapted, the condition (4.3) is replaced with

$$g(X, \phi Y) = \mathrm{d}\lambda(X, Y), \qquad \forall X, Y \in \mathcal{S}T\mathcal{M}.$$

In this paper (4.3) is used. Para-Sasakian manifolds are para-contact metric manifolds satisfying the so-called normality condition.

Coordinate expressions were given for a para-contact metric manifold (and a para-Sasakian manifold) $(\mathcal{M}, \phi, \xi, \lambda, g)$ in [20]. They are summarized here. Let (x, y, z) be coordinates on $\mathcal{M}$ with $x = \{x^1, \ldots, x^m\}$ and $y = \{y_1, \ldots, y_m\}$ such that $\lambda = \mathrm{d}z - \sum_{a=1}^m y_a \mathrm{d}x^a$. Since various dimensional manifolds will be discussed in this paper the Einstein convention is not used throughout. Introduce the pseudo-Riemannian metric tensor field, referred to as the *Mrugala metric tensor field* [30],

$$G = \frac{1}{2}\sum_{a=1}^{m}\left[\mathrm{d}x^a \otimes \mathrm{d}y_a + \mathrm{d}y_a \otimes \mathrm{d}x^a\right] + \lambda \otimes \lambda, \tag{4.4}$$

which is shown to induce a para-contact metric manifold $(\mathcal{M}, \phi, \xi, \lambda, G)$. In what follows we consider the case where $y_a > 0$ for all $a \in \{1, \ldots, m\}$. Introduce the co-frame $\{\widehat{\theta}^0, \widehat{\theta}^1_-, \widehat{\theta}^1_+, \ldots, \widehat{\theta}^m_-, \widehat{\theta}^m_+\}$ and frame $\{e_0, e_1^-, e_1^+, \ldots, e_m^-, e_m^+\}$ with

$$\widehat{\theta}^0 := \lambda, \qquad \widehat{\theta}^a_\pm := \frac{1}{2\sqrt{y_a}}\left[y_a \mathrm{d}x^a \pm \mathrm{d}y_a\right],$$

$$e_0 := \xi, \qquad e_a^\pm := \sqrt{y_a}\left[\frac{1}{y_a}\left(\frac{\partial}{\partial x^a} + y_a\frac{\partial}{\partial z}\right) \pm \frac{\partial}{\partial y_a}\right],$$

so that

$$\widehat{\theta}^0(e_0) = 1, \qquad \widehat{\theta}^a_+(e_b^+) = \widehat{\theta}^a_-(e_b^-) = \delta^a_b, \qquad \text{others vanish},$$

where δ^a_b is the Kronecker delta, giving unity for $a = b$ and zero otherwise. One can then show that

$$G = \widehat{\theta}^0 \otimes \widehat{\theta}^0 + \sum_{a=1}^{m}\left[\widehat{\theta}^a_+ \otimes \widehat{\theta}^a_+ - \widehat{\theta}^a_- \otimes \widehat{\theta}^a_-\right], \quad \xi = \frac{\partial}{\partial z},$$

$$\phi = -\sum_{a=1}^{m}\left[\widehat{\theta}^a_- \otimes e_a^+ + \widehat{\theta}^a_+ \otimes e_a^-\right].$$

Some of relations are explicitly verified below. First, one has

$$\phi(e_a^+) = -e_a^-, \quad \phi(e_a^-) = -e_a^+, \quad \text{and} \quad \phi(e_0) = 0.$$

Moreover, (4.3) is verified due to

$$G(-,\phi-) = -\sum_{a=1}^{m}[\widehat{\theta}_{+}^{a}\otimes\widehat{\theta}_{-}^{a} - \widehat{\theta}_{-}^{a}\otimes\widehat{\theta}_{+}^{a}] = -\sum_{a=1}^{m}\widehat{\theta}_{+}^{a}\wedge\widehat{\theta}_{-}^{a},$$

$$\mathrm{d}\lambda = \sum_{a=1}^{m}\mathrm{d}x^{a}\wedge\mathrm{d}y_{a} = \sum_{a=1}(\widehat{\theta}_{+}^{a}+\widehat{\theta}_{-}^{a})\wedge(\widehat{\theta}_{+}^{a}-\widehat{\theta}_{-}^{a}) = -2\sum_{a=1}^{m}\widehat{\theta}_{+}^{a}\wedge\widehat{\theta}_{-}^{a}.$$

Observe that

$$\lambda\wedge\underbrace{\mathrm{d}\lambda\wedge\cdots\wedge\mathrm{d}\lambda}_{m}\neq 0. \tag{4.5}$$

It was also shown in [20] that the Ricci tensor field $\mathbf{Ric}^{G}$ associated with the Levi-Civita connection induced from G is given such that

$$\mathbf{Ric}^{G}(X,Y) = -(2m+2)\lambda(X)\lambda(Y) + 2G(X,Y). \qquad \forall X,Y\in\mathcal{S}T\mathcal{M} \tag{4.6}$$

4.2.2 *Contact Manifold*

In the context of geometry of thermodynamics, contact manifold is identified with the so-called thermodynamic phase space [12], and is defined as follows (see also [9] for details). Let $\mathcal{C}$ be a $(2m+1)$-dimensional manifold ($m = 1, 2, \ldots$), and λ' a one-form. If λ' satisfies

$$\lambda'\wedge\underbrace{\mathrm{d}\lambda'\wedge\cdots\wedge\mathrm{d}\lambda'}_{m}\neq 0,$$

then the pair $(\mathcal{C},\lambda')$ is referred to as a *contact manifold*, and λ' a *contact one-form*. It has been known as the Darboux theorem that there exists a special set of coordinates (x,y,z) with $x=\{x^{1},\ldots,x^{m}\}$ and $y=\{y_{1},\ldots,y_{m}\}$ such that $\lambda' = \mathrm{d}z - \sum_{a=1}^{m} y_{a}\mathrm{d}x^{a}$. From the property (4.5), it follows that para-contact metric manifolds are contact manifolds. Thus, λ' is identified with λ introduced in Sect. 4.2.1, $\lambda' = \lambda$.

The *Legendre submanifold* $\mathcal{A}\subset\mathcal{C}$ of a $(2m+1)$-dimensional contact manifold $(\mathcal{C},\lambda)$ is an m-dimensional submanifold where $\iota^{*}\lambda = 0$ holds, where ι^{*} is the pull-back of the embedding $\iota:\mathcal{A}\to\mathcal{C}$. One can verify that

$$\mathcal{A}_{\varpi} = \left\{ (x,y,z) \,\middle|\, y_{a} = \frac{\partial\varpi}{\partial x^{a}},\ \text{and } z=\varpi(x),\quad a=1,\ldots,m. \right\}, \tag{4.7}$$

is a Legendre submanifold, where $\varpi:\mathcal{C}\to\mathbb{R}$ is a function of x on $\mathcal{C}$. The submanifold $\mathcal{A}_{\varpi}$ is referred to as the *Legendre submanifold generated by* ϖ, and is used for describing equilibrium thermodynamic systems [12]. It is known that a wider class of functions induce Legendre submanifolds [7]. If ϖ is convex on $\mathcal{A}_\varpi$, then the metric tensor field on $\mathcal{A}_\varpi$ in a para-contact metric manifold $(\mathcal{C}, \phi, \xi, \lambda, G)$ can be introduced as

$$g_\varpi = \sum_{a=1}^{m}\sum_{b=1}^{m} \frac{\partial^2 \varpi}{\partial x^a \partial x^b} \mathrm{d}x^a \otimes \mathrm{d}x^b.$$

Similar to the case of Symplectic manifold, one can introduce vector fields on contact manifolds so that they preserve contact structure, $\ker(\lambda)$. If a vector field X of a contact manifold $(\mathcal{C}, \lambda)$ is such that

$$\mathcal{L}_X \lambda = \rho\,\lambda, \qquad \text{i.e.,} \quad (\mathcal{L}_X \lambda) \wedge \lambda = 0,$$

with some function ρ on $\mathcal{C}$, then X is referred to as a *contact vector field*. A *contact vector field associated with a function*, or a *contact Hamiltonian vector field* X_h associated with a function h on $\mathcal{C}$ is the vector field satisfying

$$\imath_{X_h} \lambda = h, \qquad \text{and} \qquad \imath_{X_h} \mathrm{d}\lambda = -\,(\,\mathrm{d}h - (Rh)\,\lambda\,),$$

where $R \in \mathcal{S}T\mathcal{C}$ is the Reeb vector field defined such that

$$\imath_R\, \mathrm{d}\lambda = 0, \quad \text{and} \quad \imath_R \lambda = 1.$$

The function h above is referred to as a *contact Hamiltonian*. With the Cartan formula $\mathcal{L}_X \beta = (\imath_X \mathrm{d} + \mathrm{d}\imath_X)\beta$ for all $X \in \mathcal{S}T\mathcal{M}$ and $\beta \in \mathcal{S}\Lambda^p\mathcal{M}$, one can show that

$$\mathcal{L}_{X_h} \lambda = (Rh)\lambda,$$

from which contact Hamiltonian vector fields are verified to be contact vector fields. In addition, it follows that

$$\mathcal{L}_{X_h} h = (Rh)h.$$

The Reeb vector field R is the characteristic vector field ξ that has been introduced in Sect. 4.2.1, $R = \xi$.

The coordinate expression of R is $R = \partial/\partial z$, and that of $X_h \in \mathcal{S}T\mathcal{C}$ is

$$X_h = \sum_{a=1}^{m}\left[\dot{x}^a \frac{\partial}{\partial x^a} + \dot{y}_a \frac{\partial}{\partial y_a}\right] + \dot{z}\frac{\partial}{\partial z},$$

where $\{\dot{x}^a\}$, $\{\dot{y}_a\}$, $\dot{z}$ are expressed as

$$\dot{x}^a = -\frac{\partial h}{\partial y_a}, \quad \dot{y}_a = \frac{\partial h}{\partial x^a} + y_a\frac{\partial h}{\partial z}, \quad \dot{z} = h - \sum_{b=1}^{m} y_b \frac{\partial h}{\partial y_b}, \quad a = 1, \ldots, m.$$

When identifying a contact Hamiltonian vector field with a dynamical system, $\dot{x}^a$ is identified with $\mathrm{d}x^a/\mathrm{d}t$ for each a.

4.2.3 Legendre Submanifold as Dually Flat Space

As shown in [3], a Legendre submanifold can be related to a dually flat space defined in information geometry. To state this explicitly, some of notions developed in information geometry are shown below [31, 32].

1. Let $(\mathcal{M}, g)$ be a (pseudo) Riemannian manifold, and ∇ a connection on $\mathcal{M}$. If a connection ∇^* satisfies

$$X[g(Y, Z)] = g(\nabla_X Y, Z) + g(Y, \nabla^*_X Z), \qquad \forall\, X, Y, Z \in \mathcal{S}T\mathcal{M}$$

then ∇^* is referred to as the *dual connection* of ∇ with respect to g.
2. On a Riemannian manifold $(\mathcal{M}, g)$, if there exists a function ϖ such that $g = \nabla \mathrm{d}\varpi$, then (∇, g) is referred to as a *Hessian structure*, and the triplet $(\mathcal{M}, \nabla, g)$ a *Hessian manifold*.
3. On a Hessian manifold $(\mathcal{M}, \nabla, g)$, there exists a coordinate system θ such that all the connection coefficients vanish everywhere. This θ is referred to as ∇*-affine coordinates*. Moreover there exists the ∇^*-affine coordinate system denoted by η, such that

$$g\left(\frac{\partial}{\partial \theta^a}, \frac{\partial}{\partial \eta_b}\right) = \delta^b_a, \qquad a, b = 1, \dots, \dim \mathcal{M}.$$

This η is referred to as the *dual coordinate system* of θ with respect to g.
4. On a Riemannian manifold $(\mathcal{M}, g)$ with a connection ∇, if the following two conditions for all $X, Y, Z \in \mathcal{S}T\mathcal{M}$

   1. $T^\nabla(X, Y) = 0$, where $T^\nabla(X, Y) := \nabla_X Y - \nabla_Y X - [X, Y]$,
   2. $(\nabla_X g)(Y, Z) = (\nabla_Y g)(X, Z)$,

   are satisfied, then a triplet $(\mathcal{M}, \nabla, g)$ is referred to as a *statistical manifold*. The introduced $(1, 2)$-tensor field T^∇ is *torsion*.
5. On a statistical manifold $(\mathcal{M}, \nabla, g)$, if curvature tensor field R^∇ defined by

$$R^\nabla(X, Y)Z = \nabla_X \nabla_Y Z - \nabla_Y \nabla_X Z - \nabla_{[X,Y]} Z, \qquad \forall\, X, Y, Z \in \mathcal{S}T\mathcal{M}$$

vanishes, then $(\mathcal{M}, \nabla, g)$ is referred to as a flat statistical manifold. A flat statistical manifold is referred to as a *dually flat space*, and is locally a Hessian manifold. In general, given a connection ∇, if $T^\nabla \equiv 0$ and $R^\nabla \equiv 0$, then the connection ∇ is referred to as being flat.

The following is a relation between contact geometry and information geometry.

Proposition 1 *(A contact manifold and a strictly convex function induce a dually flat space, [3, 19]). Let $(\mathcal{C}, \lambda)$ be a $(2m+1)$-dimensional contact manifold, (x, y, z) a set of coordinates such that $\lambda = \mathrm{d}z - \sum_{a=1}^{m} y_a \mathrm{d}x^a$ with $x = \{x^1, \ldots, x^m\}$ and $y = \{y_1, \ldots, y_m\}$, and ϖ a strictly convex function depending only on x. Then, $((\mathcal{C}, \lambda), \varpi)$ induces an m-dimensional dually flat space.*

4.2.4 Legendre Submanifold as Attractor

As shown in [3] and [21], a class of relaxation processes, initial states approach to the equilibrium state as time develops, can be formulated as contact Hamiltonian vector fields on contact manifolds. This statement on a class of contact Hamiltonian vector fields can be summarized as follows.

Proposition 2 *(Legendre submanifold as an attractor, [3]). Let $(\mathcal{C}, \lambda)$ be a $(2m+1)$-dimensional contact manifold with λ being a contact form, (x, y, z) its coordinates so that $\lambda = \mathrm{d}z - \sum_{a=1}^{m} y_a \mathrm{d}x^a$, and ϖ a function depending only on x. Then, one has the following.*

1. *The contact Hamiltonian vector field associated with the contact Hamiltonian $h_\varpi : \mathcal{C} \to \mathbb{R}$ with*
$$h_\varpi(x, y, z) := \varpi(x) - z, \tag{4.8}$$
gives
$$\frac{\mathrm{d}}{\mathrm{d}t} x^a = 0, \qquad \frac{\mathrm{d}}{\mathrm{d}t} y_a = \frac{\partial \varpi}{\partial x^a} - y_a, \qquad \frac{\mathrm{d}}{\mathrm{d}t} z = \varpi(x) - z, \tag{4.9}$$
where $a = 1, \ldots, m$.
2. *The Legendre submanifold generated by ϖ, given by (4.7), is an invariant manifold for the contact Hamiltonian vector field.*
3. *Every point on $\mathcal{C} \setminus \mathcal{A}_\varpi$ approaches to $\mathcal{A}_\varpi$ along an integral curve as time develops. Equivalently $\mathcal{A}_\varpi$ is an attractor in $\mathcal{C}$:*
$$\lim_{t \to \infty} (\, x(t), y(t), z(t) \,) \in \mathcal{A}_\varpi. \tag{4.10}$$
4. *Let $\{x(0), y(0), z(0)\}$ be a point on $\mathcal{C} \setminus \mathcal{A}_\varpi$. Then for any $t \in \mathbb{R}$, it follows that*
$$h_\varpi(x(t), y(t), z(t)) = \exp(-t)\, h_\varpi(x(0), y(0), z(0)). \tag{4.11}$$

Points on the integral curve of the contact Hamiltonian vector field in Proposition 2 can be characterized by the Mrugala metric tensor field G. To state this, one defines the length of a curve $\gamma : \mathbb{R} \to \mathcal{M}$, $(\, t \mapsto \gamma(t) \,)$, associated with a vector field $\dot{\gamma} := (\mathrm{d}/\mathrm{d}t)\gamma$ on a (pseudo) Riemannian manifold $(\mathcal{M}, g)$ as[1]

[1] In [27, 28], the sign of the length of curves is not correct.

$$l[\,g,\dot{\gamma}\,]_t^{\infty} := \int_t^{\infty} \sqrt{g(\dot{\gamma}(t'),\dot{\gamma}(t'))}\,\mathrm{d}t'.$$

Then one has the statement below.

Proposition 3 *(Length from the Legendre submanifold generated by a function, [28]). Let G be the Mrugala metric tensor field (4.4), and X_{ϖ} the contact Hamiltonian vector field whose contact Hamiltonian is given by h_{ϖ} in (4.8). Then it follows that*

$$l[\,G,X_{\varpi}\,]_t^{\infty} = |\,z(t)-\varpi(x)\,| = |\,h_{\varpi}(x(t),y(t),z(t))\,|.$$

Proof Substituting the explicit form

$$X_{\varpi} = \sum_{a=1}^{m}\left(\frac{\partial\,\varpi}{\partial x^a} - y_a\right)\frac{\partial}{\partial y_a} + (\,\varpi(x)-z\,)\frac{\partial}{\partial z},$$

into $G(X_{\varpi},X_{\varpi})$, one has

$$G(X_{\varpi},X_{\varpi}) = \lambda(X_{\varpi})\,\lambda(X_{\varpi}) = (\,h_{\varpi}\,)^2.$$

Thus, the length is expressed as

$$l[\,G,X_{\varpi}\,]_t^{\infty} = \int_t^{\infty}\sqrt{G(X_{\varpi},X_{\varpi})}\,\mathrm{d}t' = \int_t^{\infty} |\,h_{\varpi}(x(t'),y(t'),z(t'))\,|\,\mathrm{d}t'.$$

This definite integral above is evaluated using (4.11) as

$$l[\,G,X_{\varpi}\,]_t^{\infty} = |\,h(x(0),y(0),z(0))\,|\int_t^{\infty}\exp(-\,t')\,\mathrm{d}t' = |\,h_{\varpi}(x(t),y(t),z(t))\,|,$$

which is the same as $|\varpi(x)-z|$, due to (4.8). □

The following also characterizes the relaxation process.

Proposition 4 *(Ricci curvature along relaxation process).*

$$\mathbf{Ric}^G(X_{\varpi},X_{\varpi}) = -2m(\,h_{\varpi}\,)^2, \qquad \textit{and} \qquad \mathbf{Ric}^G(X_{\varpi},R) = -2m\,h_{\varpi}.$$

Proof Substituting $\lambda(X_{\varpi}) = h_{\varpi}$,

$$G(X_{\varpi},X_{\varpi}) = (\,\lambda(X_{\varpi})\,)^2 = (h_{\varpi})^2, \quad \text{and} \quad G(X_{\varpi},R) = \lambda(X_{\varpi}) = h_{\varpi},$$

into (4.6), one completes the proof. □

From Proposition 4 and (4.11), it follows that the values $|\mathbf{Ric}^G(X_{\varpi},X_{\varpi})|$ and $|\mathbf{Ric}^G(X_{\varpi},R)|$ decrease as time develops.

One is interested in how the (1, 1)-tensor field ϕ discussed in Sect. 4.2.1 plays a role for the contact Hamiltonian vector field (4.9). To give an answer, one needs the following.

Lemma 1 *([28]). Let $\{\dot{x}_a\}, \{\dot{y}_a\}, \dot{z}$ be some functions, and X_0 the vector field*

$$X_0 = \sum_{a=1}^{m} \left[\dot{x}^a \frac{\partial}{\partial x^a} + \dot{y}_a \frac{\partial}{\partial y_a} \right] + \dot{z} \frac{\partial}{\partial z}.$$

Then, $\phi(X_0)$ and $\phi^2(X_0)$ are calculated as

$$\phi^\mu(X_0) = \sum_{a=1}^{m} \left[(-1)^\mu \dot{x}^a \left(\frac{\partial}{\partial x^a} + y_a \frac{\partial}{\partial z} \right) + \dot{y}_a \frac{\partial}{\partial y_a} \right], \quad \mu = 1, 2.$$

Proof With the local expressions shown in Sect. 4.2.1, one has

$$\widehat{\theta}^a_\pm(X_0) = \frac{\sqrt{y_a}}{2} \dot{x}^a \pm \frac{\dot{y}_a}{2\sqrt{y_a}},$$

$$e^+_a + e^-_a = \frac{2}{\sqrt{y_a}} \left(\frac{\partial}{\partial x^a} + y_a \frac{\partial}{\partial z} \right), \quad \text{and} \quad e^+_a - e^-_a = 2\sqrt{y_a} \frac{\partial}{\partial y_a}.$$

Combining these, one has

$$\begin{aligned} \phi(X_0) &= \sum_{a=1}^{m} \left[-\widehat{\theta}^a_-(X_0) e^+_a + \widehat{\theta}^a_+(X_0) e^-_a \right] \\ &= \sum_{a=1}^{m} \left[-\frac{\sqrt{y_a}}{2} \dot{x}^a (e^+_a + e^-_a) + \frac{\dot{y}_a}{2\sqrt{y_a}} (e^+_a - e^-_a) \right] \\ &= \sum_{a=1}^{m} \left[-\dot{x}^a \left(\frac{\partial}{\partial x^a} + y_a \frac{\partial}{\partial z} \right) + \dot{y}_a \frac{\partial}{\partial y_a} \right]. \end{aligned}$$

For $\phi^2(X_0)$, substituting $\lambda(X_0) = \dot{z} - \sum_{a=1}^{m} y_a \dot{x}^a$ into $\phi^2(X_0) = X_0 - \lambda(X_0)\xi$, one has the desired expression. □

Applying Lemma 1, one has the following.

Proposition 5 *(Roles of ϕ for the contact Hamiltonian system, [28]). Let X_ϖ be the contact Hamiltonian vector field in Proposition 2. Then*

$$\mathcal{L}_{\phi(X_\varpi)} h_\varpi = \mathcal{L}_{\phi^2(X_\varpi)} h_\varpi = 0.$$

Proof Substituting $\dot{x}^a = 0$ into $\phi^\mu(X_0)$ in Lemma 1, one has

$$\phi^\mu(X_\varpi) = \sum_{a=1}^{m} \dot{y}_a \frac{\partial}{\partial y_a}, \qquad \text{where} \ \ \dot{y}_a = \frac{\partial \varpi}{\partial x^a} - y_a, \qquad \mu = 1, 2.$$

Then, with $\partial h_\varpi / \partial y_a = 0$, one has

$$\mathcal{L}_{\phi^\mu(X_\varpi)} h_\varpi = \left[\phi^\mu(X_\varpi)\right] h_\varpi = 0, \qquad \mu = 1, 2.$$

□

This states that the h_ϖ is preserved along $\phi^\mu(X_\varpi) \in \mathcal{S}TC$, which should be compared with the case of $\mathcal{L}_{X_\varpi} h_\varpi$:

$$\mathcal{L}_{X_\varpi} h_\varpi = -\dot{z} = -(\varpi(x) - z) = -h_\varpi .$$

In addition to Proposition 5, one has the following.

Remark 1 The vector field $\phi^\mu(X_\varpi)$, $(\mu = 1, 2)$ is not a contact one, since

$$\mathcal{L}_{\phi^\mu(X_\varpi)} \lambda = \sum_{a=1}^{m} \left(\frac{\partial \varpi}{\partial x^a} - y_a \right) \mathrm{d}x^a \neq \rho\lambda, \qquad \mu = 1, 2,$$

where ρ is some function.

The discussions above can be generalized by taking a class of contact Hamiltonians [3]. The following can be proven straightforwardly.

Proposition 6 *(Generalization of the contact Hamiltonian system with h_ϖ). Consider the contact Hamiltonian h_Δ with some function $\widehat{h}$ so that*

$$h_\Delta(x, y, z) = \widehat{h}(\Delta), \qquad \textit{where} \ \ \Delta := \frac{\partial \varpi}{\partial x^a} - z.$$

Then, its contact Hamiltonian vector field X_Δ has the properties

$$l[G, X_\Delta]_t^\infty = |\, h_\Delta \,|, \qquad \mathcal{L}_{X_\Delta} h_\Delta = -\frac{\mathrm{d}\widehat{h}}{\mathrm{d}\Delta} h_\Delta,$$

$$\mathbf{Ric}^G(X_\Delta, X_\Delta) = -2m\,(h_\Delta)^2, \quad \mathbf{Ric}^G(X_\Delta, R) = -2m\, h_\Delta,$$

and

$$\mathcal{L}_{\phi^\mu(X_\Delta)} h_\Delta = 0, \qquad \mu = 1, 2.$$

4.3 Distribution Functions from Solvable Master Equations

In this section a set of master equations with a particular set of Markov kernels (transition matrix elements) is introduced, and then its time-development is analyzed.

Let Γ be a set of finite discrete states, $t \in \mathbb{R}$ time, and $p(j,t)\,\mathrm{d}t$ a probability distribution function that a state $j \in \Gamma$ is found in between t and $t + \mathrm{d}t$. The first objective is to realize a given distribution function p_θ^{eq} that can be written as

$$p_\theta^{\mathrm{eq}}(j) = \frac{\pi_\theta(j)}{Z(\theta)}, \qquad Z(\theta) := \sum_{j \in \Gamma} \pi_\theta(j), \tag{4.12}$$

where $\theta \in \Theta \subset \mathbb{R}^n$ is a parameter set with $\theta = \{\theta^1, \ldots, \theta^n\}$, and $Z : \Theta \to \mathbb{R}$ the so-called partition function so that p_θ^{eq} is normalized : $\sum_{j \in \Gamma} p_\theta^{\mathrm{eq}}(j) = 1$. Although it is often the case that $n \ll |\Gamma|$ when thermodynamic systems are considered, we do not assume this case, where $|\Gamma|$ is the number of elements of Γ, $|\Gamma| := \#\Gamma$.

In what follows, attention is focused on a class of master equations. Let $p : \Gamma \times \mathbb{R} \to \mathbb{R}_{\geq 0}$ be a time-dependent probability distribution function, where $t \in \mathbb{R}$ denotes time. Then, consider the set of master equations (4.1),

$$\frac{\partial}{\partial t} p(j,t) = \sum_{j' \in \Gamma \setminus \{j\}} \left[w(j|j')\, p(j',t) - w(j'|j)\, p(j,t) \right],$$

where $w : \Gamma \times \Gamma \to I, (I := [0,1] \subset \mathbb{R})$ is such that $w(j|j')$ denotes a probability that a state jumps from j' to j. This w is referred to as a transition matrix (and kernel). With (4.1), and the assumptions

$$w_\theta(j|j') = p_\theta^{\mathrm{eq}}(j),$$

together with

$$p_\theta^{\mathrm{eq}}(j) \neq 0, \quad \forall j \in \Gamma, \tag{4.13}$$

one derives the *solvable master equations*, (4.2):

$$\frac{\partial}{\partial t} p(j,t) = p_\theta^{\mathrm{eq}}(j) - p(j,t).$$

Although the term "solvable master equation" is used in this paper, such a term can be replaced with "simple master equation" for example. In addition the solvable master equations are related to another class of master equations that have interaction terms. Such a class is related to the Poisson distribution (see Appendix A).

An explicit form of $p(j,t)$ is obtained by solving (4.2). Then the following proposition can easily be shown.

Proposition 7 *(Solutions of the master equations, [27]). The solution of* (4.2) *is*

$$p(j,t) = \mathrm{e}^{-t}\, p(j,0) + (1-\mathrm{e}^{-t})p_\theta^{\mathrm{eq}}(j), \quad \textit{from which} \quad \lim_{t\to\infty} p(j,t) = p_\theta^{\mathrm{eq}}(j).$$

With this proposition, one notices the following.

1. Every solution p depends on θ.
2. The equilibrium state is realized with (4.2) as the time-asymptotic limit.
3. If $\sum_j p(j,0) = 1$, then $\sum_j p(j,t) = 1$ for any $t > 0$.

Taking into account 1, $p(j,t)$ is also denoted $p(j,t;\theta)$ in this paper. From 3, $\{p(j,t)\}$ is an element of the probability simplex,

$$\mathcal{S}^{|\Gamma|-1} := \left\{ \{p(j,t)\} \in \mathbb{R}_{\geq 0}^{|\Gamma|} \,\middle|\, \sum_{j=1}^{|\Gamma|} p(j,t) = 1 \right\}.$$

4.3.1 Denormalization

Denormalized distribution functions are often considered in the literature [33]. In this paper they are denoted $\{\widetilde{p}(j,t)\}$. For $\{\widetilde{p}(j,t)\}$, the normalization condition 3 above is not imposed, $\sum_j \widetilde{p}(j,0) \neq 1$. From the linearity of master equations and (4.2), one has

$$\frac{\partial}{\partial t}\widetilde{p}(j,t) = \widetilde{p}_\theta^{\mathrm{eq}}(j) - \widetilde{p}(j,t).$$

Discrete distribution functions belong to the exponential family, then one can write

$$p(j,t) = \exp(\theta^a \mathcal{O}_a(j) - \Psi(\theta)).$$

One realization of denormalization is to introduce a non-zero constant $w \in \mathbb{R}$ so that

$$\widetilde{p}(j,t) = w\, p(j,t).$$

For the exponential family, if $\widetilde{\Psi}$ is introduced so that

$$\widetilde{p}(j,t) = \exp\left(\theta^a \mathcal{O}_a(j) - \widetilde{\Psi}(\theta)\right),$$

then one has

$$\widetilde{\Psi}(\theta) = \Psi(\theta) - \ln w.$$

4.4 Observables with Solvable Master Equations

In this section differential equations describing time-development of observables are derived with the solvable master equations under some assumptions. Then, the time-asymptotic limit of such observables is stated. Here *observable* in this paper is defined as a function that depend on neither a random variable nor a state. Thus expectation values with respect to a probability distribution function are observables.

Let $\mathcal{O}_a : \Gamma \to \mathbb{R}$ be a function with $a \in \{1, \ldots, n\}$, and $p : \Gamma \times \mathbb{R} \to \mathbb{R}_{\geq 0}$ a distribution function that follows (4.2). Then

$$\langle \mathcal{O}_a \rangle_\theta (t) := \sum_{j\in\Gamma} \mathcal{O}_a(j)\, p(j,t;\theta), \qquad \text{and} \qquad \langle \mathcal{O}_a \rangle_\theta^{\mathrm{eq}} := \sum_{j\in\Gamma} \mathcal{O}_a(j)\, p_\theta^{\mathrm{eq}}(j),$$

are referred to as the *expectation variable* of $\mathcal{O}_a$ with respect to p, and that with respect to p_θ^{eq}, respectively.

If an equilibrium distribution function belongs to the exponential family, then the function $\Psi^{\mathrm{eq}} : \Theta \to \mathbb{R}$ with

$$\Psi^{\mathrm{eq}}(\theta) := \ln\left(\sum_{j\in\Gamma} \mathrm{e}^{\sum_{b=1}^{n}\theta^b \mathcal{O}_b(j)}\right), \tag{4.14}$$

plays various roles. Here and in what follows, (4.14) is assumed to exist. In the context of information geometry, this function is referred to as a θ*-potential.* Discrete distribution functions are considered in this paper and it has been known that such distribution functions belong to the exponential family, then Ψ^{eq} in (4.14) also plays a role throughout this paper. The value $\Psi^{\mathrm{eq}}(\theta)$ can be interpreted as the negative dimension-less free-energy, since the relation between the free-energy $\mathcal{F}$ and Ψ^{eq}

$$\mathcal{F}(\theta) = -k_{\mathrm{B}}T \ln Z(\theta) = -k_{\mathrm{B}}T\Psi^{\mathrm{eq}}(\theta),$$

holds, where k_{B} is the Boltzmann constant and T the absolute temperature, and the physical dimension of $k_{\mathrm{B}}T$ and that of $\mathcal{F}$ are energy. From (4.14), the function Ψ^{eq} relates θ^a with $\langle \mathcal{O}_a \rangle_\theta^{\mathrm{eq}}$ for each a as

$$\langle \mathcal{O}_a \rangle_\theta^{\mathrm{eq}} = \frac{\partial \Psi^{\mathrm{eq}}}{\partial \theta^a}.$$

One then can generalize Ψ^{eq} defined at equilibrium state to a function defined in nonequilibrium states as $\Psi : \Theta \times \mathbb{R} \to \mathbb{R}$,

$$\Psi(\theta, t) := \left(\frac{1}{|\Gamma|}\sum_{j\in\Gamma}\frac{p(j,t;\theta)}{p_\theta^{\mathrm{eq}}(j)}\right)\Psi^{\mathrm{eq}}(\theta).$$

Since $p_\theta^{\mathrm{eq}}(j) \neq 0$ and $\Psi^{\mathrm{eq}}(\theta) < \infty$ by assumptions, the function Ψ exists. Generalizing the idea for Ψ^{eq} in the equilibrium case, the function Ψ may be interpreted as a nonequilibrium negative dimension-less free-energy.

A set of differential equations for $\{\langle \mathcal{O}_a \rangle_\theta\}$ and Ψ can be derived as follows.

Proposition 8 *(Dynamical system obtained from the master equations, [27]). Let θ be a time-independent parameter set specifying a discrete distribution function p_θ^{eq}, and $p(j,t;\theta)$ the solution to the solvable master equations. Then $\{\langle \mathcal{O}_a \rangle_\theta\}$ and Ψ are solutions to the differential equations on $\mathbb{R}^{2n+1}$*

$$\frac{\mathrm{d}}{\mathrm{d}t}\theta^a = 0,\qquad \frac{\mathrm{d}}{\mathrm{d}t}\langle \mathcal{O}_a \rangle_\theta = -\langle \mathcal{O}_a \rangle_\theta + \frac{\partial \Psi^{\mathrm{eq}}}{\partial \theta^a},\quad \textit{and}\quad \frac{\mathrm{d}}{\mathrm{d}t}\Psi = -\Psi + \Psi^{\mathrm{eq}}.$$

Remark 2 The explicit time-dependence for this system is obtained as $\theta^a(t) = \theta^a(0)$, and $\Psi(\theta,t) = \mathrm{e}^{-t}\,[\,\Psi(0) - \Psi^{\mathrm{eq}}(\theta)\,] + \Psi^{\mathrm{eq}}(\theta)$, and

$$\langle \mathcal{O}_a \rangle_\theta(t) = \mathrm{e}^{-t}\left[\langle \mathcal{O}_a \rangle_\theta(0) - \frac{\partial \Psi^{\mathrm{eq}}}{\partial \theta^a}\right] + \frac{\partial \Psi^{\mathrm{eq}}}{\partial \theta^a}.$$

From these, one can verify that the time-asymptotic limit of these variables are those defined at equilibrium. In this paper the dynamical system in Proposition 8 is referred to as the *moment dynamical system*.

There is a symmetry between the system with $\mathcal{O}_a$ and that with $\chi\mathcal{O}_a$, where χ is a non-zero constant.

Proposition 9 *(Scale invariance). Consider the system stated in Proposition 8. Introduce a scale factor $\chi \in \mathbb{R}\setminus\{0\}$, and define $\mathcal{O}_a^\chi$ so that*

$$\mathcal{O}_a^\chi(j) := \chi\,\mathcal{O}_a(j),\qquad a = 1,\dots,n,$$

and

$$\Psi_\chi^{\mathrm{eq}}(\theta) := \ln\left(\sum_{j\in\Gamma} \mathrm{e}^{\chi\theta^b\mathcal{O}_b(j)}\right),\qquad \Psi_\chi(\theta,t) := \left(\frac{1}{|\Gamma|}\sum_{j\in\Gamma}\frac{p(j,t;\theta)}{p_\theta^{\mathrm{eq}}(j)}\right)\Psi_\chi^{\mathrm{eq}}(\theta).$$

Then one has formally the same equations in Proposition 8:

$$\frac{\mathrm{d}}{\mathrm{d}t}\theta^a = 0,\quad \frac{\mathrm{d}}{\mathrm{d}t}\left\langle \mathcal{O}_a^\chi \right\rangle_\theta = -\left\langle \mathcal{O}_a^\chi \right\rangle_\theta + \frac{\partial \Psi_\chi^{\mathrm{eq}}}{\partial \theta^a},\quad \textit{and}\quad \frac{\mathrm{d}}{\mathrm{d}t}\Psi_\chi = -\Psi_\chi + \Psi_\chi^{\mathrm{eq}}.$$

Proof The first equation is trivial. The other two equations are verified by the use of the solvable master equation (4.2) together with the definitions of $\mathcal{O}_a^\chi(j)$, $\left\langle \mathcal{O}_a^\chi \right\rangle_\theta$, and Ψ_χ. □

4.5 Geometric Description of Master Equations

There have been several attempts to geometrically describe the time-development of probability distribution functions [34–39]. In addition to these, any theoretical development of geometric description of master equations will be expected to yield several benefits. In this section, a geometrization of the solvable master equations is proposed. This is accomplished by the use of contact (metric) geometry. Before showing this, a geometrization of equilibrium states is focused.

4.5.1 *Geometry of Equilibrium States*

Equilibrium state is identified with the Legendre submanifold generated by a function on a contact manifold, as the standard contact geometric description of equilibrium thermodynamics [12]. In this subsection this aspect of the equilibrium state of the solvable master equations is explicitly shown. To this end the equilibrium distribution function p_θ^{eq} is identified with (4.12), where $p_\theta^{\mathrm{eq}}(j) > 0, (j \in \Gamma)$ holds due to (4.13).

Choose an appropriate contact manifold. If a convex function on it exists, then it follows from Proposition 1 that the corresponding dually flat space is induced. As will be explained below, since a function ϖ defining the contact Hamiltonian discussed earlier is not convex on a Legendre submanifold in Darboux coordinates, such a dually flat space is not induced for the master equations. However still a Legendre submanifold for expressing the equilibrium state can be defined. To do so, one introduces a contact manifold as an ambient manifold. After this, a possible dually flat space is discussed in another set of coordinates.

To express this contact manifold, one introduces coordinates as follows. Let $f : \mathbb{R}_{>0} \to \mathbb{R}$ be a function, define $I_f^{\mathrm{eq}} : \mathbb{R}^{|\Gamma|} \to \mathbb{R}$ and $p_{\mathrm{eq}}^{f\,j} \in \mathbb{R}$ with

$$I_f^{\mathrm{eq}}(p_{\mathrm{eq}}^f) := \sum_{j\in\Gamma} p_{\mathrm{eq}}^{f\,j}, \quad \text{where} \quad p_{\mathrm{eq}}^f = \{p_{\mathrm{eq}}^{f\,1}, \dots, p_{\mathrm{eq}}^{f\,|\Gamma|}\} \in \mathbb{R}^{|\Gamma|}$$

and

$$p_{\mathrm{eq}}^{f\,j} := p_{\mathrm{eq}}^j\, f(p_{\mathrm{eq}}^j), \quad \text{where} \quad p_{\mathrm{eq}}^j := p_\theta^{\mathrm{eq}}(j), \quad j = 1, \dots, |\Gamma|. \tag{4.15}$$

In addition, introduce $\psi_j : \mathbb{R} \to \mathbb{R}_{>0}, (j \in \Gamma)$ so that

$$p(j,t) = p_{\mathrm{eq}}^j\, \psi_j(t), \quad j = 1, \dots, |\Gamma|. \tag{4.16}$$

and I_f is such that

$$I_f(t) := \sum_{j\in\Gamma} p_{\mathrm{eq}}^{f\,j}\, \psi_j(t). \tag{4.17}$$

The set of initial values $\{\psi_j(0)\}$ is fixed with Proposition 7 or (4.16) as

$$\psi_j(0) = \frac{1}{p_{\rm eq}^j} p(j,0), \qquad j = 1, \dots, |\Gamma|,$$

where the right hand side above has been assumed to exist (see (4.13)). Then a pair $(\mathcal{C}_f^\Gamma, \lambda_f^\Gamma)$ is a $(2|\Gamma|+1)$-dimensional contact manifold, where

$$\mathcal{C}_f^\Gamma := \mathbb{R}^{|\Gamma|} \times \mathbb{R}_{>0}^{|\Gamma|} \times \mathbb{R},$$

and

$$\lambda_f^\Gamma = \mathrm{d}I_f - \sum_{j\in\Gamma} \psi_j \, \mathrm{d}\, p_{\rm eq}^{fj}. \tag{4.18}$$

Its Darboux coordinates are $(p_{\rm eq}^f, \psi, I_f)$ with $\psi = \{\psi_1, \dots, \psi_{|\Gamma|}\}$.

An example of f is

$$f(p_{\rm eq}^j) = -\ln p_{\rm eq}^j, \qquad j = 1, \dots, |\Gamma|,$$

and in this case $I_f^{\rm eq}$ is the entropy of the equilibrium distribution function.

The function $I_f^{\rm eq}$ generates the Legendre submanifold $\mathcal{A}_{I_f^{\rm eq}} \subset \mathcal{C}_f^\Gamma$ as in (4.7) with $\varpi = I_f^{\rm eq}$, which is explicitly written as

$$\mathcal{A}_{I_f^{\rm eq}} = \left\{ (p_{\rm eq}^f, \psi, I_f) \in \mathcal{C}_f^\Gamma \;\middle|\; \psi_j = \frac{\partial I_f^{\rm eq}}{\partial p_{\rm eq}^{fj}}, \quad I_f = I_f^{\rm eq}(p_{\rm eq}^f), \quad j = 1, \dots, |\Gamma| \right\},$$

which reduces to

$$\mathcal{A}_{I_f^{\rm eq}} = \left\{ (p_{\rm eq}^f, \psi, I_f) \in \mathcal{C}_f^\Gamma \;\middle|\; \psi_j = 1, \quad I_f = I_f^{\rm eq}(p_{\rm eq}^f), \quad j = 1, \dots, |\Gamma| \right\},$$

due to

$$\frac{\partial I_f^{\rm eq}}{\partial p_{\rm eq}^{fj}} = 1, \qquad j = 1, \dots, |\Gamma|.$$

Notice that the state where $\psi_j = 1$ for all $j \in \Gamma$ is the equilibrium state due to (4.16).

Then combining the discussions so far and a part of Proposition 1, one has the following.

Proposition 10 *The equilibrium state of the solvable master equations is expressed as the Legendre submanifold $\mathcal{A}_{I_f^{\rm eq}}$ of the contact manifold $(\mathcal{C}_f^\Gamma, \lambda_f^\Gamma)$.*

Remark 3 Since

$$\frac{\partial^2 I_f^{\rm eq}}{\partial p_{\rm eq}^{fj} \partial p_{\rm eq}^{fk}} = 0, \qquad j,k = 1, \dots, |\Gamma|,$$

the function I_f^{eq} does not induce a metric tensor field with respect to the coordinates p_{eq}^f.

Although I_f^{eq} does not induce a metric tensor field with respect to p_{eq}^f, the function I_f^{eq} induces a metric tensor field with respect to p_{eq}. Then its dual coordinate system is induced. For the sake of completeness, this is briefly discussed below.

On the manifold $\mathbb{R}^{|\Gamma|}$, let p_{eq} be a coordinate system, I_{eq}^f a function of p_{eq}, and define

$$g_f^\Gamma := \sum_{j\in\Gamma}\sum_{k\in\Gamma} \frac{\partial^2 I_f^{\mathrm{eq}}}{\partial p_{\mathrm{eq}}^j \partial p_{\mathrm{eq}}^k} \mathrm{d}p_{\mathrm{eq}}^j \otimes \mathrm{d}p_{\mathrm{eq}}^k.$$

If f is chosen as $f(p_{\mathrm{eq}}^j) = \ln p_{\mathrm{eq}}^j$, $(j \in \Gamma)$ so that

$$I_f^{\mathrm{eq}}(p_{\mathrm{eq}}) = \sum_{j\in\Gamma} p_{\mathrm{eq}}^j \ln p_{\mathrm{eq}}^j,$$

then one has the *Shahshahani metric tensor field* [33, 40]:

$$g^{\mathrm{S}} := \sum_{j\in\Gamma} \frac{1}{p_{\mathrm{eq}}^j} \mathrm{d}p_{\mathrm{eq}}^j \otimes \mathrm{d}p_{\mathrm{eq}}^j.$$

The dual coordinates for this choice of f, $f(p_{\mathrm{eq}}^j) = \ln p_{\mathrm{eq}}^j$, are

$$P^j := p_{\mathrm{eq}}^j, \quad \text{and} \quad P_j^* := \frac{\partial I_f^{\mathrm{eq}}(p_{\mathrm{eq}})}{\partial p_{\mathrm{eq}}^j} = 1 + \ln p_{\mathrm{eq}}^j, \qquad j = 1, \ldots, |\Gamma|,$$

in the sense that

$$g^{\mathrm{S}}\left(\frac{\partial}{\partial P^j}, \frac{\partial}{\partial P_k^*}\right) = \delta_j^k.$$

Notice that if $f \equiv 1$, then I_f^{eq} is not convex. Thus in this case Riemannian metric tensor field is not induced. In addition, in this case $p_{\mathrm{eq}}^f = p_{\mathrm{eq}} \in \mathcal{S}^{|\Gamma|-1}$.

4.5.2 Geometry of Nonequilibrium States

So far geometry of equilibrium states has been discussed. One remaining issue is how to give the physical meaning of the set outside $\mathcal{A}_{I_f^{\mathrm{eq}}}$, $\mathcal{C}_f^\Gamma \setminus \mathcal{A}_{I_f^{\mathrm{eq}}}$. A natural interpretation of $\mathcal{C}_f^\Gamma \setminus \mathcal{A}_{I_f^{\mathrm{eq}}}$ would be some set of nonequilibrium states. We make this interpretation in this paper (see also [21]).

In the contact geometric framework of thermodynamics, the equilibrium state is identified with a Legendre submanifold. Then, as found in [3] and [21], some dynamical systems expressing nonequilibrium process of thermodynamic variables can be identified with a class of contact Hamiltonian vector fields on a contact manifold. The above claim also holds on para-contact metric manifolds.

To clarify how time-development of probability distribution functions, that are not thermodynamic variables, can be described on contact and para-contact metric manifolds, in the following a geometric description of the time-development of the master equations is shown. Our basic strategy is the same as that of thermodynamic variables as explained below. As shown in Proposition 7, initial states approach to the equilibrium state as time develops. This time-development can be reformulated on contact manifolds and para-contact metric manifolds with Proposition 2. The details are as follows.

Given a function $f : \mathbb{R}_{>0} \to \mathbb{R}$, the functions $\{p_{\text{eq}}^{f\,j}\}$, $\{\psi_j\}$, and I_f have been defined as (4.15), (4.16), and (4.17), respectively. Then the dynamic equations for $\{p_{\text{eq}}^{j}\}$, $\{\psi_j\}$ and I_f are obtained from (4.2) as

$$\frac{\mathrm{d}}{\mathrm{d}t} p_{\text{eq}}^{f\,j} = 0, \quad \frac{\mathrm{d}}{\mathrm{d}t}\psi_j = 1 - \psi_j, \quad \frac{\mathrm{d}}{\mathrm{d}t} I_f = I_f^{\text{eq}} - I_f, \quad j = 1, \dots, |\Gamma|. \tag{4.19}$$

The dynamical system (4.19) is a contact Hamiltonian system as stated below.

Proposition 11 *(Master equations as contact Hamiltonian system). Let* $(\mathcal{C}_f^{\Gamma}, \lambda_f^{\Gamma})$ *be a* $(2|\Gamma| + 1)$*-dimensional contact manifold, and* $(p_{\text{eq}}^{f}, \psi, I_f)$ *the Darboux coordinates so that* λ_f^{Γ} *is given by* (4.18).

Then (4.19) *can be written as a contact Hamiltonian system with the contact Hamiltonian*

$$h_{I_f^{\text{eq}}}^{\Gamma}(p_{\text{eq}}^{f}, \psi, I_f) = I_f^{\text{eq}}(p_{\text{eq}}) - I_f. \tag{4.20}$$

In addition, it follows that

$$\lim_{t\to\infty} (\, p_{\text{eq}}^{f}, \psi(t), I_f(t)\,) \in \mathcal{A}_{I_f^{\text{eq}}}. \tag{4.21}$$

Proof Identify $p_{\text{eq}}^{f}, \psi, I_f$ with x, y, z in Proposition 2, respectively. In addition, identify I_f^{eq} with ϖ. Then it follows that (4.9) is equivalent to (4.19). Thus (4.19) is the contact Hamiltonian system with the contact Hamiltonian (4.20). From the general discussion stated as (4.10), the property (4.21) holds. □

For the sake of completeness, the solution to (4.19) is given. This is immediately obtained as

$$\begin{aligned} p_{\text{eq}}^{j}(t)\, f(p_{\text{eq}}^{j}(t)) &= p_{\text{eq}}^{j}(0) f(p_{\text{eq}}^{j}(0)), \\ \psi_j(t) &= \mathrm{e}^{-t}\psi_j(0) + (1 - \mathrm{e}^{-t}), \\ I_f(t) &= \mathrm{e}^{-t} I_f(0) + (1 - \mathrm{e}^{-t}) I_f^{\text{eq}}. \end{aligned}$$

To characterize points on an integral curve of $X_{I_f^{\rm eq}}$ that is the contact Hamiltonian vector field generated by $h^{\Gamma}_{I_f^{\rm eq}}$ in (4.20), introduce the Mrugla metric tensor field adapted to the contact manifold $(\mathcal{C}^{\Gamma}_f, \lambda^{\Gamma}_f)$,

$$G^{\Gamma} := \frac{1}{2}\sum_{j\in\Gamma}\left[\mathrm{d}p_{\rm eq}^{f\,j}\otimes \mathrm{d}\psi_j + \mathrm{d}\psi_j\otimes \mathrm{d}p_{\rm eq}^{f\,j}\right] + \lambda^{\Gamma}_f\otimes\lambda^{\Gamma}_f.$$

Then from discussions in Sect. 4.2.4, one has the following.

Proposition 12 *The length between a state and the equilibrium state for the master equations calculated with (4.4) is*

$$l[\,G^{\Gamma}, X_{I_f^{\rm eq}}\,]_t^{\infty} = \int_t^{\infty}\sqrt{G^{\Gamma}(X_{I_f^{\rm eq}}, X_{I_f^{\rm eq}})}\,\mathrm{d}t' = |\,h^{\Gamma}_{I_f^{\rm eq}}(p^f_{\rm eq}, \psi, I_f)\,|, \qquad (4.22)$$

where $X_{I_f^{\rm eq}}$ is the contact Hamiltonian vector field associated with $h^{\Gamma}_{I_f^{\rm eq}}(p^f_{\rm eq}, \psi, I_f)$. Then the convergence rate is exponential. In addition, it follows that

$$\mathbf{Ric}^{G^{\Gamma}}\left(X_{I_f^{\rm eq}}, X_{I_f^{\rm eq}}\right) = -2|\Gamma|\left(h^{\Gamma}_{I_f^{\rm eq}}\right)^2,$$

and

$$\mathbf{Ric}^{G^{\Gamma}}\left(X_{I_f^{\rm eq}}, \frac{\partial}{\partial I_f}\right) = -2|\Gamma|\,h^{\Gamma}_{I_f^{\rm eq}}.$$

4.5.2.1 Denormalization

In below, denormalized distribution functions $\{\widetilde{p}(j,t)\}$ are notationally distinguished from $\{p(j,t)\}$. Introduce $\{\widetilde{\psi}_j\}$ and $\widetilde{I}_f$ such that

$$\widetilde{p}(j,t) = p^j_{\rm eq}\,\widetilde{\psi}_j(t), \qquad \widetilde{I}_f(t) := \sum_{j\in\Gamma} p^{f\,j}_{\rm eq}\,\widetilde{\psi}_j(t),$$

where f has been a given function, and $\{p^{f\,j}_{\rm eq}\}$ have been defined in (4.15). The dynamic equations for $\{p^j_{\rm eq}\}$, $\{\widetilde{\psi}_j\}$ and $\widetilde{I}_f$ are obtained from (4.2), from which

$$\frac{\mathrm{d}}{\mathrm{d}t}p^{f\,j}_{\rm eq} = 0, \quad \frac{\mathrm{d}}{\mathrm{d}t}\widetilde{\psi}_j = 1 - \widetilde{\psi}_j, \quad \frac{\mathrm{d}}{\mathrm{d}t}\widetilde{I}_f = I^{\rm eq}_f(p_{\rm eq}) - \widetilde{I}_f, \quad j = 1,\ldots,|\Gamma|. \qquad (4.23)$$

It should be noted that the dynamical system (4.23) is formally same as that of (4.19).

4.6 Geometric Description of Expectation Variables

Several geometrization of expectation variables and thermodynamic variables in nonequilibrium states for some models and methods have been proposed. Yet, suffice to say that there remains no general consensus on how best to extend a geometry of equilibrium states to a geometry of nonequilibrium states. In this section, a geometrization of nonequilibrium states is proposed for the observables associated with the moment dynamical system defined in Proposition 8.

4.6.1 Geometry of Equilibrium States

Equilibrium states are identified with the Legendre submanifolds generated by functions in the context of geometric thermodynamics [11, 12]. Besides, in the context of information geometry, equilibrium states are identified with dually flat spaces [1]. Combining these identifications, one can employ Proposition 1 to discuss information geometric aspects of equilibrium states. To apply Proposition 1 to physical systems, the coordinate sets x and y are chosen such that x^a and y_a form a thermodynamic conjugate pair for each a. Here it is assumed that such thermodynamic variables can be defined even for nonequilibrium states, and that they are consistent with those variables defined at equilibrium. In addition to this, the physical dimension of ϖ should be equal to that of $y_a \, \mathrm{d}x^a$. Moreover, Ψ and its Legendre transform are chosen as ϖ.

Choose an appropriate contact manifold and a convex function on it. Then, it follows from Proposition 1 that the corresponding dually flat space is induced. To have such a space, a contact manifold is specified first.

This appropriate contact manifold is the pair $(\mathcal{C}^{\mathcal{O}}, \lambda^{\mathcal{O}})$, where

$$\mathcal{C}^{\mathcal{O}} := \mathbb{R}^n \times \mathbb{R}^n \times \mathbb{R},$$

and

$$\lambda^{\mathcal{O}} = \mathrm{d}\,\Psi - \sum_{a=1}^{n} \langle \mathcal{O}_a \rangle_\theta \, \mathrm{d}\theta^a .$$

To have a dually flat space, the function Ψ^{eq} in (4.14) is used as a convex function. This convex function generates the Legendre submanifold $\mathcal{A}_{\Psi^{\mathrm{eq}}} \subset \mathcal{C}^{\mathcal{O}}$ as in (4.7) with $\varpi = \Psi^{\mathrm{eq}}$, which is explicitly written in coordinates as

$$\mathcal{A}_{\Psi^{\mathrm{eq}}} = \left\{ (\theta, \langle \mathcal{O} \rangle, \Psi) \in \mathcal{C}^{\mathcal{O}} \;\middle|\; \langle \mathcal{O}_a \rangle_\theta = \frac{\partial\, \Psi^{\mathrm{eq}}}{\partial \theta^a}, \; \Psi = \Psi^{\mathrm{eq}}(\theta), \; j = 1, \ldots, |\Gamma| \right\}.$$

Combining the discussions so far and Proposition 1, one has the following.

Proposition 13 *The equilibrium state of the moment dynamical system is expressed as the Legendre submanifold $\mathcal{A}_{\Psi^{\rm eq}}$ of the contact manifold $(\mathcal{C}^{\mathcal{O}}, \lambda^{\mathcal{O}})$. Then such equilibrium state induces a dually flat space.*

The induced dually flat space is denoted by $(\mathcal{A}_{\Psi^{\rm eq}}, \nabla^{\Psi^{\rm eq}}, g^{\Psi^{\rm eq}})$, where the Riemannian metric tensor field $g^{\Psi^{\rm eq}}$ is

$$g^{\Psi^{\rm eq}} = \sum_{a=1}^{n}\sum_{b=1}^{n} \frac{\partial^2 \Psi^{\rm eq}}{\partial\theta^a \partial\theta^b} \mathrm{d}\theta^a \otimes \mathrm{d}\theta^b.$$

The dual coordinates are

$$\theta^a \qquad \text{and} \qquad \eta_a = \frac{\partial \Psi^{\rm eq}}{\partial \theta^a}.$$

4.6.2 Geometry of Nonequilibrium States

So far geometry of equilibrium states has been discussed. In this section our interpretation of outside the Legendre submanifold is same as that in Sect. 4.5.2.

Proposition 8 is written in a contact geometric language here. In what follows phase space is identified with a $(2n+1)$-dimensional para-contact metric manifold $(\mathcal{C}, \phi, \xi, \lambda, G)$.

The moment dynamical system is a contact Hamiltonian system as stated below.

Proposition 14 *(Moment dynamical system as a contact Hamiltonian system, [27]). The dynamical system in Proposition 8 can be written as a contact Hamiltonian system.*

Proof Identify x, y, z and ϖ in Proposition 2 with $(\theta, \langle\mathcal{O}\rangle, \Psi^{\rm eq}, \Psi)$ in Proposition 8 as

$$x^a = \theta^a, \quad y_a = \langle \mathcal{O}_a\rangle_\theta, \quad \varpi(x) = \Psi^{\rm eq}(\theta), \quad z = \Psi.$$

This set of identifications yields the proof. □

In nonequilibrium statistical physics, attention is often concentrated on how far a state is close to the equilibrium state. To characterize points on an integral curve of the contact Hamiltonian vector field $X_{\Psi^{\rm eq}}$ associated with the contact Hamiltonian

$$h^{\mathcal{O}}_{\Psi^{\rm eq}}(\theta, \langle\mathcal{O}\rangle_\theta, \Psi) = \Psi^{\rm eq}(\theta) - \Psi,$$

introduce the Mrugala metric tensor field adapted to the contact manifold $(\mathcal{C}^{\mathcal{O}}, \lambda^{\mathcal{O}})$,

$$G^{\mathcal{O}} := \frac{1}{2}\sum_{a=1}^{n}\left[\mathrm{d}\theta^a \otimes \mathrm{d}\langle\mathcal{O}_a\rangle_\theta + \mathrm{d}\langle\mathcal{O}_a\rangle_\theta \otimes \mathrm{d}\theta^a\right] + \lambda_{\mathcal{O}} \otimes \lambda_{\mathcal{O}}.$$

The relation between $G^{\mathcal{O}}$ and $g^{\Psi_{\rm eq}}$ is

$$g^{\Psi_{\rm eq}} = \iota^{\mathcal{O}*} G^{\mathcal{O}},$$

where $\iota^{\mathcal{O}} : \mathcal{A}_{\Psi_{\rm eq}} \to \mathcal{C}^{\mathcal{O}}$ is the embedding.

Then from discussions in Sect. 4.2.4, one has the following.

Proposition 15 *The length between a state and the equilibrium state for the moment dynamical system calculated with (4.4) is*

$$l[G^{\mathcal{O}}, X_{\Psi_{\rm eq}}]_t^{\infty} = \int_t^{\infty} \sqrt{G^{\mathcal{O}}(X_{\Psi_{\rm eq}}, X_{\Psi_{\rm eq}})}\, {\rm d}t' = |\, h^{\mathcal{O}}_{\Psi_{\rm eq}}(\theta, \langle \mathcal{O} \rangle_\theta, \Psi)\,|, \quad (4.24)$$

where $\langle \mathcal{O} \rangle_\theta = \{\langle \mathcal{O}_1 \rangle_\theta, \ldots, \langle \mathcal{O}_n \rangle_\theta\}$, and $X_{\Psi_{\rm eq}}$ the contact Hamiltonian vector field associated with $h^{\mathcal{O}}_{\Psi_{\rm eq}}$. Then the convergence rate for (4.24) is exponential. In addition, it follows that

$$\mathbf{Ric}^{G^{\mathcal{O}}}(X_{\Psi_{\rm eq}}, X_{\Psi_{\rm eq}}) = -2n \left(h^{\mathcal{O}}_{\Psi_{\rm eq}}\right)^2,$$

and

$$\mathbf{Ric}^{G^{\mathcal{O}}}\left(X_{\Psi_{\rm eq}}, \frac{\partial}{\partial \Psi}\right) = -2n\, h^{\mathcal{O}}_{\Psi_{\rm eq}}.$$

Combining Propositions 10–15, one arrives at the main theorem in this paper.

Theorem 1 *(Geometric descriptions of the solvable master equations and moment dynamical systems). The solvable master equations and moment dynamical system derived from the solvable master equations are described on para-contact metric manifolds, and its convergence to the equilibrium states are characterized by the Mrugala metric fields and the Ricci tensor fields associated with the Levi-Civita connections.*

4.7 Beyond the Toy Model

So far as a toy model, called the solvable master equations in this paper, the master equations with a particular choice of transition matrix have mainly been discussed. Although various mathematical structures have been clarified due to its simplicity, this toy model lacks generality. In what follows, how much the present contact geometric approach and its variant can possibly be applied to the case with general transition matrices is briefly discussed.

4.7.1 Equilibrium States

By recalling the claims in Sect. 4.5.1, one realizes that a peculiarity of the choice of transition matrix for master equations has never been used in the discussion of equilibrium states. Therefore claims on geometric description of equilibrium states for master equations can be extended to general master equations. For example, from Proposition 10, one immediately has the following.

Proposition 16 *Given master equations, assume that there exists an equilibrium state. The equilibrium state of the master equations is then expressed as the Legendre submanifold of the contact manifold.*

4.7.2 Nonequilibrium States

For the toy model studied in this paper there is a simple relaxation process, that is, $p_{\theta}^{\mathrm{eq}}(j)$ is realized in the asymptotic limit, $t \to \infty$. This model with the carefully chosen transition matrix enables the process to be described in the contact geometric language. Since some systems with other complicated transition matrices show relaxation process, one possible extension of the present toy model could be obtained by choosing such a transition matrix carefully. In addition, as shown in Appendix A, the use of the probability generating function simplifies a class of master equations. However it is unclear which class of master equations can be described by contact Hamiltonian systems, and such a direction should be pursued as a future work.

Another direction to study master equations in the contact geometric language is to focus on the Hamilton-Jacobi equation obtained from master equations under some approximation [41], since contact geometric descriptions of the Hamilton-Jacobi equation have been well-recognized [42].

4.8 Conclusions

This paper has offered a viewpoint that the solvable master equations and expectation variables of the moment dynamical system derived from the master equations can be described on para-contact metric manifolds. To give a geometric description of these, contact Hamiltonian vector fields have been introduced on para-contact metric manifolds. Then relaxation processes have been characterized by the use of metric and the Ricci tensor fields. Moreover possible information geometric structures for moment dynamical systems have been clarified.

The significance of these descriptions given in this paper is mentioned here. Although a number of pure mathematical studies of para-contact metric manifolds exist in the literature, there were a few applications to geometric thermodynamics. Together with this, Riemannian metric tensor fields are often considered in contact geometric thermodynamics, applications of facts found in pure mathematical studies are thus expected to yield beneficial statements in thermodynamics. The present study has given some of such applications. In addition, although there have been a variety of studies of master equations as dynamical systems, any contact geometric approach to these equations had not been shown. Thus the present study is a first step to build a complete contact geometric theory of master equations.

There remain unsolved problems that have not been addressed in this article. These include how this analysis is applied to a wider class of master equations and corresponding moment dynamical systems. In addition, how theorems found in the study of para-contact metric geometry can be applied to the master equations more, and how other geometric formulations of thermodynamics [43, 44] can be related to the present study should be addressed. By addressing these questions together with this study, it is expected that a relevant and sophisticated geometric methodology will be established for dealing with master equations and their applications.

Acknowledgements The author S.G. is partially supported by JSPS (KAKENHI) grant number JP19K03635. The other author H.H. is partially supported by JSPS (KAKENHI) grant number JP17H01793. In addition, both of the authors are partially supported by JST CREST JPMJCR1761.

Disclaimer: Views and opinions expressed are those of the authors and do not necessarily represent official positions of their respective companies.

A. Link Between the Solvable Model and One-Step Process

In this section it is shown that there is a link between the solvable model (4.2) and a model described by master equations with non-trivial transition matrix elements.

The solvable model (or the toy model) (4.2) is rewritten by introducing the set of the variables

$$q_\theta(j,t) = p(j,t) - p_\theta^{\mathrm{eq}}(j),$$

as

$$\frac{\mathrm{d}}{\mathrm{d}t} q_\theta(j,t) = -\, q_\theta(j,t), \qquad j = 1, \ldots, |\Gamma|. \tag{4.25}$$

It is shown below that this set of equations, (4.25), is related to another class of master equations.

First, consider the master equations with some parameter $\theta \in \Theta \subset \mathbb{R}^n$

$$\frac{\mathrm{d}}{\mathrm{d}t} \wp_\theta(j,t) = \gamma\, [\, \wp_\theta(j-1,t) - \wp_\theta(j,t)\,], \qquad j = 0, \ldots, \infty \tag{4.26}$$

where $\gamma > 0$ is constant, and $\wp_\theta(j,t) = 0$ for all t and $j < 0$. This model, (4.26), belongs to a class of one-step processes [22], and is obtained by choosing $\Gamma = \{0, 1, \ldots\}$ and $\{w(j\,|\,j')\}$ appropriately in (4.1).

Second, to see a link between (4.25) and (4.26), introduce the probability generating function

$$\widehat{\wp}_{\theta}(s,\tau)=\sum_{j=0}^{\infty}\wp_{\theta}(j,\tau)\,s^{j}, \tag{4.27}$$

where the domain of $s\in\mathbb{C}$ is defined so that the series converges, $|s|\leq 1$, and $\tau\in\mathbb{R}$ a scaled time whose scale is determined later. From (4.27), the normalization condition $\sum_{j=0}^{\infty}\wp_{\theta}(j,\tau)=1$ is equivalent to

$$\widehat{\wp}_{\theta}(1,\tau)=1. \qquad \tau\in\mathbb{R} \tag{4.28}$$

Equation (4.26) is written in terms of $\widehat{\wp}_{\theta}$ as follows. It follows from (4.26) that

$$\frac{\mathrm{d}}{\mathrm{d}t}\widehat{\wp}_{\theta}(s,t)=\sum_{j=0}^{\infty}\left[\frac{\mathrm{d}}{\mathrm{d}t}\wp_{\theta}(j,t)\right]s^{j}=\gamma\left[\sum_{j=-1}^{\infty}\wp_{\theta}(j,t)\,s^{j+1}-\widehat{\wp}_{\theta}(s,t)\right].$$

Rewriting the the most right hand side of the equation above in terms of $\widehat{\wp}_{\theta}$ with $\wp_{\theta}(-1,t)=0$, one has

$$\frac{\mathrm{d}}{\mathrm{d}t}\widehat{\wp}_{\theta}(s,t)=-\gamma(1-s)\,\widehat{\wp}_{\theta}(s,t).$$

Then putting $\tau=\gamma(1-s)\,t$ and restricting $0\leq s<1$ with $s\in\mathbb{R}$, one has that $\widehat{\wp}_{\theta}(s,\tau)\in\mathbb{R}$, and

$$\frac{\mathrm{d}}{\mathrm{d}\tau}\widehat{\wp}_{\theta}(s,\tau)=-\,\widehat{\wp}_{\theta}(s,\tau), \qquad \tau\in\mathbb{R} \tag{4.29}$$

where $s=1$ has been excluded so that $\tau=c\,t$ is satisfied with some $c>0$. Notice that

$$\lim_{\tau\to\infty}\widehat{\wp}_{\theta}(s,\tau)=0,$$

for all s in $[0,1)$. In addition, it follows from (4.28) that $\widehat{\wp}_{\theta}(1,\tau)$ is constant, from which (4.29) does not hold for $s=1$, at which τ vanishes. Thus, in what follows the domain of s is chosen to be $[0,1)$ for (4.29).

To see a relation between (4.2) and (4.29) more directly, (4.29) is rewritten below. First, introduce $\widehat{\wp}_{\theta}^{\mathrm{eq}}(s)$ as a prescribed equilibrium distribution function of s with some parameter $\theta\in\Theta$, and the set of variables $\widehat{\wp}$, where

$$\widehat{\wp}(s,\tau)=\widehat{\wp}_{\theta}(s,\tau)+\widehat{\wp}_{\theta}^{\mathrm{eq}}(s), \qquad 0\leq s<1,$$

so that

$$\lim_{\tau\to\infty}\widehat{\wp}(s,\tau)=\widehat{\wp}_{\theta}^{\mathrm{eq}}(s), \qquad 0\leq s<1.$$

By assuming the existence of $\wp_{\theta}^{\mathrm{eq}}$ and that of $\wp$, the normalization condition is written with (4.27) as

$$\widehat{\wp}_{\theta}^{\mathrm{eq}}(1) = \sum_{j=0}^{\infty} \wp_{\theta}^{\mathrm{eq}}(j) = 1, \qquad \text{and} \qquad \widehat{\wp}(1,\tau) = \sum_{j=0}^{\infty} \wp(j,\tau) = 0, \qquad \tau \in \mathbb{R}.$$

Then, the system (4.29) is immediately written in terms of $\widehat{\wp}$ as

$$\frac{\mathrm{d}}{\mathrm{d}\tau}\widehat{\wp}(s,\tau) = \widehat{\wp}_{\theta}^{\mathrm{eq}}(s) - \widehat{\wp}(s,\tau), \qquad 0 \le s < 1. \tag{4.30}$$

Thus, one has arrived at the following:

Proposition 17 *The system* (4.30) *that is obtained from* (4.26) *is formally same as the solvable master equations* (4.2)*, where the systems* (4.30) *and* (4.26) *are linked by the probability generating function* (4.27) *and a change of variables.*

Remark 4 Although (4.30) is formally same as (4.2), there is a significant difference. The totality of the label s for (4.30) is $[0,1)$, and that of j for (4.2) is the discrete set Γ. This continuous label s does not allow us to discuss (4.30) in terms of the standard contact geometry employed in the main text of this paper, however, we could formally proceed with a careful analysis. Such an analysis is left for future work.

For the sake of completeness, the solutions to (4.29), (4.30), and (4.26) are expressed as follows. First, the solution to (4.29) is immediately obtained as

$$\widehat{\wp}_{\theta}(s,\tau) = \mathrm{e}^{-\tau}\,\widehat{\wp}_{\theta}(s,0),$$

or equivalently,

$$\widehat{\wp}_{\theta}(s,t) = \mathrm{e}^{-\gamma(1-s)t}\,\widehat{\wp}_{\theta}(s,0).$$

Second, the solution to (4.30) is then

$$\widehat{\wp}(s,\tau) = \mathrm{e}^{-\tau}\,\widehat{\wp}_{\theta}(s,0) + (1-\mathrm{e}^{-\tau})\,\widehat{\wp}_{\theta}^{\mathrm{eq}}(s), \qquad 0 \le s < 1.$$

In what follows the initial conditions for $\wp_{\theta}$ are imposed as

$$\wp_{\theta}(j,0) = \delta_{j,0}, \qquad j = 0,\ldots,\infty,$$

from which one has the relation:

$$\widehat{\wp}_{\theta}(s,0) = \sum_{j=0}^{\infty} \delta_{j,0}\, s^{j} = 1.$$

Finally, to obtain the solution to (4.26) for $\wp_\theta$, one rewrites $\widehat{\wp}_\theta(s,t)$ with $\widehat{p}_\theta(s,0)=1$ as

$$\widehat{\wp}_\theta(s,t)=\mathrm{e}^{\gamma(s-1)t}\,\widehat{\wp}_\theta(s,0)=\sum_{j=0}^{\infty}\frac{(\gamma\,s\,t)^j}{j\,!}\mathrm{e}^{-\gamma\,t}=\sum_{j=0}^{\infty}\left[\frac{(\gamma\,t)^j}{j\,!}\,\mathrm{e}^{-\gamma\,t}\right]s^j.$$

Since the term $[\cdots]$ of the equation above should be equal to $\wp_\theta(j,t)$, one has

$$\wp_\theta(j,t)=\frac{(\gamma\,t)^j}{j\,!}\,\mathrm{e}^{-\gamma\,t},\qquad j=0,\ldots,\infty,$$

where this distribution function is known as the Poisson distribution with the parameter $\gamma\,t$.

References

1. Amari, S., Nagaoka, H.: Methods of Information Geometry. Oxford University Press, AMS (2000)
2. Ay, N., et al.: Information Geometry. Springer (2017)
3. Goto, S.: Legendre submanifolds in contact manifolds as attractors and geometric nonequilibrium thermodynamics. J. Math. Phys. **56**, 073301, 30 pp. (2015)
4. Nakamura, Y.: Gradient systems associated with probability distribution. Jpn. J. Ind. App. Math. **11**, 21–30 (1994)
5. Fujiwara, A., Amari, S.I.: Gradient systems in view of information geometry. Physica D **80**, 317–327 (1995)
6. Pistone, G.: Information geometry of the probability simplex: A short course arXiv:1911.01876
7. Arnold, V.I.: Methods of Classical Mechanics. Springer (1976)
8. Libermann, P., Marle, C.-M.: Symplectic Geometry and Analytical Mechanics. Springer (1987)
9. da Silva, A.C.: Lectures on Symplectic Geometry, 2nd edn. Springer (2008)
10. Hermann, R.: Geometry. Dekker, Systems and Physics (1973)
11. Mrugala, R.: Geometrical formulation of equilibrium phenomenological thermodynamics. Rep. Math. Phys. **14**, 419–427 (1978)
12. Mrugala, R.: On contact and metric structures on thermodynamic spaces. Suken kokyuroku **1142**, 167–181 (2000)
13. Etnyre, J., Ghrist, R.: Contact topology and hydrodynamics I: Beltrami fields and the Seifert conjecture. Nonlinearity **13**, 441–458 (2000)
14. Ohsawa, T.: contact geometry of the Pontryagin maximum principle. Automatica **55**, 1–5 (2015)
15. Bravetti, A., Cruz, H., Tapias, D.: Contact Hamiltonian mechanics. Ann. Phys. **376**, 17–39 (2017)
16. Leon, M., Valcazar, M.L.: Contact Hamiltonian systems. J. Math. Phys. **60**, 102902 (2019)
17. Grmela, M.: Contact geometry of mesoscopic thermodynamics and dynamics. Entropy **16**, 1652–1686 (2014)
18. van der Schaft, A.J., Maschke, B.: Geometry of thermodynamic process. Entropy, **20**, 925 [23 pp.], (2018)
19. Goto, S.: Contact geometric descriptions of vector fields on dually flat spaces and their applications in electric circuit models and nonequilibrium thermodynamics. J. Math. Phys. **57**, 102702 [40 pp.], (2016)

20. Bravetti, A., Lopez-Monsalvo, C.S.: Para-Sasakian geometry in thermodynamic fluctuation theory. J. Phys. A: Math. Theor. **48**, 125206 [21 pp.], (2015)
21. Bravetti, A., Lopez-Monsalve, C.S., Nettel, F.: Contact symmetries and Hamiltonian thermodynamics. Ann. Phys. **361**, 377–400 (2015)
22. van Kampen: *Stochastic Processes in Physics and Chemistry*. North Holland, (1981)
23. Landau, D., Binder, K.: A guide to Monte-Carlo Simulations in Statistical Physics. Cambridge Univ Press (2005)
24. Gelman, A. et al.: Beysian Data Analysis 3rd Ed. Chapman and Hal/CRC (2013)
25. Richey, M.: The evolution of Markov Chain Monte Carlo methods. Am. Math. Month. **117**, 383–413 (2010)
26. Goto, S., Hino, H.: Diffusion equations from master equations – A discrete geometric approach. J. Math. Phys. **61**, 113301 [27 pp.], (2020). https://aip.scitation.org/doi/10.1063/5.0003656
27. Goto, S., Hino, H.: Information and contact geometric description of expectation variables exactly derived from master equations. Physica Scripta **95**, 015207 [14 pp.], (2020)
28. Goto, S., Hino, H.: Expectation variables on a para-contact metric manifold exactly derived from master equations. Geom. Sci. Inf. **239–247**, (2019)
29. Zamkovoy, S.: Canonical connections on paracontact manifolds. Ann. Glob. Geom. **36**, 37–60 (2009)
30. Mrugala, R.: Statistical approach to the geometric structure of thermodynamics. Phys. Rev. A **41**, 3156–3160 (1990)
31. Matsuzoe, H., Henmi, M.: Hessian structures on deformed exponential families, GSI 2013. LNCS **8085**, 275–282 (2013)
32. Henmi, M., Matsuzoe, H.: Statistical manifolds admitting torsion and partially flat spaces. Springer, Geometric Structures of Information. Signal and Communication Technology (2019)
33. Harper, M.: Information geometry and evolutionary game theory. arXiv:0911.1383
34. Ezra, G.S.: Geometric approach to response theory in non-Hamiltonian systems. J. Math. Chem. **32**, 339–360 (2002)
35. Ezra, G.S.: On the statistical mechanics of non-Hamiltonian systems: the generalized Liouville equation, entropy, and time-dependent metrics. J. Math. Chem. **35**, 29–53 (2004)
36. Sergi, A., Giaquinta, P.V.: On the geometry and entropy of non-Hamiltonian phase space. J. Stat. Mech. **2007**, PO2013, (2007)
37. Ohara, A., Wada, T.: Information geometry of q-Gaussian densities and behaviors of solutions to related diffusion equations. J. Phys. A **43**, 035002 [18 pp.], (2010)
38. Bravetti, A., Tapias, D.: Liouville's theorem and the canonical measure for nonconservative system for contact geometry. J. Phys. A **48**, 245001 [11 pp.], (2015)
39. Goto, S., Umeno, K.: Maps on statistical manifolds exactly reduced from the Peron-Frobenius equations for solvable chaotic maps. J. Math. Phys. **59**, 032701 [13 pp.], (2018)
40. Shahshahani, S.: A new mathematical framework for the study of linkage and selection, Memories of AMS, (1979)
41. Suzuki, M.: Statistical mechanics of non-equilibrium systems II Prog. Theo. Phys. **55**, 383–399 (1976)
42. Rajeev, S.G.: A Hamilton-Jacobi formalism for thermodynamics. Ann. Phys. **323**, 2265–2285 (2008)
43. Balian, R., Valentin, P.: Hamiltonian structure of thermodynamics with guage. Eur. Phys. J. B **21**, 269–282 (2001)
44. Gay-Balmaz, F., Yoshimura, H.: Dirac structures in nonequilibrium thermodynamics. J. Math. Phys. **59**, 012701 [29 pp.], (2018)

Chapter 5
Invariant Koszul Form of Homogeneous Bounded Domains and Information Geometry Structures

Frédéric Barbaresco

Abstract In 1955, Jean-Louis Koszul has written a seminal paper entitled "*Sur la forme hermitienne canonique des espaces homogènes complexes*". Let G be a Lie group and let B be a closed subgroup of G, Koszul introduced a hermitian form that he called canonical hermitian form of the complex homogeneous space G/B with an invariant volume form. In this seminal paper, Koszul has investigated homogeneous spaces and necessary conditions to carry a nondegenerate 2-form derived from the invariant volume element. The advantage of this form for the determination of homogeneous bounded domains was underlined by Elie Cartan in last sentence of his 1932 paper "*Sur les domaines bornés de l'espace de n variables complexes*", observing that a necessary condition for G/B to be a bounded domain is that this form is positive definite. Hirohiko Shima has first observed that this geometry of Koszul Hessian structures is linked with Information Geometry. Jean-Louis Koszul also developed these structures in a Lecture "*Exposés sur les Espaces Homogènes Symétriques*" given at Sao-Paulo in 1958. We will synthetize these Jean-Louis Koszul works of 1955 and 1958 on this invariant form. We will put these elements in the context of more recent researches as a recent work of Della Vedova on covering spaces of symplectic manifolds (co-adjoint orbits, endowed with their canonical Kirillov-Kostant-Souriau symplectic structure) admitting homogeneous non Chern-Ricci flat special compatible almost complex structures, or Biquard extension of Kostant-Sekiguchi-Vergne correspondance. In last part, we develop the role of homogeneous bounded domains and invariant Koszul form in the framework of Information Geometry with use-case of Gaussian laws.

Keywords Invariant koszul form · Homogeneous bounded domains · Information geometry

---

F. Barbaresco (✉)
Thales Land & Air Systems, Voie Pierre-Gilles de Gennes, F91470 Limours, France
e-mail: frederic.barbaresco@thalesgroup.com

F. Nielsen (ed.), *Progress in Information Geometry*,
Signals and Communication Technology,
https://doi.org/10.1007/978-3-030-65459-7_5

5.1 Preamble

«C'est au problème de la détermination des domaines bornés homogènes posé par E. Cartan vers 1935 que se rattache [mes travaux]…Ce sont les travaux de Piatetskii Shapiro sur les domaines de Siegel, puis ceux de E.B. Vinberg sur les cônes homogènes qui m'ont amené à l'étude des groupes de transformation affines des variétés localement plates et en particulier aux critères de convexité liés aux formes invariantes.»- Jean-Louis Koszul, 1995

Jean-Louis Koszul was influenced by Elie Cartan who determined all symmetric bounded domains in C^n, $\forall n$, and all homogeneous bounded domains in C^n, $n = 2, 3$.

As soon as 1955, Koszul studied the class of homogeneous complex manifolds *G/B* (not only for Kähler manifold) with an invariant volume form such that the Ricci tensor is non-degenerate and proved that any bounded domain which is homogeneous with respect to a semi-simple group of automorphisms is symmetric, followed by Hano who proved the same for a unimodular group of matrices (if *G/B* is a homogeneous Kähler manifold with *G* unimodular and non-degenerate Ricci tensor, then *G* is semi-simple, and then every homogeneous bounded domain *G/B* with G unimodular is symmetric). In the following, we will synthetize Koszul approach for this result in his 1955 paper "*Sur la forme hermitienne canonique des espaces homogènes complexes*" but also detailed in his Lecture given at Sao Paulo "*Exposés sur les espaces homogènes symétriques*" in September and October 1958. More especially, we will underline importance of the left invariant form of degree 1 introduced by Koszul in these documents to deduce invariant metric in homogeneous bounded domains, that plays a fundamental role in the framework of Information Geometry. In Sao Paulo Lecture, Koszul illustrated these results with use-cases for the Poincaré upper-half plane and for the half-space of Siegel, and showed that with its trace formula of endomorphism *g/b*, he verify that with the canonical Kähler hermitian form and the associated metrics, we can recover metrics introduced by Henri Poincaré and Carl-Ludwig Siegel in these bounded domains. Based on seminal work of Elie Cartan [1], Jean-Louis Koszul has developed his model [2–4] of homogeneous bounded domains. This topic has been studied by authors [5–12]. On information Geometry, based on first papers [13–15], other works have been developed [16–30], with extension of Fisher matrix by Souriau on Symplectic Manifolds [31, 32], with links with other hessian structures as Kähler ones [33–35]. Biography and testimony on Jean-Louis Loszul are available in [36–38]. For more classical introduction to information geometry, we make reference to S.I. Amari papers [39, 40]. Koszul seminal papers [41, 42] and their developments are given in [43–50]. For references on moment map on homogeneous spaces, we make reference to [51–53] and other extension [54–56].

Henri Cartan was Jean-Louis Koszul PhD supervisor and in a letter from André Weil to Henri Cartan, cited in the proceedings of the conference "*Elie Cartan and today's mathematics*" in 1984, we can read "*As to the symmetrical spaces, and more particularly to the symmetric bounded domains at the birth of which you contributed, I have kept alive the memory of the satisfaction I felt in finding some incarnations in Siegel from his first works on quadratic forms, and later to convince Siegel of the*

Fig. 5.1 Henri Cartan lecture on homogeneous bounded domains, in Freiburg, March 13, 1987

value of your father's ideas on the subject". This citation makes reference to Elie Cartan's seminal work on symmetric bounded domains, exploited by carl-Ludwig Siegel (Fig. 5.1).

During 50's, Jean-Louis Koszul gave 3 Lectures in Sao Paulo entitled «*Faisceaux et Cohomologie*»; «*Variétés Kählériennes*» and «*Exposés sur les espaces homogènes symétriques*». Koszul visits to Brazil have been co-organized with a first visit to USP in 1956 to give a course in August and September (lecture notes taken by José de Barros Neto) on "*Faisceaux et cohomologie*" (general treatise on Cech cohomology with coefficients on a sheaf), published in 1957. A 2nd Course was given in September and October of 1956 (lecture notes taken by Chaim Samuel Honig and Carlos Benjamin de Lyra) on "*Variétés Kâhlériennes*", published in 1957, and also course on Multilinear Algebra, with notes by L. H. Jacy Monteiro, published in 1956. Last course on "*Exposés sur les espaces homogènes symétriques* " was given in September and October of 1958 with lecture notes published in 1959. R. Bott commented on these seminars "*very pleasant. The pace is fast, and the considerable material is covered elegantly. In addition to the more or less standard theorems on symmetric spaces, the author discusses the geometry of geodesics, Bergmann's metrics, and finally studies the bounded domains with many details*".

Elected to Academia de Ciencias do Estado de Sao Paulo in 1981, Jean-Louis Koszul visited USP again in 1986 for an inaugural talk at Instituto de Estudos Avançados, hosted by A.A. M. Rodrigues on "*The genesis of Bourbaki*" (Fig. 5.2).

In the book "*Selected papers of JL Koszul*" [57], Koszul summarizes the work, I will detail in the following: "*It is with the problem of the determination of the homogeneous bounded domains posed by E. Cartan around 1935 that are related [my papers]. The idea of approaching the question through invariant Hermitian forms already appears explicitly in Cartan. This leads to an algebraic approach which constitutes the essence of Cartan's work and which, with the Lie J-algebras,*

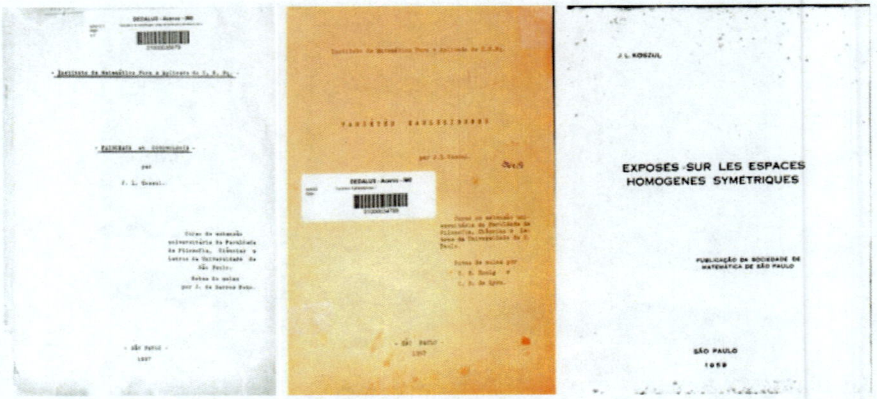

Fig. 5.2 Three courses given by Jean-Louis Koszul at Sao-Paulo during 50's on «Faisceaux et Cohomologie»;«Variétés Kählériennes» and «Exposés sur les espaces homogènes symétriques»

was pushed much further by the Russian School [58–71]. *It is the work of Piatetskii Shapiro on the Siegel domains, then those of E.B. Vinberg on the homogeneous cones that led me to the study of the affine transformation groups of the locally flat manifolds and in particular to the convexity criteria related to invariant forms"*. In particular, J.L. Koszul main source of inspiration is given in this last sentence of Elie Cartan's 1935 paper [72]: *"It is clear that if one could demonstrate that all homogeneous domains whose form* $\Phi = \sum_{i,j} \frac{\partial^2 \log K(z,z^*)}{\partial z_i \partial z_j^*} dz_i dz_j^*$ *is positive definite are symmetric, the whole theory of homogeneous bounded domains would be elucidated. This is a problem of Hermitian geometry certainly very interesting*". The work of Koszul has also been extended and deepened by one of his student Jacques Vey in [73] and [74]. Jacques Vey has transposed the notion of hyperbolicity, developed by W. Kaup for Riemann surfaces, into the category of differentiable manifolds with flat linear connection (locally flat manifolds), which makes it possible to completely characterize the locally flat manifolds admitting as universal covering a convex open sharp cone of R^n, which had been studied by Koszul in [75]. The links between Koszul's work and those of Ernest B. Vinberg [58–65] were recently developed at the conference "*Transformation groups 2017*" in Moscow dedicated to the 80th anniversary of Professor EB Vinberg, in Dmitri Alekseevsky's talk on "*Vinberg's theory of homogeneous convex cones: developments and applications*" [76].

In the paper *«Sur la forme hermitienne canonique des espaces homogènes complexes»* [77] of 1955: Koszul considers the Hermitian structure of a homogeneous *G/B* manifold (*G* related Lie group and *B* a closed subgroup of *G*, associated, up to a constant factor, to the single invariant *G*, and to the invariant complex structure by the operations of *G)*. Koszul says "*The interest of this form for the determination of homogeneous bounded domains has been emphasized by Elie Cartan: a necessary condition for G/B to be a bounded domain is indeed that this form is positive definite*". Koszul calculated this canonical form from infinitesimal data Lie algebra of *G*, the sub-algebra corresponding to *B* and an endomorphism algebra defining the invariant complex structure of *G/B*. The results obtained by Koszul proved that the homogeneous bounded domains whose group of automorphisms is semi-simple are bounded symmetric domains in the sense of Elie Cartan. In this seminal paper, Koszul introduced a left invariant form of degree 1 on *G* that is given by the following: $\Psi(X) = Tr_{g/b}[ad(JX) - J.ad(X)] \quad \forall X \in g$ with *J* an endomorphism of the Lie algebra space and the trace $Tr_{g/b}[.]$ corresponding to that of the endomorphism *g/b*. The Kähler form of the canonical Hermitian form is given by the differential of $-\Psi(X)$ of this form of degree 1.

These results were deepened by Koszul in a Lecture *«Exposés sur les espaces homogènes symétriques»* [78] published in 1959 after a seminar held in September and October 1958 at the University of Sao Paulo, which details the determination of homogeneous bounded domains. He returned to [77] and showed that any symmetric bounded domain is a direct product of irreducible symmetric bounded domains, determined by Elie Cartan (4 classes corresponding to classical groups and 2 exceptional domains). For the study of irreducible symmetric bounded domains, Koszul refered to Elie Cartan, Carl-Ludwig Siegel and Loo-Keng Hua. Koszul illustrated the subject with two particular cases, the upper half-plane of Poincaré and the upper half-space of Siegel, and showed that with its trace formula of endomorphism *g/h*, he found that the canonical Kähler hermitian form and the associated metrics allow to recover those introduced by Henri Poincaré and Carl-Ludwig Siegel [6] (who introduced them as invariant metric under action of the automorphisms of these spaces). Koszul has extended these results in four other papers published until 1970.

In the paper *«Domaines bornées homogènes et orbites de groupes de transformations affines»* [79] of 1961 written by Koszul at the Institute for Advanced Study at Princeton during a stay funded by the National Science Foundation, Koszul demonstrated the reciprocal of its 1955 result for a class of complex homogeneous spaces. This class consists of some open orbits of complex affine transformation groups and contains all homogeneous bounded domains. Koszul addressed again the problem of knowing if a complex homogeneous space, whose canonical Hermitian form is positive definite is isomorphic to a bounded domain, but via the study of the invariant bilinear form defined on a real homogeneous space by an invariant volume and an invariant flat connection. Koszul demonstrated that if this bilinear form is positive definite then the homogeneous space with its flat connection is isomorphic to a convex open domain containing no straight line in a real vector space and extended it to the initial problem for the complex homogeneous spaces obtained in defining a complex structure in the variety of vectors of a real homogeneous space provided with an invariant flat connection. It is in this article that Koszul used the affine representation of Lie groups and algebras.

In the paper *«Ouverts convexes homogènes des espaces affines»* [80] of 1962, Koszul was interested by the structure of the convex open non-degenerate Ω (with no straight line) and homogeneous (the group of affine transformations of *E* leaving stable Ω operates transitively in Ω) in a real affine space of finite dimension. Koszul demonstrated that they can be all deduced from non-degenerate and homogeneous convex open cones built in [79]. He used for this the properties of the group of affine transformations leaving stable a non-degenerate convex open domain and a homogeneous domain.

In the paper *«Variétés localement plates et convexité»* [75] of 1965, Koszul established the following theorem: let M *be* a locally related differentiable manifold. If the universal covering of M is isomorphic as a flat manifold with a convex open domain containing no straight line in a real affine space, then there exists on M a closed differential form α such that $D\alpha$ *(D* linear covariant derivative of zero torsion) is defined positive in all respects and which is invariant under every automorphism of *M.* If G is a group of automorphisms of M such that $G\backslash M$ is quasi-compact and if there exists on M a closed 1-differential form α invariant by G and such that $D\alpha$ is definite positive at any point, then the universal covering of M is isomorphic as a flat manifold with a convex open domain that does not contain a straight line in a real affine space.

In the paper *«Déformations des variétés localement plates»* [81] of 1968, Koszul provided other proofs of theorems introduced in [75]. Koszul considered related differentiable manifolds of dimension n and TM the fibered space of M. The linear connections on M constitute a subspace of the space of the differentiable applications of the TMxTM fiber *product* in the space *T(TM)* of the TM vectors. Any locally flat connection D (the curvature and the torsion are zero) defines a locally flat connection on the covering of *M,* and is hyperbolic when universal covering of *M,* with this connection, is isomorphic to a sharp convex open domain (without straight lines) in R^n. Koszul showed that, if M is a compact manifold, for a locally flat connection on M to be hyperbolic, it is necessary and sufficient that there exists a closed differential form of degree 1 on M whose covariant differential is positive definite.

In the paper *«Trajectoires convexes de groupes affines unimodulaires»* [82] in 1970, Koszul demonstrated that a convex sharp open domain in R^n that admits a unimodular transitive group of affine automorphisms is an auto-dual cone. This is a more geometric demonstration of the results shown by Ernest Vinberg [64] on the automorphisms of convex cones.

We will also introduce in the following more recent results as a recent work of Della Vedova on covering spaces of symplectic manifolds (co-adjoint orbits, endowed with their canonical Kirillov-Kostant-Souriau symplectic structure) admitting homogeneous non Chern-Ricci flat special compatible almost complex structures, or Biquard extension of Kostant-Sekiguchi-Vergne correspondance.

5.2 Invariant Koszul Form for Homogeneous Bounded Domains

Koszul considered on *G/B* an invariant complex structure tensor *I*. All the invariant volumes on *G/B*, equal up to a constant factor, define with the complex structure the same invariant Hermitian form on *G/B*, called Hermitian canonical form, denoted *h.* Let E be a differentiable fiber space of base M and let p be the projection of E

on M, such that $p^*((pX).f) = X.(p^*f)$. The projection $p : E \to M$ defines an injective homomorphism p^* of the space of differential forms of M in the space of the differential forms of E such that for any form α of degree n on M and any sequence of n projectable vectors fields, we have $p^*(\alpha(pX_1, pX_2, \ldots, pX_n)) = (p^*\alpha)(X_1, X_2, \ldots, X_n)$. Let I be the tensor of an almost complex structure on the basis M, there exists on E a tensor J of type *(1,1)* and only one which possesses the following properties $p(JX) = I(pX)$ and $J^2X = -X \mod h, X \in g$ for any vector field X on E. Let G be a connected Lie group and B a closed subgroup of G, we note g the Lie algebra left invariant vector fields on G and b sub-algebra of g corresponding to B. The canonical mapping of G on G/B is denoted p (defining E as before). We assume that there exists on G/B an invariant volume by G, which consist in assuming that, for all $s \in B$, the automorphism $X \to Xs$ of g defines by passing to the quotient an automorsphism of determinant *1* in *g/b* . Let r be the dimension of G/B and $(X_i)_{1 \le i \le m}$ a base of g such that $X_i \in b$, for $r \le i \le m$. Let $(\xi_i)_{1 \le i \le m}$ the base of the space of differential forms of degree *1* left invariant on G such that $\xi_i(X_j) = \delta_{ij}$. If ω is an invariant volume on G/B, then $\Omega = p^*\omega$ is equal, up to a constant factor, to $\xi_1 \wedge \xi_2 \wedge \ldots \wedge \xi_r$. We will assume the base (X_j) chosen so that this factor is equal to *1*, let $\Omega = \xi_1 \wedge \xi_2 \wedge \ldots \wedge \xi_r$. For any vector field that can be projected X on G, we have:

$$\begin{aligned} p^*(div(pX))\Omega &= p^*((div(pX))\omega) = p^*((pX)\omega) \\ &= X\Omega = \sum_{j=1}^{r} \xi_j([X_j, X])\Omega \end{aligned} \tag{5.1}$$

$$p^*(div(pX)) = \sum_{j=1}^{r} \xi_j([X_j, X]) \tag{5.2}$$

These elements being defined, Koszul calculates the Hermitian canonical form of G/B, denoted h, more particularly $\eta = p^*h$ on G. Let X and Y both right invariant vector fields on G. They are projectable and the fields pX and pY are conformal vector fields on G/B such that $div(pX) = div(pY) = 0$, because the volume and the complex structure of G/B are invariant under G. As a result, if κ is the Kähler form of h and if $\alpha = p^*\kappa$, then:

$$4\alpha(X, Y) = 4p^*(\kappa(pX, pY)) = p^*div(I[pX, pY]) \tag{5.3}$$

and as $p(J[X, Y]) = I[pX, pY]$, we obtain:

$$4\alpha(X, Y) = p^*div(J[X, Y]) = \sum_{i=1}^{2n} \xi_i([X_i, J[X, Y]]) \tag{5.4}$$

X and Y are two left invariant vectors fields on G. X' and Y' right invariant vectors fields coinciding with X and Y at the point e, neutral element of G. If $T = [X', Y']$ is tight invariant vectors fields which coincide with $-[X, Y]$ on e, then:

$$[X, JT] = J[X, [X, Y]] - [X, J[X, Y]] \text{ at point } e \tag{5.5}$$

At point e, we have the equality:

$$4\alpha(X, Y) = \sum_{i=1}^{2n} \xi_i([J[X, Y], X_i] - J[[X, Y], X_i]) \tag{5.6}$$

As the form α is invariant on the left by G, this equality is verified for all points. For any endomorphism Θ of the space g such that $\Theta b \subset b$, we denote by $Tr_b\Theta$ the trace of the restriction of Θ to b and by $Tr_{g/b}\Theta$ the trace of the endomorphism of g/b deduced from Θ by passage to the quotient, with $Tr\Theta = Tr_b\Theta + Tr_{g/b}\Theta$. We have:

$$Tr_{g/b}\Theta = \sum_{i=1}^{2n} \xi_i(\Theta X_i) \tag{5.7}$$

Whatever $X \in g$ and $s \in B$, we have $J(Xs) - (JX)s \in b$. If $ad(Y)$ is the endomorphism of g defined by $ad(Y).Z = [Y, Z]$, we have $(Jad(Y) - ad(Y)J)g \subset b$ for all $Y \in b$. We can deduce, for all $X \in g$, the endomorphism $ad(JX) - Jad(X)$ leaves steady the subspace b. Koszul defines a linear form Ψ on the space g by defining:

$$\Psi(X) = Tr_{g/b}(ad(JX) - Jad(X)), \quad \forall X \in g \tag{5.8}$$

Koszul has finally obtained the following fundamental theorem:

Theorem of Koszul [77] The Kähler form of the Hermitian canonical form has for image by p^* the differential of the form $-\frac{1}{4}\Psi(X)$ with $\Psi(X) = Tr_{g/b}(ad(JX) - Jad(X)), \quad \forall X \in g$

Koszul note that the form Ψ is independent of the choice of the tensor J. It is determined by the invariant complex structure of G/B. The form Ψ is right invariant by B. For all $s \in B$, note the endomorphism $r(s) : X \to Xs$ of g . Since $J(Xs) = (JX)s \mod b$ and that $Tr_{g/b}ad(Y) = 0$, we have:

$$\Psi(Xs) = Tr_{g/b}(ad((JX)s) - Jad(Xs)), \quad \forall X \in g, \forall Y \in b \tag{5.9}$$

$$\Psi(Xs) = Tr_{g/b}\left(r(s)ad(JX)r(s)^{-1} - Jr(s)ad(X)r(s)^{-1}\right) \tag{5.10}$$

$$\Psi(Xs) = \Psi(X) + Tr_{g/b}\big((J - r(s)^{-1} Jr(s))ad(X)\big), \quad \forall X \in g, s \in B \tag{5.11}$$

As $\big(J - r(s)^{-1} Jr(s)\big)$ maps g in b, we get $\Psi(Xs) = \Psi(X)$. The form Ψ is not zero on b. This is not the image by p^* of a differential form of G/B. However, the right invariance of Ψ on B is translated, infinitesimally by the relation:

$$\Psi([b, g]) = (0) \tag{5.12}$$

Koszul proved that the canonical hermitian form h of a homogeneous Kähler manifold G/B has the following expression:

$$\eta(X, Y) = \frac{1}{2}\Psi([JX, Y])$$
$$\text{with} \begin{cases} \Psi([X, Y]) = \Psi([JX, JY]) \\ \eta([JX, JY]) = \eta(X, Y) \end{cases} \quad \forall X, Y \in g \tag{5.13}$$

To do, the link with the first chapters, I can summarize the main result of Koszul that there is an integrable structure almost complex J on g, and for $l \in g^*$ defined by a positive J -invariant inner product on g:

$$\langle X, Y\rangle_i = \langle [JX, Y], l\rangle \tag{5.14}$$

Koszul has proposed as admissible form, $l \in g^*$, the form ξ :

$$\Psi(X) = \langle X, \xi\rangle = \text{Tr}[ad(JX) - Jad(X)] \quad \forall X \in g \tag{5.15}$$

Koszul proved that $\langle X, Y\rangle_\xi$ coincides, up to a positive multiplicative constant; with the real part of the Hermitian inner product obtained by the Bergman metric of symmetric homogeneous bounded domains by identifying g with the tangent space. $\Psi(X)$ is the restriction to g of a differential form Ψ of degree 1, with left invariance on G. This form is fully defined by the invariant complex structure of G/B. This form is invariant to the choice of J. This form is invariant on the right by B. We have $\Psi([X, Y]) = 0$ with $X \in g, Y \in b$. The exterior differential $d\Psi$ of Ψ is the inverse image by the projection $G \to G/B$ of degree 2 form Ω. This form Ω is, up to a constant, the Kähler form h, defined by the canonical Hermitian form of G/B: $h(\pi.X, \pi.Y) = \frac{1}{2}(d\Psi)(X, J.Y), \quad \forall X, Y \in G$ as it is proved in Bourbaki seminar by Koszul in [129].

The 1st Koszul form is then given by:

$$\alpha = -\frac{1}{4} d\Psi(X) \tag{5.16}$$

Koszul has illustrated this structure for the simplest example of *Siegel Domains*. First, the Poincaré upper half-plane $V = \{z = x + iy / y > 0\}$ which is isomorphic to the open $zz^* < 1$, which is a bounded domain. The group G of transformations

$z \rightarrow az+b$ with a and b real values with $a > 0$ is simply transitive in V. We identify G and V by the application passing from $s \in G$ an element to the image $i = \sqrt{-1}$ by s.

Let's define vector fields $X = y\frac{d}{dx}$ and $Y = y\frac{d}{dy}$ which generate the vector space of left invariant vectors fields on G, and J an almost complex structure on V defined by $JX = Y$. As $[X, Y] = -Y$ and $ad(Y).Z = [Y, Z]$ then:

$$\begin{cases} Tr[ad(JX) - Jad(X)] = 2 \\ Tr[ad(JY) - Jad(Y)] = 0 \end{cases} \tag{5.17}$$

The Koszul forms and the Koszul metric are respectively given by:

$$\Psi(X) = 2\frac{dx}{y} \Rightarrow \alpha = -\frac{1}{4}d\Psi = -\frac{1}{2}\frac{dx \wedge dy}{y^2} \Rightarrow ds^2 = \frac{dx^2 + dy^2}{2y^2} \tag{5.18}$$

We note that $\alpha = -\frac{1}{4}d\Psi(X)$ is indeed the Kähler form of Poincaré's metric, which is invariant by the automorphisms of the upper half-plane.

The next example used by Koszul concerns $V = \{Z = X + iY/X, Y \in Sym(p), Y > 0\}$ the upper half-space of Siegel (which is the most natural extension of the Poincaré half-plane) with:

$$\begin{cases} SZ = (AZ + B)D^{-1} \\ A^T D = I, B^T D = D^T B \end{cases} \text{with } S = \begin{pmatrix} A & B \\ 0 & D \end{pmatrix} \text{ and } J = \begin{pmatrix} 0 & I \\ -I & 0 \end{pmatrix} \tag{5.19}$$

We can then compute Koszul forms and the metric:

$$\Psi(dX + idY) = \frac{3p+1}{2} Tr\left(Y^{-1}dX\right)$$
$$\Rightarrow \begin{cases} \alpha = -\frac{1}{4}d\Psi = \frac{3p+1}{8} Tr\left(Y^{-1}dZ \wedge Y^{-1}d\bar{Z}\right) \\ ds^2 = \frac{(3p+1)}{8} Tr\left(Y^{-1}dZY^{-1}d\bar{Z}\right) \end{cases} \tag{5.20}$$

We recover Carl-Ludwig Siegel metric for the upper half space.

Jean-Louis Koszul [77, 78, 79, 80, 75, 83, 81, 82] and his student Jacques Vey [73, 74] introduced new theorems:

Koszul Theorem [82]: Let Ω be a sharp convex open in an affine space of E of finite dimension on R. If a unimodular Lie group of affine transformations operates transitively on Ω, Ω is a cone.

Koszul-Vey Theorem [74]: Let M a hessian connected manifold associated with the hessian metric g. Assume that M has a closed 1-form α such that $D\alpha = g$ and that there is a group G of affine automorphisms of M preserving α, then:

- If M/G is almost compact, then the manifold, universal covering of M, is affinely isomorphic to a convex domain of an affine space containing no straight line.
- If M/G is compact, then Ω is a sharp convex cone.

Koszul studied symmetric homogeneous spaces and defines the relation between invariant flat affine connections and the affine representations of Lie algebras and invariant Hessian metrics characterized by affine representations of Lie algebras. Koszul provides a correspondence between symmetric homogeneous spaces with invariant Hessian structures using affine representations of Lie algebras, and proves that a symmetric homogeneous space simply connected with an invariant Hessian structure is a direct product of a Euclidean space and of a homogeneous dual-cone. Let *G* be a connected Lie group and *G/K* a homogeneous space over which *G* acts effectively. Koszul gives a bijective correspondence between all planar *G* -invariantes connections on *G/K* and all of a certain class of affine representations of the Lie algebra of *G*. The main theorem of Koszul is:

Koszul's Theorem: Let *G/K* be a homogeneous space of a connected Lie group *G* and be $\mathfrak{g}$ and $\mathfrak{k}$ the Lie algebras of *G* and *K*, assuming that *G/K* has G-invariant connection, then admits an affine representation *(f, q)* on the vector space *E*. Conversely, assume that *G* is simply connected and has an affine representation, then *G/K* admits a flat *G*-invariant connection.

5.3 Contextualization with Last Advanced Works

Every Kähler manifold has a symplectic structure, but the converse that every closed symplectic manifold has also a Kähler structure is not true. Based on an observation of Paulette Liberman, W. P. Thurston [84] has produced some counter-examples of symplectic manifolds which are not Kähler (Kodaira-Thurston manifold). But, recently Alberto Della Vedova has observed that co-adjoint orbits, endowed with their canonical Kirillov-Kostant-Souriau symplectic structure [85–87], serve as covering spaces of symplectic manifolds admitting homogeneous non Chern-Ricci flat special compatible almost complex structures. This result has been developed in PhD of Alice Gatti, supervised by Alberto Della Vedova that adjoint orbits of non-compact semisimple Lie groups turn out to be naturally almost-Kähler manifolds endowed with the Kirillov-Kostant-Souriau symplectic form and a canonically defined almost-complex structure, providing explicit formulae for the Chern-Ricci form.

Let us denote by ρ the Ricci curvature form, then we have $\Psi([X, Y]) = \rho(X, Y)$. This relation imply that a symplectic manifold admitting a homogeneous special compatible almost complex structure with non-zero Hermitian scalar curvature is, up to coverings, a coadjoint orbit equipped with the KKS (Kirillov-Kostant-Souriau) symplectic form [88–90].

Theorem [84, 88] Let (M, ω) be a symplectic manifold admitting a homogeneous compatible almost complex structure satisfying $\rho = \lambda\omega$ for some $\lambda \neq 0$. Then *M* is a covering space of a coadjoint orbit and ω is the pull-back via the covering map of the canonical symplectic form:

$\Psi([X, Y]) = \lambda\sigma(X, Y) = \lambda B(v, [X, Y])$ with $B(X, Y) = Tr(ad_X ad_Y)$ killing form where *K* is the isotropy subgroup of v with respect to the adjoint representation of G, and $[v, X] = 0$. v has compact isotropy subgroup $V \subset G$. The canonical symplectic form $\rho = \lambda\omega$ on the orbit is determined by the exact two-cocycle σ. The canonical almost complex structure J on G/V is compatibile with the Kirillov-Kostant-Souriau symplectic form ω. The author in [88] proves also the converse of previous theorem that a coadjoint orbit having the first Chern class equal to a non-zero multiple of the class of the canonical symplectic form admits a homogeneous special compatible almost complex structure, under the additional assumptions that G is semi-simple and the coadjoint orbit has compact isotropy.

In a preprint [91], O. Biquard has extended the Kostant-Sekiguchi-Vergne [92] correspondence. This classical Kostant-Sekiguchi-Vergne correspondence concerns G/H a symmetric space of noncompact type, with H a maximal compact subgroup of G, associated to a Cartan decomposition of the Lie algebra $\mathfrak{g} = \mathfrak{h} \oplus \mathfrak{m}$, and gives a diffeomorphism between the nilpotent G-orbits in $\mathfrak{g}$ and the nilpotent H^C-orbits in $\mathfrak{m}^C$. Olivier Biquard has extend this correspondence to all G-orbits in $\mathfrak{g}$, turns out that each G-orbit in $\mathfrak{g}$ is diffeomorphic to each orbit in a family of H^C-orbits in $\mathfrak{m}^C$, providing a set of H-invariant Kähler metrics on any G-orbit in $\mathfrak{g}$, such that the Kähler form equals the Kirillov-Kostant-Souriau symplectic form of the orbit. $\mathfrak{t}_{\mathfrak{g}} = \mathfrak{t} \oplus \mathfrak{a}$, and any semi-simple element in $\mathfrak{g}$ is G-conjugate to $\tau = \tau_1 + i\tau_2$, where $\tau_1 \in \mathfrak{t}$ and $\tau_2 \in i\mathfrak{a}$, and $\mathfrak{t}_{\mathfrak{g}} = \mathfrak{t} \oplus \mathfrak{a}$ is a Cartan subalgebra of $\mathfrak{g}$. Any element of $\mathfrak{g}$ is conjugate to $\tau = \tau_1 + i\tau_2 + \sigma_1 + i\sigma_2$, and the representation $\sigma = (\sigma_1, \sigma_2, \sigma_3)$ satisfying $\sigma_1 \in \mathfrak{h}$ and $\sigma_2, \sigma_3 \in i\mathfrak{m}$, and commutes with the τ_i, such that $\left[\sigma_i, \tau_j\right] = 0$.The Cartan subalgebra of $\mathfrak{g}$ is one dimensional, so either it is compact ($\mathfrak{t}_{\mathfrak{g}} = \mathfrak{h}$), either it is noncompact ($\mathfrak{t}_{\mathfrak{g}} = \mathfrak{a} \subset \mathfrak{m}$). For all cases, the diffeomorphism with a complex H^C-space gives a H-invariant Kähler structure on the orbit.

Theorem [91]

(1) For any $\tau_3 \in i\mathfrak{a}$ such that $[\tau_3, \sigma_i] = 0$ and the regularity assumptions $C_{\mathfrak{g}}(\tau_1, \tau_2) = C_{\mathfrak{g}}(\tau_2, \tau_3) = C_{\mathfrak{g}}(\tau_1, \tau_2, \tau_3)$ are satisfied, there exist an H-invariant diffeomorphism from the G-orbit of $\tau_1 + i\tau_2 + \sigma_1 + i\sigma_2$ in $\mathfrak{g}$ to the H^C-orbit of $\tau_2 + i\tau_3 + \sigma_2 + i\sigma_3$ in $\mathfrak{m}^C$. This diffeomorphism gives a H-invariant kähler metric on the G-orbit of $\tau_1 + i\tau_2 + \sigma_1 + i\sigma_2$ in $\mathfrak{g}$, whose kähler form is the Krillov-Kostant-Souriau Symplectic form.

(2) If $\tau_3 \in i\mathfrak{a}$ satisfies $[\tau_3, \sigma_i] = 0$ but we have only the regularity $C_{\mathfrak{g}}(\tau_1, \tau_2) = C_{\mathfrak{g}}(\tau_1, \tau_2, \tau_3)$, then there still exists a H-invariant diffeomorphism from the G-orbit of $\tau_1 + i\tau_2 + \sigma_1 + i\sigma_2$ in $\mathfrak{g}$ to a H^C-space, which gives a H-invariant Kähler structure on the orbit.

Other results on this extension have been proven by Roger Bielawski [93], with a natural interpretation in terms of Nahm's equations with reference to Vergne [92].

Other elements on links between Kähler form and Symplectic KKS 2-form are considered in [94–96]. In [94], author has considered the regular coadjoint orbits of G, a noncompact real semi-simple Lie group, on which, the Iwasawa decomposition induces a left- invariant foliation which is isotropic with respect to the Kirillov symplectic form. In [95], author has given a new proof that semi-simple co-adjoint orbits through real hyperbolic elements are symplectomorphic to cotangent bundles, establishing a new connection between the Iwasawa horospherical projection and the symplectic geometry of real hyperbolic co-adjoint orbits. In [96] authors have studied orbits of coadjoint representations of classical compact Lie groups by an explicit parameterization of the orbit by means of a generalized stereographic projection, obtaining a Kählerian structure on the orbit, introducing basis two-forms for the cohomology group of the orbit.

5.4 Koszul Hessian Geometric Structure of Information Geometry

The elementary geometric structures discovered by Jean-Louis Koszul are the foundations of Information Geometry. These links were first established by Professor Hirohiko Shima [97–102] in his 2007 book entitled "*The Geometry of Hessian Structures*" [103], which is dedicated to Professor Koszul. The origin of this work followed the visit of Koszul in Japan in 1964, for a mission coordinated with the French government. Koszul taught lectures on the theory of flat manifolds at Osaka University. Hirohiko Shima was then a student and attended these lectures with the teachers Matsushima and Murakami. This lecture was at the origin of the notion of Hessian structures and the beginning of the works of Hirohiko Shima. Henri Cartan noted concerning Koszul's ties with Japan, "*Koszul has attracted eminent mathematicians from abroad to Strasbourg and Grenoble. I would like to mention in particular the links he has established with representatives of the Japanese School of Differential Geometry*". Shima's book [103] is a systematic introduction to the theory of Hessian structures (provided by a pair of a flat connection D and an Hessian metric g). Koszul studied flat manifolds with a closed 1-form α, such that $D\alpha$ be positive definite, where $D\alpha$ is a hessian metric . However, not all Hessian metrics are globally of the form $g = D\alpha$. Shima introduces the notion of Codazzi structure for a pair (D,g), with D a torsion-free connection, which verifies the Codazzi equation $(D_X g)(Y, Z) = (D_Y g)(X, Z)$. A Hessian structure is a Codazzi structure for which connection D is flat. This is an extension of Riemannian geometry. It is then possible to define a connection D' and a dual Codazzi structure (D',g) with $D' = \nabla - D$ where ∇ is the Levi-Civita connection. For a hessian structure (D, g) with $g = Dd\varphi$, the dual Codazzi structure $\left(D', g\right)$ is also a Hessian structure and $g = D'd\varphi'$, where φ' is the Legendre transform of φ: $\varphi' = \sum_i x^i \frac{\partial \varphi}{\partial x^i} - \varphi$. Shima observed that Information

Geometry framework could be introduced by dual connections, and not only founded on Fréchet, Rao and Chentsov works [103]. A hessian structure *(D, g)* is of Koszul type, if there is a closed 1-form ω as $g = D\omega$. Using D and the volume element of g, Koszul introduced a 2nd form, which plays a similar role to the Ricci tensor for a Kählerian metric. Let υ be the volume element of g, we define a closed 1-form α such that $D_X\upsilon = \alpha(X)\upsilon$ and a symmetric bilinear form $\gamma = D\alpha$. In the following, α and γ forms are called 1st and 2nd form of Koszul for Hessian structure (D, g). We can consider the forms associated with the Hessian dual structure (D', g) by $\alpha' = -\alpha$ and $\gamma' = \gamma - 2\nabla\alpha$. In the case of a homogeneous regular convex cone Ω, with D the canonical flat connection of the ambient vector space, the Koszul forms α and γ for the canonical Hessian structure $(D, g = Dd\psi)$ are given by $\alpha = d\log\psi$ and $\gamma = g$. The volume element υ determined by g is invariant under the action of the group of automorphisms G of Ω.

Jean-Louis Koszul, invited by the intermediary of Professor Boyom, attended the 1st GSI "*Geometric Science of Information*" conference in August 2013 at the Ecole des Mines in Paris, where he attended the presentation of Hirohiko Shima, given for his honor on the topic "*Geometry of Hessian Structures* " [104]. In the photo below, we can see from left to right, Jean-Louis Koszul and Hirohiko Shima. Professor Michel Boyom has extensively studied and developed, at the University of Montpellier, Koszul models [105, 106, 107, 108, 109, 85, 86, 87] in relation to symplectic flat affine manifolds and to the cohomology of Koszul-Vinberg algebras (KV Cohomology). Professor Boyom with his PhD student Byande [90, 110] have explored other links with Information Geometry.

Links with Koszul-Vinberg characteristic function could be found in [111, 110] (Fig. 5.3).

The main object studied by Information Geometry is invariant distance between probability densities in the space of their parameters. The most natural metric has been introduced by Rao to be the Fisher metric ds_θ^2 that is invariant by changes of non-singular parametrization:

$$w = W(\theta) \Rightarrow ds_w^2 = ds_\theta^2 \text{ si en posant } [I(\theta)]_{i,j} = -E\left[\frac{\partial^2 \log p_\theta(w)}{\partial\theta_i \partial\theta_j}\right]$$

$$ds_\theta^2 = -\int p_\theta(w) \log \frac{p_{\theta+d\theta}(w)}{p_\theta(w)} dw$$

$$ds_\theta^2 \underset{Taylor}{\approx} \sum_{i,j} g_{ij} d\theta_i d\theta_j = \sum_{i,j} [I(\theta)]_{i,j} d\theta_i d\theta_j = d\theta^T . I(\theta) . d\theta$$

linked with Fr échet-Darmois-Cramer-Rao Bound:

$$E\left[\left(\theta - \hat{\theta}\right)\left(\theta - \hat{\theta}\right)^T\right] \geq I(\theta)^{-1} \tag{5.21}$$

For families of exponential densities, these parameters belong to homogeneous bounded domains, we illustrate this property for Gaussian laws. For monovariate

Fig. 5.3 Jean-Louis Koszul and Hirihiko Shima at GSI'13 "Geometric Science of Information" conference in Ecole des Mines ParisTech in Paris, October 2013

gaussian density, parameters are given by $\theta = \begin{pmatrix} m \\ \sigma \end{pmatrix}$ with m the mean and σ the standard deviation of $p_\theta(w) = \frac{1}{\sqrt{2\pi\sigma^2}} e^{-\frac{1}{2}\frac{(w-m)^2}{\sigma^2}}$. Fisher metric computation provides:

$$I(\theta) = \begin{bmatrix} \frac{1}{\sigma^2} & 0 \\ 0 & \frac{2}{\sigma^2} \end{bmatrix} \Rightarrow ds_\theta^2 = d\theta^T.I(\theta).d\theta = \frac{dm^2}{\sigma^2} + 2.\frac{d\sigma^2}{\sigma^2}$$

$$ds_\theta^2 = 2.\frac{|dz|^2}{(\mathrm{Im}(z))^2} \text{ with } z = \frac{m}{\sqrt{2}} + i\sigma \qquad (5.22)$$

We recover the metric of the Poincaré upper-half plan $H = \{z = x + iy/y > 0\}$, the most simple example of homogeneous bounded domain. It is obvious to recover this geometry for the Fisher metric of Gaussian densities, because their space of parameters $\theta = \begin{pmatrix} m \\ \sigma \end{pmatrix}$ lie in $H_{Gaussienne} = \left\{ z = m + i\sqrt{2}\sigma/m, \sigma \in R, \sigma > 0 \right\}$, the upper half plan. Fisher metric invariance by reparametrization inherits this property by a more richer invariance given by invariance by automorphisms of this homogeneous bounded domain. Poincaré metric $ds^2 = \frac{dx^2}{y^2} + \frac{dy^2}{y^2} = \frac{|dz|^2}{(\mathrm{Im}(z))^2} = y^{-1}dzy^{-1}dz^*$ for upper-half plan is invariant by its automorphisms given by $M(z) = \frac{az+b}{c.z+d}$ with $ad - bc = 1$. Then, a Gaussian density is coded by 1 point in the upper

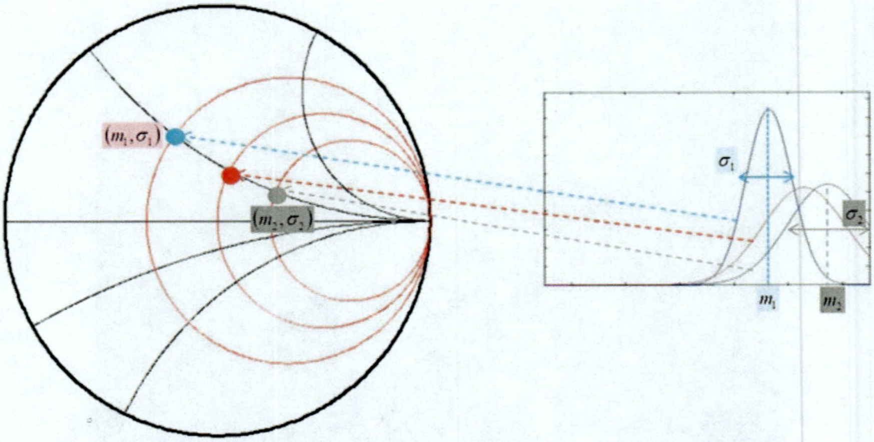

Fig. 5.4 Information geometry of monovariate Gaussian densities and their parametrization in the complex Poincaré unit disk. A Gaussian parameterized by (m,σ) is represented by 1 point in the Poincaré unit disc. The distance between 2 Gaussians is given by the geodesic distance in the unit disk

half plan, and by Cayley transform to a point in the Poincaré unit disk:

$$\chi = \frac{z-i}{z+i} \; (|\chi| < 1) \Rightarrow ds^2 = 8.\frac{|d\chi|^2}{\left(1-|\chi|^2\right)^2} \tag{5.23}$$

We can then define a distance between gaussian laws using Poincaré formula in unit disk, by integrating along a radial from the center $r = |\chi| \Rightarrow ds = 2\sqrt{2}.\frac{dr}{1-r^2}$ that make appear the primitive an hyperbolic tangent. For the distance between two arbitrary points in the disk, we have to use isometry of the disk $\phi_\tau(\chi) = \frac{\chi-\tau}{1-\chi\tau^*}$ (Fig. 5.4):

$$d^2(\{m_1,\sigma_1\},\{m_2,\sigma_2\}) = 2.\left(\log\frac{1+\delta(\chi^{(1)},\chi^{(2)})}{1-\delta(\chi^{(1)},\chi^{(2)})}\right)^2$$
$$\text{with} \quad \delta(\chi^{(1)},\chi^{(2)}) = \left|\frac{\chi^{(1)}-\chi^{(2)}}{1-\chi^{(1)}\chi^{(2)*}}\right|, \chi = \frac{z-i}{z+i} \text{ and } z = \frac{m}{\sqrt{2}} + i\sigma \tag{5.24}$$

A Generalization of Poincaré upper half plan is given by Siegel upper half space $H_{Siegel} = \{Z = X + iY/X, Y \in Sym(n), Y > 0\}$ where real variables have been replaced by symmetric matrices. Carl-Ludwig Siegel has looked for the metric which is invariant by the automorphisms of this space given by $M(Z) = (AZ+B)(CZ+D)^{-1}$ with $A^TD - B^TC = I_d$. He proved that this invariant metric is given:

$ds^2 = Trace\left(Y^{-1}dZY^{-1}d\bar{Z}\right)$ with $\bar{Z}$ the transpose and conjugate of the matrix.

We can observe that we are able to recover the Fisher metric for multivariate gaussian densities (M, R) with R the covariance matrix, in the case where the mean vector $M = 0$. If we use previous metric in the Siegel upper half space:

$$X = 0, Y = R \Rightarrow Z = iR \Rightarrow ds^2 = Trace\left(\left(R^{-1}dR\right)^2\right) \tag{5.25}$$

In the Siegel unit disk $SH_{Siegel} = \left\{W / WW^{+} < I_d\right\}$, extension of Poincaré unit disk, obtained by Cayley transform $W = (Z - iI_d)(Z + iI_d)^{-1}$, Siegel metric is given by (Fig. 5.5):

$$ds^2 = Trace\left(\left(1 - W\bar{W}\right)dW\left(1 - \bar{W}W\right)d\bar{W}\right) \tag{5.26}$$

In next chapter, we will consider the Poincaré unit disk as an Homogeneous Symplectic Manifolds associated to SU(1,1) co-adjoint orbits by Kirillov-Kostant-Souriau 2-form. We will then deduced how to define Gaussian density parametrized in Poincaré unit disk using Jean-Marie Souriau Lie Groups Thermodynamics.

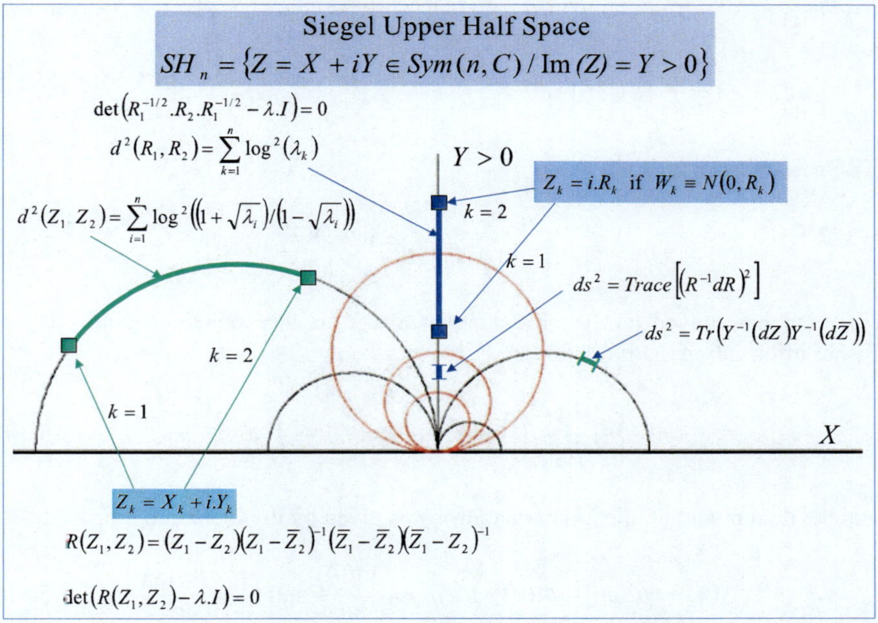

Fig. 5.5 Siegel Upper-Half Space, its metric and distance along geodesics

5.5 Homogeneous Symplectic Manifold as Co-Adjoint Orbits and Their Density of Probability: Poincaré Unit Disk and Its Gaussian Density by Moment Map of SU(1,1) Lie Group

Classically, to optimize the parameter θ of a probabilistic model, based on a sequence of observations y_t, is an online gradient descent:

$$\theta_t \leftarrow \theta_{t-1} - \eta_t \frac{\partial l_t(y_t)^T}{\partial \theta} \tag{5.27}$$

with learning rate η_t, and the loss function $l_t = -\log p(y_t/\hat{y}_t)$. This simple gradient descent has a first drawback of using the same non-adaptive learning rate for all parameter components, and a second drawback of non invariance with respect to parameter re-encoding inducing different learning rates. Amari has introduced the natural gradient to preserve this invariance to be insensitive to the characteristic scale of each parameter direction. The gradient descent could be corrected by $I(\theta)^{-1}$ where I is the Fisher information matrix with respect to parameter θ, given by:

$$I(\theta) = [g_{\bar{j}}] \text{ with } g_{ij} = \left[-E_{y \approx p(y/\theta)} \left[\frac{\partial^2 \log p(y/\theta)}{\partial \theta_i \partial \theta_j} \right] \right]_{ij} \tag{5.28}$$

with natural gradient:

$$\theta_t \leftarrow \theta_{t-1} - \eta_t I(\theta)^{-1} \frac{\partial l_t(y_t)^T}{\partial \theta} \tag{5.29}$$

Amari has proved that the Riemannian metric in an exponential family is the Fisher information matrix defined by:

$$g_{ij} = -\left[\frac{\partial^2 \Phi}{\partial \theta_i \partial \theta_j} \right]_{ij} \text{ with } \Phi(\theta) = -\log \int_R e^{-\langle \theta, y \rangle} dy \tag{5.30}$$

and the dual potential, the Shannon entropy, is given by the Legendre transform:

$$S(\eta) = \langle \theta, \eta \rangle - \Phi(\theta) \text{ with } \eta_i = \frac{\partial \Phi(\theta)}{\partial \theta_i} \text{ and } \theta_t = \frac{\partial S(\eta)}{\partial \eta_i} \tag{5.31}$$

We can observe that $\Phi(\theta) = -\log \int_R e^{-\langle \theta, y \rangle} dy = -\log \psi(\theta)$ is related to the classical cumulant generating function.

J. L. Koszul and E. Vinberg have introduced an affinely invariant Hessian metric on a sharp convex cone through its characteristic function:

$$\Phi_\Omega(\theta) = -\log \int_{\Omega^*} e^{-\langle\theta,y\rangle} dy = -\log \psi_\Omega(\theta) \text{ with } \theta \in \Omega \text{ sharp convex cone}$$

$$\psi_\Omega(\theta) = \int e^{-\langle\theta,y\rangle} dy \text{ with Koszul-Vinberg Characteristic function} \quad (5.32)$$

Jean-Louis Koszul has introduced the following forms

$$\text{1st Koszul form } \alpha : \alpha = d\Phi_\Omega(\theta) = -d\log\psi_\Omega(\theta) \quad (5.33)$$

$$\text{2nd Koszul form } \gamma : \gamma = D\alpha = Dd\log\psi_\Omega(\theta) \quad (5.34)$$

with the following property of positive definitiveness:

$$(Dd\log\psi_\Omega(x))(u) = \frac{1}{\psi_\Omega(u)^2}\left[\int_{\Omega^*} F(\xi)^2 d\xi . \int_{\Omega^*} G(\xi)^2 d\xi - \left(\int_{\Omega^*} F(\xi).G(\xi)d\xi\right)^2\right]$$

$$(Dd\log\psi_\Omega(x))(u) > 0$$

$$\text{with } F(\xi) = e^{-\frac{1}{2}\langle x,\xi\rangle} \text{ and } G(\xi) = e^{-\frac{1}{2}\langle x,\xi\rangle}\langle u,\xi\rangle \quad (5.35)$$

Koszul has defined the following Diffeomorphism:

$$\eta = \alpha = -d\log\psi_\Omega(\theta) = \int_{\Omega^*} \xi p_\theta(\xi) d\xi \text{ with } p_\theta(\xi) = \frac{e^{-\langle\xi,\theta\rangle}}{\int_\Omega e^{-\langle\xi,\theta\rangle} d\xi} \quad (5.36)$$

with preservation of Legendre transform:

$$S_\Omega(\eta) = \langle\theta,\eta\rangle - \Phi_\Omega(\theta) \text{ with } \eta = d\Phi_\Omega(\theta) \text{ and } \theta = dS_\Omega(\eta) \quad (5.37)$$

This relations have been extended by Jean-Marie Souriau in geometric statistical mechanics, where he developed a "Lie groups thermodynamics" of dynamical systems where the (maximum entropy) Gibbs density is covariant with respect to the action of the Lie group [112–116]. In the Souriau model, previous structures of information geometry are preserved:

$$I(\beta) = -\frac{\partial^2\Phi}{\partial\beta^2} \text{ with } \Phi(\beta) = -\log\int_M e^{-\langle U(\xi),\beta\rangle} d\lambda_\omega$$
$$\text{and } U : M \to \mathfrak{g}^* \quad (5.38)$$

We preserve the Legendre transform:

$$S(Q) = \langle Q, \beta \rangle - \Phi(\beta)$$
$$\text{with } Q = \frac{\partial \Phi(\beta)}{\partial \beta} \in \mathfrak{g}^* \text{ and } \beta = \frac{\partial S(Q)}{\partial Q} \in \mathfrak{g} \tag{5.39}$$

In the Souriau Lie groups thermodynamics model, β is a "geometric" (Planck) temperature, element of Lie algebra $\mathfrak{g}$of the group, and Q is a "geometric" heat, element of the dual space of the Lie algebra $\mathfrak{g}^*$of the group. Souriau has proposed a Riemannian metric that we have identified as a generalization of the Fisher metric:

$$I(\beta) = [g_\beta] \text{ with } g_\beta([\beta, Z_1], [\beta, Z_2]) = \tilde{\Theta}_\beta(Z_1, [\beta, Z_2]) \tag{25}$$

$$\text{with } \tilde{\Theta}_\beta(Z_1, Z_2) = \tilde{\Theta}(Z_1, Z_2) + \langle Q, ad_{Z_1}(Z_2) \rangle$$
$$\text{where } ad_{Z_1}(Z_2) = [Z_1, Z_2] \tag{5.40}$$

Souriau has proved that all co-adjoint orbit of a Lie Group given by $O_F = \{ Ad_g^* F, g \in G \}$ subset of $\mathfrak{g}^*, F \in \mathfrak{g}^*$carries a natural homogeneous symplectic structure by a closed G-invariant 2-form. If we define $K = Ad_g^* = (Ad_{g^{-1}})^*$ and $K_*(X) = -(ad_X)^*$ with:

$$K_*(X) = -(ad_X)^* \text{ with:}$$
$$\langle Ad_g^* F, Y \rangle = \langle F, Ad_{g^{-1}} Y \rangle, \forall g \in G, Y \in \mathfrak{g}, F \in \mathfrak{g}^* \tag{5.41}$$

where if $X \in \mathfrak{g}$, $Ad_g(X) = gXg^{-1} \in \mathfrak{g}$, the G-invariant 2-form is given by the following expression:

$$\sigma_\Omega(ad_X F, ad_Y F) = B_F(X, Y) = \langle F, [X, Y] \rangle, X, Y \in \mathfrak{g} \tag{5.42}$$

Souriau Foundamental Theorem is that «Every symplectic manifold on which a Lie group acts transitively by a Hamiltonian action is a covering space of a coadjoint orbit». We can observe that for Souriau model, Fisher metric is an extension of this 2-form in non-equivariant case:

$$g_\beta([\beta, Z_1], [\beta, Z_2]) = \tilde{\Theta}(Z_1, [\beta, Z_2]) + \langle Q, [Z_1, [\beta, Z_2]] \rangle \tag{5.43}$$

The Souriau additional term $\tilde{\Theta}(Z_1, [\beta, Z_2])$ is generated by non-equivariance through Symplectic cocycle. The tensor $\tilde{\Theta}$ used to define this extended Fisher metric is defined by the moment map $J(x)$, application from M(homogeneous symplectic manifold) to the dual space of the Lie algebra $\mathfrak{g}^*$, given by:

$$\tilde{\Theta}(X,Y)=J_{[X,Y]}-\{J_X,J_Y\} \quad \text{with } J(x):M\rightarrow\mathfrak{g}^* \text{ such that } J_X(x)=\langle J(x),X\rangle, X\in\mathfrak{g} \tag{5.44}$$

This tensor $\tilde{\Theta}$ is also defined in tangent space of the cocycle $\theta(g)\in\mathfrak{g}^*$(this cocycle appears due to the non-equivariance of the coadjoint operator Ad_g^*, action of the group on the dual space of the lie algebra; the action of the group on the dual space of the Lie algebra is modified with a cocycle so that the momentu map becomes equivariant relative to this new affine action):

$$Q\big(Ad_g(\beta)\big)=Ad_g^*(Q)+\theta(g) \tag{5.45}$$

$\theta(g)\in\mathfrak{g}^*$is called nonequivariance one-cocycle, and it is a measure of the lack of equivariance of the moment map.

$$\tilde{\Theta}(X,Y):\mathfrak{g}\times\mathfrak{g}\rightarrow\Re \quad \text{with } \Theta(X)=T_e\theta\big(X(e)\big) \quad X,Y:\ \langle\Theta(X),Y\rangle \tag{5.46}$$

Souriau has then defined a Gibbs density that is covariant under the action of the group:

$$p_{Gibbs}(\xi)=e^{\Phi(\beta)-\langle U(\xi),\beta\rangle}=\frac{e^{-\langle U(\xi),\beta\rangle}}{\int_M e^{-\langle U(\xi),\beta\rangle}d\lambda_\omega} \quad \text{with } \Phi(\beta)=-\log\int_M e^{-\langle U(\xi),\beta\rangle}d\lambda_\omega \tag{5.47}$$

$$Q=\frac{\partial\Phi(\beta)}{\partial\beta}=\frac{\int\limits_M U(\xi)e^{-\langle U(\xi),\beta\rangle}d\lambda_\omega}{\int\limits_M e^{-\langle U(\xi),\beta\rangle}d\lambda_\omega}=\int_M U(\xi)p(\xi)d\lambda_\omega \tag{5.48}$$

We can express the Gibbs density with respect to Q by inverting the relation $Q=\frac{\partial\Phi(\beta)}{\partial\beta}=\Theta(\beta)$.

Then $p_{Gibbs,Q}(\xi)=e^{\Phi(\beta)-\langle U(\xi),\Theta^{-1}(Q)\rangle}$ with $\beta=\Theta^{-1}(Q)$.

This domain is very active and close approaches are explored by Koichi Tojo to build harmonic exponential families on homogeneous spaces [117, 118, 115] (Figs. 5.6 and 5.7).

We will introduce Souriau moment map for SU(1,1)/U(1) group that acts transitively on Poincaré Unit Disk, based on moment map. Considering the Lie group

$$SU(1,1)=\left\{\begin{pmatrix} a & b\\ b^* & a^*\end{pmatrix}/a,b\in C,|a|^2-|b|^2=1\right\} \tag{5.35}$$

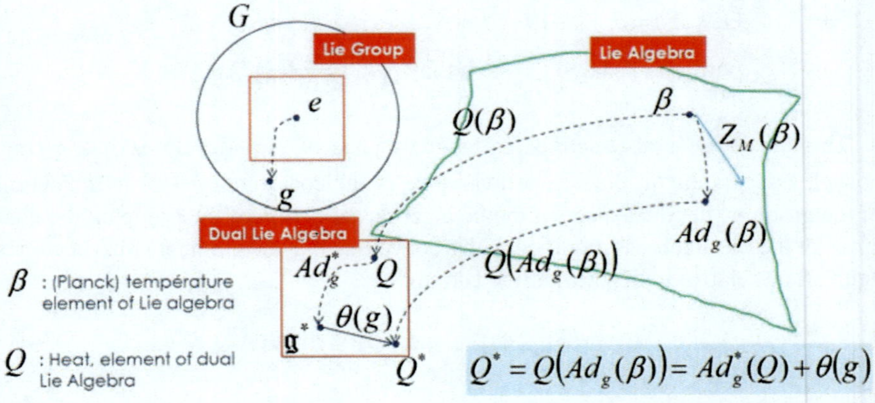

Fig. 5.6 Fondamental equation of Souriau Lie Groups Thermodynamics. Q is the geometric heat in dual Lie algebra, β is the geometric temperature in Lie algebra

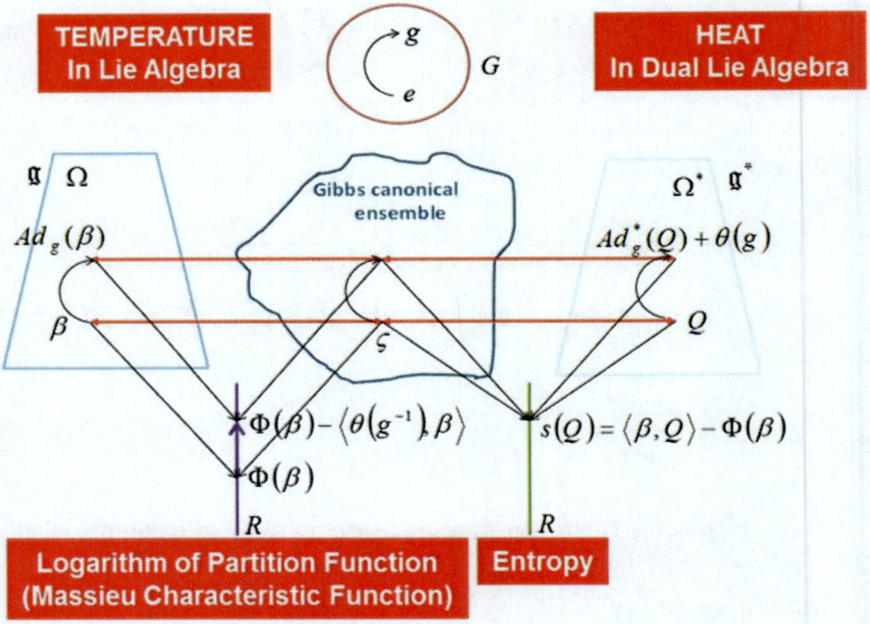

Fig. 5.7 Souriau model of Lie groups thermodynamics

and its Lie algebra given by elements

$$su(1,1) = \left\{ \begin{pmatrix} ir & \eta \\ \eta^* & -ir \end{pmatrix} / r \in R, \eta \in C \right\} \tag{5.36}$$

A basis for this Lie algebra $su(1,1)$ is $(u_1, u_2, u_3) \in \mathfrak{g}$ with

$$u_1 = \frac{i}{2}\begin{pmatrix} 1 & 0 \\ 0 & -1 \end{pmatrix}, u_2 = -\frac{1}{2}\begin{pmatrix} 0 & 1 \\ 1 & 0 \end{pmatrix} \text{ and } u_3 = \frac{1}{2}\begin{pmatrix} 0 & -i \\ i & 0 \end{pmatrix} \tag{5.37}$$

with $[u_1, u_3] = -u_2$, $[u_1, u_2] = u_3$, $[u_2, u_3] = -u_1$.

The compact subgroup is generated by u_1, while u_2 and u_3 generate a hyperbolic subgroup. The dual space of the Lie algebra is given by:

$$su(1,1)^* = \left\{ \begin{pmatrix} z & x+iy \\ -x+iy & -z \end{pmatrix} / x, y, z \in R \right\} \tag{5.38}$$

with the basis $(u_1^*, u_2^*, u_3^*) \in \mathfrak{g}^*$ with

$$u_1^* = \begin{pmatrix} 1 & 0 \\ 0 & -1 \end{pmatrix}, u_2^* = \begin{pmatrix} 0 & i \\ j & 0 \end{pmatrix} \text{ and } u_3^* = \begin{pmatrix} 0 & 1 \\ -1 & 0 \end{pmatrix} \tag{5.39}$$

Let consider $D = \{z \in C / |z| < 1\}$ be the open unit disk of Poincaré. For each $\rho > 0$, the pair (D, ω_ρ) is a symplectic homogeneous manifold with $\omega_\rho = 2i\rho \frac{dz \wedge dz^*}{(1-|z|^2)^2}$, where ω_ρ is invariant under the action:

$$\begin{gathered} SU(1,1) \times D \to D \\ (g, z) \mapsto g \cdot Z = \frac{az+b}{b^*z+a^*} \end{gathered} \tag{5.40}$$

This action is transitive and is globally and strongly Hamiltonian. Its generators are the hamiltonian vector fields associated to the functions:

$$J_1(z, z^*) = \rho \frac{1+|z|^2}{1-|z|^2}, J_2(z, z^*) = \frac{\rho}{i} \frac{z-z^*}{1-|z|^2}, J_3(z, z^*) = -\rho \frac{z+z^*}{1-|z|^2}$$

The associated moment map [119, 120] $J : D \to su^*(1,1)$ defined by $J(z).u_i = J_i(z, z^*)$, maps D into a coadjoint orbit in $SU^*(1,1)$. Then, we can write the moment map as a matrix element of $SU^*(1,1)$:

$$J(z) = J_1(z,z^*)u_1^* + J_2(z,z^*)u_2^* + J_3(z,z^*)u_3^*$$

$$J(z) = \rho \begin{pmatrix} \frac{1+|z|^2}{1-|z|^2} & -2\frac{z^*}{1-|z|^2} \\ 2\frac{z}{1-|z|^2} & -\frac{1+|z|^2}{1-|z|^2} \end{pmatrix} \in \mathfrak{g}^* \tag{5.41}$$

The moment map J is a diffeomorphism of D onto one sheet of the two-sheeted hyperboloid in $SU^*(1,1)$, determined bythe following equation $J_1^2 - J_2^2 - J_3^2 = \rho^2$, $J_1 \geq \rho$ with $J_1 u_1^* + J_2 u_2^* + J_3 u_3^* \in su^*(1,1)$. We note O_ρ^+ the coadjoint orbit $Ad^*_{SU(1,1)}$ of $SU(1,1)$, given by the upper sheet of the two-sheeted hyperboloid given by previous equation. The orbit method of Kostant-Kirillov-Souriau associates to each of these coadjoint orbits a representation of the discrete series of $SU(1,1)$, provided that ρ is a half integer greater or equal than 1 ($\rho = \frac{k}{2}$, $k \in N$ and $\rho \geq 1$). When explicitly executing the Kostant-Kirillov construction, the representation Hilbert spaces H_ρ are realized as closed reproducing kernel subspaces of $L^2(D, \omega_\rho)$. The Kostant-Kirillov-Souriau orbit method shows that to each coadjoint orbit of a connected Lie group is associated a unitary irreducible representation of G acting in a Hilbert space H.

Souriau has oberved that action of the full Galilean group on the space of motions of an isolated mechanical system is not related to any equilibrium Gibbs state (the open subset of the Lie algebra, associated to this Gibbs state is empty). The main Souriau idea was to define the Gibbs states for one-parameter subgroups of the Galilean group. We will use the same approach, in this case We will consider action of the Lie group $SU(1,1)$ on the symplectic manifold (M,ω) (Poincaré unit disk) and its momentum map J are such that the open subset $\Lambda_\beta = \left\{ \beta \in \mathfrak{g} / \int_D e^{-\langle J(z), \beta \rangle} d\lambda(z) < +\infty \right\}$ is not empty. This condition is not always satisfied when (M, ω) is a cotangent bundle, but of course it is satisfied when it is a compact manifold. The idea of Souriau is to consider a one parameter subgroup of $SU(1,1)$. To parametrize elements of $SU(1,1)$ is through its Lie algebra. In the neighborhood of the identity element, the elements of $g \in SU(1,1)$ can be written as the exponential of an element β of its Lie algebra:

$$g = \exp(\varepsilon\beta) \text{ with } \beta \in \mathfrak{g} \tag{5.42}$$

The condition $g^+ M g = M$ for $M = \begin{pmatrix} 1 & 0 \\ 0 & -1 \end{pmatrix}$ can be expanded for $\varepsilon << 1$ and is equivalent to $\beta^+ M + M\beta = 0$ which then implies $\beta = \begin{pmatrix} \dot{u}^* & \eta \\ \eta^* & -\dot{u}^* \end{pmatrix}$, $r \in R$, $\eta \in C$. We can observe that r and $\eta = \eta_R + i\eta_I$ contain 3 degrees of freedom, as required. Also because det $g = 1$, we get $Tr(\beta) = 0$. We can then exponentiate β

with exponential map to get :

$$g = \exp(\varepsilon\beta) = \sum_{k=0}^{\infty} \frac{(\varepsilon\beta)^k}{k!} = \begin{pmatrix} a_\varepsilon(\beta) & b_\varepsilon(\beta) \\ b_\varepsilon^*(\beta) & a_\varepsilon^*(\beta) \end{pmatrix} \tag{5.43}$$

If we make the remark that we have the following relation $\beta^2 = \begin{pmatrix} ir & r \\ \eta^* & -ir \end{pmatrix}\begin{pmatrix} i^* & \eta \\ \eta^* & -i^* \end{pmatrix} = \left(|\eta|^2 - r^2\right)I$, we can developed the exponential map :

$$g = \exp(\varepsilon\beta)$$
$$g = \begin{pmatrix} \cosh(\varepsilon R) + ir\frac{\sinh(\varepsilon R)}{R} & \eta\frac{\sinh(\varepsilon R)}{R} \\ \eta^*\frac{\sinh(\varepsilon R)}{R} & \cosh(\varepsilon R) - ir\frac{\sinh(\varepsilon R)}{R} \end{pmatrix}$$
$$\text{with } R^2 = |\eta|^2 - r^2 \tag{5.44}$$

We can observe that one condition is that $|\eta|^2 - r^2 > 0$ then the subset to consider is given by the subset $\Lambda_\beta = \left\{\beta = \begin{pmatrix} ir & \eta \\ \eta^* & -ir \end{pmatrix}, r \in R, \eta \in C / |\eta|^2 - r^2 > 0\right\}$ such that $\int_D e^{-\langle J(z),\beta\rangle} d\lambda(z) < +\infty$. The generalized Gibbs states of the full $SU(1,1)$ group do not exist. However, generalized Gibbs states for the one-parameter subgroups exp $(\alpha\beta)$, $\beta \in \Lambda_\beta$, of the $SU(1,1)$ group do exist. The generalized Gibbs state associated to β remains invariant under the restriction of the action to the one-parameter subgroup of $SU(1,1)$ generated by exp $(\varepsilon\beta)$.

To go futher, we will develop the Souriau Gibbs density from the Souriau moment map $J(z)$ and the Souriau temperature $\beta \in \Lambda_\beta$. If we note $b = \frac{1}{1-|z|^2}\begin{bmatrix} 1 \\ -z \end{bmatrix}$, we can write the moment map:

$$J(z) = \rho\left(2Mbb^+ - \mathrm{Tr}\left(Mbb^+\right)I\right) \text{ with } M = \begin{bmatrix} 1 & 0 \\ 0 & -1 \end{bmatrix} \tag{5.45}$$

We can the write the covariant Gibbs density in the unit disk given by moment map of the Lie group $SU(1,1)$ and geometric temperature in its Lie algebra $\beta \in \Lambda_\beta$:

$$p_{Gibbs}(z) = \frac{e^{-\langle J(z),\beta\rangle}}{\int_D e^{-\langle J(z),\beta\rangle} d\lambda(z)} \text{ with } d\lambda(z) = 2i\rho\frac{dz \wedge dz^*}{\left(1 - |z|^2\right)^2} \tag{5.46}$$

$$p_{Gibbs}(z) = \frac{\left\langle e^{-\{\rho(2Mbb^+ - Tr(Mbb^+)I),\beta}\right.}{\int_D e^{-\langle J(z),\beta\rangle} d\lambda(z)} = \frac{e^{\rho\left\langle\begin{pmatrix} \frac{1+|z|^2}{(1-|z|^2)} & \frac{-2z^*}{(1-|z|^2)} \\ \frac{2z}{(1-|z|^2)} & -\frac{1+|z|^2}{(1-|z|^2)} \end{pmatrix}, \begin{pmatrix} tr & \eta \\ \eta^* & -ir \end{pmatrix}\right\rangle}}{\int_D e^{-\langle J(z),\beta\rangle} d\lambda(z)}$$

To write the Gibbs density with respect to its statistical moments, we have to express the density with respect to $Q = E[J(z)]$. Then, we have to invert the relation between Q and β, to replace this last variable $\beta = \begin{pmatrix} ir & \eta \\ \eta^* & -ir \end{pmatrix} \in \Lambda_\beta$ by $\beta = \Theta^{-1}(Q) \in \mathfrak{g}$ where $Q = \frac{\partial \Phi(\beta)}{\partial \beta} = \Theta(\beta) \in \mathfrak{g}^*$ with $\Phi(\beta) = -\log \int_D e^{-\langle J(z)\beta \rangle} d\lambda(z)$, deduce from Legendre tranform. The mean moment map is given by:

$$Q = E[J(z)] = E\left[\rho \begin{bmatrix} \frac{1+|w|^2}{(1-|w|^2)} & \frac{-2w^*}{(1-|w|^2)} \\ \frac{2w}{(1-|w|^2)} & -\frac{1+|w|^2}{(1-|w|^2)} \end{bmatrix}\right] \text{ where } w \in D$$

Moment map interpretation by stereographic projection is well explained in a paper of Charles-Michel Marle [121].

The extension of Kostant-Sekiguchi-Vergne correspondence is illustrated by Olivier Biquard also with the simplest case $G = SU(1, 1)$. From its Lie algebra, we have the Cartan decomposition:

$$\mathfrak{g} = \left\{ \begin{pmatrix} ir & \eta^* \\ \eta & -ir \end{pmatrix} / r \in R, \eta \in C \right\} = \mathfrak{h} \oplus \mathfrak{m}$$
$$\text{with } \mathfrak{h} = \begin{pmatrix} ir & 0 \\ 0 & -ir \end{pmatrix} \text{ and } \mathfrak{m} = \begin{pmatrix} 0 & \eta^* \\ \eta & 0 \end{pmatrix} \tag{5.49}$$

The complexified action of $H = SO(2, R)$ on $\mathfrak{m} = R^2$ is that of $SO(2, C)$ on C^2, so the nonzero H^C-orbits in $\mathfrak{m}^C$ are copies of C^*. The nonzero G-orbits in $\mathfrak{g}$ are the connected components of:

$$\det \begin{pmatrix} ir & \eta^* \\ \eta & -ir \end{pmatrix} = r^2 - |\eta|^2 = \lambda \tag{5.50}$$

More especially for $\lambda < 0$, $\mathfrak{t}_{\mathfrak{g}} \subset \mathfrak{m}$ and the orbit is a hyperboloid, diffeomorphic to the semi-simple orbits in $\mathfrak{m}^C$ (so the complex structure is that of C^*); the corresponding orbits in $\mathfrak{m}^C$ depend on a parameter $\tau_3 \in i\mathfrak{a} = it$.

5.6 Extension to Siegel Disk and Moment Map of *SU(P,Q)* Lie Group

To address computation of covariant Gibbs density for Siegel Unit Disk, we will consider in this section $SU(p,q)$ Unitary Group:

$$\begin{aligned} G &= SU(p,q) \text{ and} \\ K &= S(U(p)\times U(q)) = \left\{ \begin{pmatrix} A & 0 \\ 0 & D \end{pmatrix} / A \in U(p), \right. \\ & \left. D \in U(q), \det(A)\det(D) = 1 \right\} \end{aligned} \tag{5.51}$$

We can use the following decomposition for $g \in G^C$:

$$g = \begin{pmatrix} A & B \\ C & D \end{pmatrix} \in G^C, g = \begin{pmatrix} I_p & BD^{-1} \\ 0 & I_q \end{pmatrix} \begin{pmatrix} A - BD^{-1}C & 0 \\ 0 & D \end{pmatrix} \begin{pmatrix} I_p & 0 \\ D^{-1}C & I_q \end{pmatrix} \tag{5.52}$$

and consider the action of $g \in G^C$ on Siegel Unit Disk $SD = \{ Z \in M_{pq}(C)/I_p - ZZ^+ > 0 \}$ given by:

$$g = \begin{pmatrix} A & B \\ C & D \end{pmatrix} \in G^C, g = \begin{pmatrix} I_p & BD^{-1} \\ 0 & I_q \end{pmatrix} \begin{pmatrix} A - BD^{-1}C & 0 \\ 0 & D \end{pmatrix} \begin{pmatrix} I_p & 0 \\ D^{-1}C & I_q \end{pmatrix} \tag{5.53}$$

Benjamin Cahen has study this case and introduced the moment map by identifing G-equivariantly $\mathfrak{g}^*$ with $\mathfrak{g}$ by means of the Killing form β on $\mathfrak{g}^C$:

$\mathfrak{z}^*$ G-equivariant with $\mathfrak{g}$ by Killing form $\beta(X,Y) = 2(p+q)Tr(XY)$

The set of all elements of $\mathfrak{g}$ fixed by K is $\mathfrak{h}$:

$$\mathfrak{h} = \{ \text{element of } G \text{ fixed by } K \} \;, \xi_0 \in \mathfrak{h}, \xi_0 = i\lambda \begin{pmatrix} -qI_p & 0 \\ 0 & pI_q \end{pmatrix} \tag{5.54}$$

Then, we the equivatiant moment map is given by:

$$\begin{aligned} & \forall X \in g^C, Z \in D, \psi(Z) = Ad^*(\exp(-Z^+)\zeta(\exp Z^+ \exp Z))\xi_0 \\ & \forall g \in G, Z \in D \text{ then } \psi(g.Z) = Ad_g^* \psi(Z) \\ & \psi \text{ is a diffeomorphism from } SD \text{ onto orbit } O(\xi_0) \end{aligned} \tag{5.55}$$

with:

$$\psi(Z) = i\lambda \begin{pmatrix} (I_p - ZZ^+)^{-1}(-pZZ^+ - qI_p) & (p+q)Z(I_q - Z^+Z)^{-1} \\ -(p+q)(I_q - Z^+Z)^{-1}Z^+ & (pI_q + qZ^+Z)(I_q - Z^+Z)^{-1} \end{pmatrix} \quad (5.56)$$

$$\text{and } \zeta\left(\exp Z^+ \exp Z\right) = \begin{pmatrix} I_p & Z(I_q - Z^+Z)^{-1} \\ 0 & I_q \end{pmatrix} \quad (5.57)$$

Same developments as used for Poincaré unit disk and SU(1,1) moment map could be easily extended for SU(p,q) its moment map to define density of probability for Siegel Disk.

5.7 Conclusion

The work of Professor Koszul is a proof of fidelity to his masters and in the first place to Prof. Elie Cartan, who inspired him throughout his life. Henri Cartan writes on this subject "*I do not forget the homage he paid to Elie Cartan's work in Differential Geometry during the celebration, in Bucharest, in 1969, of the centenary of his birth. It is not a coincidence that this centenary was also celebrated in Grenoble the same year. As always, Koszul spoke with the discretion and tact that we know him, and that we love so much at home*". Our generation and previous one have forgotten or misunderstood the depth of the work of Jean-Louis Koszul and Elie Cartan on the study of bounded homogeneous domains. It is our responsibility to correct this omission, and to make it the new inspiration for the Geometric Science of Information (Fig. 5.8).

Jean-Louis Koszul was the cousin of the composer Henri Dutilleux, two grandsons of Julien Koszul, musician, director of the Roubaix conservatory, student of Camille Saint-Saëns, fellow student of Gabriel Fauré and Eugène Gigout at the Ecole Niedermeyer, and professor of Albert Roussel (who was Professor of Jean Cartan, brother of Henri Cartan and son of Elie Cartan). As told by his grandfather, Julien Koszul in [122], the Koszul family is French thanks to the Emperor Napoleon. Matheus Koziet, then Kosziel, then Mathieu Koszul (Wola-Radziszowska, near Krakow, Poland 10/14/1784, Niedermorschwiller 11/18/1858) is indeed the great grandfather of Jean-Louis Koszul (Mathieu Koszul was grandfather of Julien Koszul, who was the grandfather of Jean-Louis Koszul). Mathieu Koszul participated from 1800 to 1818, in many campaigns. Incorporated at the age of sixteen in the Austrian army, which he probably deserted in 1806 following the great French victories of Iena and Auerstaedt, he joined the Poniatowski army, took part in many battles against the Prussians, the Austrians and the Russians with the Grand Army. He traveled to Moscow and returned unscathed, no doubt thanks to his iron health. In 1813, he participated in the Battle of the Nations in Leipzig. Marshal Poniatowski, who was responsible for covering the Emperor's retreat, drowns in the Ester by his side. Taken prisoner, he is asked where he comes from "... from Wola-Radziszowska,

Fig. 5.8 Top from left to right, Jean-Louis Koszul at ENS Paris in 1942 at a Bourbaki Seminar in 1956, at Strasbourg University around 1959 and at Grenoble in 1993. Bottom from left to right, Jean-Louis Koszul for CIRM 30th anniversary at Luminy in 2011, at Mines ParisTech for SEE GSI'13 conference in 2013 and last interview of Jean-Louis Koszul in 2016 for 50th birthday of Institut Joseph Fourier in Grenoble

in Poland", to which the victors reply "… there is no Poland, Wola-Radziszowska is in Galicia, near Krakow, you are Austrian. » But confidence in these Polish-Austrian recruits was moderate and Mathieu only arrived in France (in Alsace) with the rearguard in 1815. His unit was stationed at Niedermorschwiller, near Colmar, where he became the order of an Austrian colonel. He finds Alsace more pleasant and hospitable than Poland, and asks for permission to marry Marie-Anne Rimelen (Niedermorschwiller, 05/13/1799 - Niedermorschwiller, 06/11/1847). He became French by marrying her on 08/18/1818. Not knowing how to read or write, he signs his marriage certificate with a cross. We can only retrospectively thank the Emperor Napoleon for making the Koszul a great French family with descent also refined in music and mathematics.

Disclaimer: Views and opinions expressed are those of the authors and do not necessarily represent official positions of their respective companies.

Appendix

Bargman Parameterization of SU(1,1)

$SU(1,1)$ is isomorphic to $SL(2,R) = Sp(2,R)$ through the complex unitary matrix W:

$$SL(2,R) = \left\{ g = \begin{pmatrix} a & b \\ c & d \end{pmatrix} / \det g = ad - bc = 1 \right\} \quad (5.A1)$$

$$Sp(2,R) = \left\{ g = \begin{pmatrix} a & b \\ c & d \end{pmatrix} / gJg^T = J,\ J = \begin{pmatrix} 0 & +1 \\ -1 & 0 \end{pmatrix} \right\} \quad (5.A2)$$

$$W = \frac{1}{\sqrt{2}} \begin{pmatrix} \omega^{-1} & \omega^{-1} \\ -\omega & \omega \end{pmatrix} = \left(W^{+}\right)^{-1} \text{ with } \omega = e^{i\pi/4} = \frac{1}{\sqrt{2}}(1+i) \quad (5.A3)$$

If we observe that $W^{-1}JW = -iM$, the isomorphism is given explicitely by:

$$\begin{pmatrix} a & b \\ c & d \end{pmatrix} = g(u) = WuW^{-1} = \begin{pmatrix} \mathrm{Re}(\alpha+\beta) & -\mathrm{Im}(\alpha-\beta) \\ \mathrm{Im}(\alpha+\beta) & \mathrm{Re}(\alpha-\beta) \end{pmatrix} \quad (5.A30)$$

$$\begin{pmatrix} \alpha & \beta \\ \beta^* & \alpha^* \end{pmatrix} = u(g) = W^{-1}gW = \frac{1}{2} \begin{pmatrix} (a+d) - i(b-c) & (a-d) + i(b+c) \\ (a-d) - i(b+c) & (a+d) + i(b-c) \end{pmatrix} \quad (5.A4)$$

We can also make also a link with $SO(2,1)$ of "1 + 2" pseudo-orthogonal matrices:

$$SO(2,1) = \left\{ \Gamma \in GL(3,3) / \det(\Gamma) = 1,\ \Gamma K \Gamma^T = \Gamma,\ K = \begin{pmatrix} +1 & 0 & 0 \\ 0 & -1 & 0 \\ 0 & 0 & -1 \end{pmatrix} \right\} \quad (5.A5)$$

$$\Gamma(g) = \begin{pmatrix} \frac{1}{2}\left(a^2+b^2+c^2+d^2\right) & \frac{1}{2}\left(a^2-b^2+c^2-d^2\right) & -cd-ab \\ \frac{1}{2}\left(a^2+b^2-c^2-d^2\right) & \frac{1}{2}\left(a^2-b^2-c^2+d^2\right) & cd-ab \\ -bd-ac & bd-ac & ad+bc \end{pmatrix} \quad (5.A6)$$

with $\Gamma(g_1)\Gamma(g_2) = \Gamma(g_1 g_2)$, $\Gamma(I) = I$, $\Gamma\left(g^{-1}\right) = \Gamma(g)^{-1}$

The $SO(2,1)$ matrix corresponds to any $SU(1,1)$:

$$\Gamma(u) = \begin{pmatrix} |\alpha|^2 + |\beta|^2 & 2\mathrm{Re}\alpha\beta^* & 2\mathrm{Im}\alpha\beta^* \\ 2\mathrm{Re}\alpha\beta & \mathrm{Re}\left(\alpha^2+\beta^2\right) & \mathrm{Im}\left(\alpha^2-\beta^2\right) \\ -2\mathrm{Im}\alpha\beta & -\mathrm{Im}\left(\alpha^2+\beta^2\right) & \mathrm{Re}\left(\alpha^2-\beta^2\right) \end{pmatrix} \quad (5.A7)$$

and $\alpha = \pm\sqrt{\frac{1}{2}(\Gamma_{11}+\Gamma_{12}) + i(\Gamma_{12}-\Gamma_{21})}$, $\beta = \frac{1}{2\alpha}(\Gamma_{10} - i\Gamma_{20})$

The properties of connectivity of $Sp(2, R)$ is described by its isomorphy with $SU(1, 1)$. Using unimodular condition:

$$|\alpha|^2 - |\beta|^2 = 1 \Rightarrow \alpha_R^2 + \alpha_I^2 - \beta_R^2 = 1 + \beta_I^2 \geq 1$$
$$\text{with } \alpha = \alpha_R + i\alpha_I \text{ and } \beta = \beta_R + i\beta_I$$

If β_I is fixed, $(\alpha_R, \alpha_I, \beta_R)$ are constrained to define a one-sheeted revolution hyperboloid, with its circular waist in the α plane.

To $SU(1, 1)$, we can associate the simply-connected universal covering group, using the maximal compact subgroup $U(1)$ and corresponding to the Iwasawa decomposition (factorization of a noncompact semisimple group into its maximal compact subgroup times a solvable subgroup).

$$\begin{pmatrix} \alpha & \beta \\ \beta^* & \alpha^* \end{pmatrix} = \begin{pmatrix} e^{i\omega} & 0 \\ 0 & e^{i\omega} \end{pmatrix} \begin{pmatrix} \lambda & \mu \\ \mu^* & \lambda \end{pmatrix} \text{with} \begin{cases} \omega = \arg \alpha = \frac{1}{2} i \ln(\alpha^* \alpha^{-1}) \\ \lambda = |\alpha| > 0 \\ \mu = e^{-i\omega}\beta = \sqrt{\frac{\alpha^*}{\alpha}}\beta \end{cases} \tag{5.A8}$$

$$\beta = e^{i\omega}\mu, |\alpha|^2 - |\beta|^2 = \lambda^2 - |\mu|^2 = 1 \text{ so } |\mu| < \lambda \tag{5.A9}$$

Bargmann has generalized this parameterization for $Sp(2N, R)$, more convenient but difficult to generalize to N dimensions. For $SU(1, 1)$:, Bargmann has used (ω, γ):

$$\gamma = \frac{\mu}{\lambda} = \frac{\beta}{\alpha}(|\gamma| < 1), \lambda = \frac{1}{\sqrt{1 - |\gamma|^2}}, \mu = \frac{\gamma}{\sqrt{1 - |\gamma|^2}} \tag{5.A10}$$

For $SL(2, R) = Sp(2, R)$, the Bargman, parameterization is given by this decomposition of a non-singular matrix into the product of an orthogonal and a positive definite symmetric matrix:

$$\begin{pmatrix} a & b \\ c & d \end{pmatrix} = \begin{pmatrix} \cos\omega & -\sin\omega \\ \sin\omega & \cos\omega \end{pmatrix} \begin{pmatrix} \lambda + \text{Re}\mu & \text{Im}\mu \\ \text{Im}\mu & \lambda - \text{Re}\mu \end{pmatrix} \tag{5.A11}$$

Conversely: $\omega = \arg[(a + d) - i(b - c)]$, $\mu = e^{-i\omega}[(a - d) + i(b + c)]$
ω is counted modulo 2π, $\omega \equiv \omega(\text{mod} 2\pi)$.

$SU(1, 1)$ and $SL(2, R) = Sp(2, R)$ are described when ω is counted modulo 2π, $\omega \equiv \omega(\text{mod} 2\pi)$.

Valentine Bargmann has proposed the covering of the general symplectic group $Sp(2N, R)$:

$$Sp(2N, R) = \left\{ g = \begin{pmatrix} A & B \\ C & D \end{pmatrix} / g J_{2N} g^T = J_{2N}, J_{2N}^T = -J_{2N}, J_{2N} = \begin{pmatrix} 0 & I_N \\ -I_N & 0 \end{pmatrix} \right\} \tag{5.A12}$$

with relations:

$$AB^T = BA^T, AC^T = CA^T, BD^T = DB^T, CD^T = DC^T, AD^T - BC^T = I_N \tag{5.A13}$$

$$g \in Sp(2N, R) \Rightarrow g^{-1} = M_{2N} g^T M_{2N} = \begin{pmatrix} D^T & -B^T \\ -C^T & A^T \end{pmatrix} \tag{5.A14}$$

Bargmann has observed that although $Sp(2N, R)$ is not isomorphic to any pseudo-unitary group, its inclusion in $U(N, N)$ will display the connectivity properties through its unitary $U(N)$ maximal compact subgroup, generalizing the role of $U(1) = SO(2)$ in $Sp(2, R)$.

$W_N = W \otimes I_N$ a $2N \times 2N$ matrix where $W = W_1 = \frac{1}{\sqrt{2}}\begin{pmatrix} \omega_{\pi/4}^{-1} & \omega_{\pi/4}^{-1} \\ -\omega_{\pi/4} & \omega_{\pi/4} \end{pmatrix}$ with $\omega = e^{i\pi/4} = \frac{1}{\sqrt{2}}(1+i)$, which gives the $N \times N$ block coefficients.

$$u(g) = W_N^{-1} g W_N = \frac{1}{2}\begin{pmatrix} [A+D] - i[B-C] & [A-D] + i[B+C] \\ [A-D] - i[B+C] & [A+D] + i[B-C] \end{pmatrix} = \begin{pmatrix} \alpha & \beta \\ \beta^* & \alpha^* \end{pmatrix} \tag{5.A15}$$

$$\text{With} \begin{cases} \alpha\alpha^+ - \beta\beta^+ = I_N, \alpha^+\alpha - \beta^T\beta^* = I_N \\ \alpha\beta^T - \beta\alpha^T = 0, \alpha^T\beta^* - \beta^+\alpha = 0 \end{cases} \tag{5.A16}$$

$$\text{and } u^{-1} = M_{2N} u^+ M_{2N}^{-1} = \begin{pmatrix} \alpha^+ & -\beta^T \\ -\beta^+ & \alpha^T \end{pmatrix} \tag{5.A17}$$

The symplecticity property of g becomes:

$$u M_{2N} u^+ = M_{2N}, M_{2N} = i W_N^{-1} J_{2N} W_N = \begin{pmatrix} I_N & 0 \\ 0 & -I_N \end{pmatrix} \tag{5.A18}$$

$$\begin{pmatrix} A & B \\ C & D \end{pmatrix} = g(u) = W_N u W_N^{-1} = \begin{pmatrix} \mathrm{Re}(\alpha+\beta) & -\mathrm{Im}(\alpha-\beta) \\ \mathrm{Im}(\alpha+\beta) & \mathrm{Re}(\alpha-\beta) \end{pmatrix} \tag{5.A19}$$

Valentine Bargmann has extended the well-know theorem that any real matrix R may be decomposed into the product of an orthogonal Q and a symmetric positive definite matrix S, uniquely as $R = QS$. Bargmann has shown that if $R \in Sp(2N, R)$, then $R = QS$ with $Q, S \in Sp(2N, R)$ where Q maps onto unitary matrix and S maps onto Hermitian positive definite matrix:

$$u(Q) = \begin{pmatrix} \alpha & 0 \\ 0 & \alpha^* \end{pmatrix}, \alpha\alpha^+ = I_N, \alpha \in U(N) \text{ and } u(S) = \exp\begin{pmatrix} 0 & \xi \\ \xi^* & 0 \end{pmatrix}, \xi = \xi^T \tag{5.A20}$$

We can generalize Bargmann parameterization of $SU(1,1)$ to $Sp(2N, R)$:

$$u\{\omega,\lambda,\mu\} = \begin{pmatrix} e^{i\omega} I_N & 0 \\ 0 & e^{-i\omega} I_N \end{pmatrix} \begin{pmatrix} \lambda & \mu \\ \mu^* & \lambda^* \end{pmatrix} \oplus, \det\lambda > 0 \tag{5.A21}$$

Then the Bargmann parameters are:

$$\omega = \frac{1}{N}\arg\det\alpha, \lambda = e^{-i\omega}\alpha, \mu = e^{-i\omega}\beta, e^{iN\omega} = \frac{\det\alpha}{|\det\alpha|}, \det\lambda = |\det\alpha| > 0 \tag{5.A22}$$

The $Sp(2N, R)$ matrices in terms of the Bargmann parameters are:

$$g\{\omega,\lambda,\mu\} = \begin{pmatrix} \cos\omega I_N & -\sin\omega I_N \\ \sin\omega I_N & \cos\omega I_N \end{pmatrix} \begin{pmatrix} \mathrm{Re}(\lambda+\mu) & -\mathrm{Im}(\lambda-\mu) \\ \mathrm{Im}(\lambda+\mu) & \mathrm{Re}(\lambda-\mu) \end{pmatrix} \tag{5.A23}$$

V. Bargmann has proposed the covering of the general symplectic group $Sp(2N, R)$:

$$Sp(2N, R) = \left\{ g = \begin{pmatrix} A & B \\ C & D \end{pmatrix} / g J_{2N} g^T = J_{2N}, J_{2N}^T = -J_{2N}, J_{2N} = \begin{pmatrix} 0 & I_N \\ -I_N & 0 \end{pmatrix} \right\} \tag{5.A24}$$

$$AB^T = BA^T, AC^T = CA^T, BD^T = DB^T, CD^T = DC^T, AD^T - BC^T = I_N \tag{5.A25}$$

Bargmann has observed that although $Sp(2N, R)$ is not isomorphic to any pseudo-unitary group, its inclusion in $U(N, N)$ will display the connectivity properties through its unitary $U(N)$ maximal compact subgroup, generalizing the role of $U(1) = SO(2)$ in $Sp(2, R)$: $W_N = W \otimes I_N$, $2N \times 2N$ matrix where $W = W_1 = \frac{1}{\sqrt{2}}\begin{pmatrix} \omega_{\pi/4}^{-1} & \omega_{\pi/4}^{-1} \\ -\omega_{\pi/4} & \omega_{\pi/4} \end{pmatrix}$ with $\omega = e^{i\pi/4} = \frac{1}{\sqrt{2}}(1+i)$.

$$u(g) = W_N^{-1} g W_N = \frac{1}{2}\begin{pmatrix} [A+D] - i[B-C] & [A-D] + i[B+C] \\ [A-D] - i[B+C] & [A+D] + i[B-C] \end{pmatrix} = \begin{pmatrix} \alpha & \beta \\ \beta^* & \alpha^* \end{pmatrix} \tag{5.A26}$$

with $\alpha\alpha^+ - \beta\beta^+ = I_N, \alpha^+\alpha - \beta^T\beta^* = I_N$ and $\alpha\beta^T - \beta\alpha^T = 0, \alpha^T\beta^* - \beta^+\alpha = 0$ (5.A27)

The symplecticity property of g becomes:

$$u M_{2N} u^+ = M_{2N}, M_{2N} = i W_N^{-1} J_{2N} W_N = \begin{pmatrix} I_N & 0 \\ 0 & -I_N \end{pmatrix} \tag{5.A28}$$

$$\begin{pmatrix} A & B \\ C & D \end{pmatrix} = g(u) = W_N u W_N^{-1} = \begin{pmatrix} \mathrm{Re}(\alpha+\beta) & -\mathrm{Im}(\alpha-\beta) \\ \mathrm{Im}(\alpha+\beta) & \mathrm{Re}(\alpha-\beta) \end{pmatrix} \tag{5.A29}$$

[123, 37, 124]

References

1. Cartan, E.: Sur les invariants intégraux de certains espaces homogènes clos et les propriétés topologiques de ces espaces. Ann. Soc. Pol. De Math. **8**, 181–225 (1929)
2. Cartan, H.: Allocution de Monsieur Henri Cartan, colloques Jean-Louis Koszul. Annales de l'Institut Fourier, tome **37**(4), 1–4 (1987)
3. Koszul, J.L.: L'œuvre d'Élie Cartan en géométrie différentielle, in Élie Cartan, 1869-1951. Hommage de l'Académie de la République Socialiste de Roumanie à l'occasion du centenaire de sa naissance. Comprenant les communications faites aux séances du 4e Congrès du Groupement des Mathématiciens d'Expression Latine, tenu à Bucarest en 1969, pp. 39–45. Editura Academiei Republicii Socialiste Romania, Bucharest (1975)
4. Koszul, J.L.: Interview for "Institut Joseph Fourier" 50th birthday in 2016: video. https://www.youtube.com/watch?v=AzK5K7Q05sw
5. Siegel, C.L.: Über der analytische Theorie der quadratischen Formen. Ann. Math. **36**, 527–606 (1935)
6. Siegel, C.L.: Symplectic geometry. Am. J. Math. **65**, 1–86 (1943)
7. Hua, L.K.: Harmonic Analysis of Functions of Several Complex Variables in the Classical Domains. American Mathematical Society, Providence, RI, USA (1963)
8. Lichnerowicz, A.: Espaces homogènes Kähleriens. In: Colloque de Géométrie Différentielle; Publication du CNRSP, Paris, France, pp. 171–184 (1953)
9. Berezin, F.: Quantization in complex symmetric spaces. Izv. Akad. Nauk SSSR Ser. Math. **9**, 363–402 (1975)
10. Vesentini, E.: Geometry of homogeneous bounded domains, Springer-Verlag 2011, reprint of the 1rst Ed. C.IM.E., Ed. Cremonese, Roma 1968
11. Sampieri, U.: A generalized exponential map for an affinely homogeneous cone, Atti della Accademia Nazionale dei Lincei. Classe di Scienze Fisiche, Matematiche e Naturali. Rendiconti Lincei. Matematica e Applicazioni, Serie 8, Vol. 75, n.6, pp. 320–330 (1983)
12. Sampieri, U.: Lie group structures and reproducing kernels on homogeneous siegel domains. Annali di Matematica **152**(1), 1–19 (1988)
13. Fréchet, M.: Sur l'extension de certaines évaluations statistiques au cas de petits échantillons. Revue de l'Institut International de Statistique **11**(3/4), 182–205 (1943)
14. Rao, C.R.: Information and the accuracy attainable in the estimation of statistical parameters. Bull. Calcutta Math. Soc. **37**, 81–89 (1945)
15. Chentsov, N.N.: Statistical Decision Rules and Optimal Inferences, Transactions of Mathematics Monograph, American Mathematical Society, Providence, RI, USA, Vol. 53 (1982)
16. Balian, R., Alhassid, Y., Reinhardt, H.: Dissipation in many-body systems: A geometric approach based on information theory. Phys. Rep. **131**, 1–146 (1986)
17. Balian, R.: The entropy-based quantum metric. Entropy **16**, 3878–3888 (2014)
18. Gromov, M.: Convex sets and Kähler manifolds, In: Advances in Differential Geometry and Topology; Tricerri, F. (Ed.), pp. 1–38. World Scientific, Singapore, Singapore (1990)
19. Gromov, M.: In a search for a structure, Part 1: On entropy, 23 June 2012 . http://www.ihes.fr/~gromov/PDF/structre-serch-entropy-july5-2012.pdf
20. Gromov, M.: Gromov Six Lectures on Probability, Symmetry, Linearity. October 2014, Jussieu, November 6th, 2014; Lecture Slides & video of Gromov lectures on youtube: http://www.ihes.fr/~gromov/PDF/probability-huge-Lecture-Nov-2014.pdf; https://www.youtube.com/watch?v=hb4D8yMdov4

21. Gromov, M.: Gromov Four Lectures on Mathematical Structures arising from Genetics and Molecular Biology, IHES, October 2013; video of Lectures on youtube: https://www.youtube.com/watch?v=v7QuYuoyLQc&t=5935s
22. Nguiffo Boyom, M.: Sur les structures affines homotopes à zéro des groupes de Lie. J. Differ. Geom. **31**, 859–911 (1990)
23. Nguiffo Boyom, M.: Structures localement plates dans certaines variétés symplectiques. Math. Scand. **76**, 61–84 (1995)
24. Nguiffo Boyom, M.: Métriques kählériennes affinement plates de certaines variétés symplectiques. I., Proc. London Math. Soc. (3), no 66, no 2, pp. 358-380 (1993)
25. Nguiffo Boyom, M.: The cohomology of Koszul-Vinberg algebras. Pacific J. Math. **225**, 119–153 (2006)
26. Nguiffo Boyom, M.: Some Lagrangian Invariants of Symplectic Manifolds, Geometry and Topology of Manifolds. Banach Center Institute of Mathematics, Polish Academy of Sciences, Warsaw **76**, 515–525 (2007)
27. Nguiffo Boyom, M., Byande, P.M.: KV Cohomology in Information Geometry Matrix Information Geometry, pp. 69–92. Springer, Heidelberg, Germany (2013)
28. Nguiffo Boyom, M.: Transversally Hessian foliations and information geometry I. Am. Inst. Phys. Proc. **1641**, 82–89 (2014)
29. Nguiffo Boyom, M., Wolak, R.: Transverse Hessian metrics information geometry MaxEnt 2014. AIP. Conf. Proc. Am. Inst. Phys. (2015)
30. Barbaresco, F.: Les densités de probabilité «distinguées» et l'équation d'Alexis Clairaut: regards croisés de Maurice Fréchet et de Jean-Louis Koszul, GRETSI'17. Juan-Les-Pins, Sept (2017)
31. Souriau, J.-M.: Structures des systèmes dynamiques. Dunod, Paris (1969)
32. Souriau, J.-M.: Mécanique statistique, groupes de Lie et cosmologie, Colloques int. du CNRS numéro 237, Géométrie symplectique et physique mathématique, pp. 59–113 (1974)
33. https://www.academia.edu/42630654/Statistical_Mechanics_Lie_Group_and_Cosmology_1_st_part_Symplectic_Model_of_Statistical_Mechanics
34. Bourguignon, J.P.: La géométrie kählérienne, domaines de la géométrie différentielle, séminaire Histoires de géométries, FMSH, 2005, video. https://youtu.be/SDmMo4a1vbk
35. Gauduchon, P.: Calabi's extremal Kähler metrics: An elementary introduction; preprint: germanio.math.unifi.it/wp-content/uploads/2015/03/dercalabi.pdf
36. Lichnerowicz, A.: Groupes de Lie à structures symplectiques ou Kähleriennes invariantes. Albert C. (eds) Géométrie Symplectique et Mécanique. Lecture Notes in Mathematics, Vol. 1416. Springer (1990)
37. Barbaresco, F.: Jean-Louis Koszul et les structures élémentaires de la géométrie de l'information, revue MATAPLI 2018, SMAI (2018)
38. Malgrange, B.: Quelques souvenirs de Jean-Louis KOSZUL, Gazette des Mathématiciens **156**, 63–64, Avril (2018)
39. Cartier, P.: In memoriam Jean-Louis KOSZUL, Gazette des Mathématiciens **156**, 64–66, Avril (2018)
40. Amari, S.I.: Differential Geometry of Statistical Models, SPRINGER Series Lecture Notes in Statistics, Vol. 28 (1985)
41. Amari,, S.I.: Information Geometry and Its Applications, SPRINGER series Applied Mathematical Sciences, Vol. 194 (2016)
42. Koszul, J.M.: Homologie et cohomologie des algèbres de Lie. Bulletin de la Société Mathématique de France, Tome **78**, 65–127 (1950)
43. Cartan, H.: Les travaux de Koszul, I, Séminaire Bourbaki, Tome 1, Exposé no. 1, pp. 7–12 (1948–1951)
44. Cartan, H.: Les travaux de Koszul, II, Séminaire Bourbaki, Tome 1, Exposé no. 8, pp. 45–52 (1948–1951)
45. Cartan, H.: Les travaux de Koszul, III, Séminaire Bourbaki, Tome 1, Exposé no. 12, pp. 71–74 (1948–1951)

46. Haefliger, A.: Des espaces homogènes à la résolution de Koszul Annales de l'institut Fourier. Tome **37**(4), 5–13 (1987)
47. Vey, J.: Travaux de Jacques Vey. Editions du CNRS, Paris (1983)
48. Gorodski, C.: 2nd Workshop of the São Paulo Journal of Mathematical Sciences: Jean-Louis Koszul in São Paulo, His Work and Legacy, Institute of Mathematics and Statistics, University of São Paulo, November 13-14 (2019). https://www.ime.usp.br/~2wspjm/
49. Barbaresco, F.: Les structures géométriques de l'Information de Jean-Louis Koszul, Colloque GRETSI, August (2019)
50. Colloque "FGSI'19 Cartan-Koszul Souriau—Foundation of Geometric Structures of Information": https://fgsi2019.sciencesconf.org/
51. Guichardet, A.: La méthode des orbites: historiques, principes, résultats. Leçons de mathématiques d'aujourd'hui, vol. 4, Cassini, pp. 33–59 (2010)
52. Cishahayo, C., de Bièvre, S.: On the contraction of the discrete series of SU(1;1). Annales de l'institut Fourier, tome **43**(2), 551–567 (1993)
53. Cahen, B.: Contraction de SU(1,1) vers le groupe de Heisenberg, pp. 19–43. Travaux mathématiques, Fascicule XV (2004)
54. McDuff, D.: The symplectic structure of Kähler manifolds of nonpositive curvature, J. Differential Geom. 3) 28 (1988), 467–475, MR 0965224, Zbl 0632.53058
55. Biquard, O.: Métriques extrémales sur les surfaces toriques [d'après S. Donaldson] [pdf]. Séminaire Bourbaki no 1018, mars 2010
56. Tojo, K., Yoshino, T.: Harmonic exponential families on homogeneous spaces, preprint (2020)
57. Selected Papers of J L Koszul, Series in Pure Mathematics, Vol. 17, World Scientific Publishing (1994)
58. Vinberg, E.B.: Homogeneous cones, Dokl. Akad. Nauk SSSR., no 133, pp. 9–12, 1960; Soviet Math. Dokl., no 1, pp. 787–790, 1961
59. Vinberg, E.B.: The Morozov-Borel theorem for real Lie groups, Dokl. Akad. Nauk SSSR., no 141, pp. 270–273, 1961; Soviet Math. Dokl., no 2, pp.1416–1419, 1962
60. Vinberg, E.B.: Convex homogeneous domains, Dokl. Akad. Nauk SSSR., 141 1961, 521–524; Soviet Math. Dokl., no 2, pp. 1470–1473, 1962
61. Vinberg, E.B.: Automorphisms of homogeneous convex cones, Dokl. Akad. Nauk SSSR., no 143, pp. 265–268 (1962); Soviet Math. Dokl., no 3, pp. 371–374, 1963
62. Vinberg, E.B.: The Theory of Homogeneous Convex Cones, Trudy Moskovskogo Matematicheskogo Obshchestva, Vol. 12, pp. 303–358 (1963); Trans. Moscow Math. Soc., no 12, pp. 340–403, 1963
63. Vinberg, E.B., Gindikin S.G., Pyatetskii-Shapiro I.I.: On the classification and canonical realization of bounded homogeneous domains, Trudy Moskov. Mat. Obshch., no 12, pp. 359–388 (1963); Trans. Moscow Math. Soc., n 12, pp.404–437, 1963
64. Vinberg, E.B.: The structure of the group of automorphisms of a convex cone, Trudy Moscov. Mat. Obshch., no 13, pp. 56–83 (1964); Trans. Moscow Math. Soc., n°13, 1964
65. Vinberg, E.B.: Structure of the group of automorphisms of a homogeneous convex cone. Trudy Moskovskogo Matematicheskogo Obshchestva **13**, 56–83 (1965)
66. Pyatetskii-Shapiro, I.I.: Certain problems of harmonic analysis in homogeneous cones, Dokl. Akad. Nauk SSSR., pp. 181–184 (1957)
67. Pyatetskii-Shapiro, I.I.: On a problem of E. Cartan, Dokl. Akad. Nauk SSSR., no 124, pp. 272–273 (1959)
68. Pyatetskii-Shapiro, I.I.: The geometry of homogeneous domains and the theory of automorphic functions, The solution of a problem of E. Cartan, Uspekhi Mat. Nauk. **14**(3), 190–192 (1959)
69. Pyatetskii-Shapiro, I.I.: On the classification of bounded homogeneous domains in n-dimensional complex space, Dokl. Akad. Nauk SSSR., n°o 141, pp. 316–319 (1961); Soviet Math. Dokl., n°, pp. 1460–1463, 1962
70. Pyatetskii-Shapiro, I.I.: On bounded homogeneous domains in n-dimensional complex space. Izv. Akad. Nauk SSSR. Ser. Mat. **26**, 107–124 (1962)

71. Gindikin, S.G.: Analysis in homogeneous domains", Uspekhi Mat. Nauk. **19**(4), 3–92 (1964); Russian Math. Surveys, vol.19, n°4, pp.1–89, 1964
72. Cartan, E.: Sur les domaines bornés de l'espace de n variables complexes. Abh. Math. Seminar Hamburg **1**, 116–162 (1935)
73. Vey, J.: Sur une notion d'hyperbolicité des variétés localement plates. Faculté des sciences de l'université de Grenoble, Thèse de troisième cycle de mathématiques pures (1969)
74. Vey, J.,:Sur les automorphismes affines des ouverts convexes saillants, Annali della Scuola Normale Superiore di Pisa, Classe di Science, 3e série, tome 24(4), pp. 641–665 (1970)
75. Koszul, J.L.: Variétés localement plates et convexité. Osaka. J. Math. **2**, 285–290 (1965)
76. Alekseevsky, D.: Vinberg's theory of homogeneous convex cones: developments and applications, Transformation groups 2017. Conference dedicated to Prof. Ernest B. Vinberg on the occasion of his 80th birthday, Moscou, December 2017; https://www.mccme.ru/tg2017/slides/alexeevsky.pdf; vidéo: http://www.mathnet.ru/present19121
77. Koszul, J.L.: Sur la forme hermitienne canonique des espaces homogènes complexes. Can. J. Math. **7**, 562–576 (1955)
78. Koszul, J.L.: Exposés sur les Espaces Homogènes Symétriques. Publicação da Sociedade de Matematica de São Paulo, São Paulo, Brazil (1959)
79. Koszul, J.L.: Domaines bornées homogènes et orbites de groupes de transformations affines. Bull. Soc. Math. France **89**, 515–533 (1961)
80. Koszul, J.L.: Ouverts convexes homogènes des espaces affines. Math. Z. **79**, 254–259 (1962)
81. Koszul, J.L.: Déformations des variétés localement plates. Ann. Inst. Fourier **18**, 103–114 (1968)
82. Koszul, J.L.: Trajectoires Convexes de Groupes Affines Unimodulaires. In: Essays on Topology and Related Topics, pp. 105–110. Springer, Berlin, Germany (1970)
83. Koszul, J.L.: Lectures on Groups of Transformations. Tata Institute of Fundamental Research, Bombay (1965)
84. Thurston, W.P.: Some simple examples of symplectic manifolds. Proc. Am. Math. Soc. **55**(2), 467–468 (1976)
85. Kirillov A.A.: Elements of the Theory of Representations. Springer (1976)
86. Kostant, B: Quantization and unitary representations. Springer (1970)
87. Souriau, J.-M.: Structure des systèmes dynamiques. Dunod, Paris (1969)
88. Della Vedova, A.: Special homogeneous almost complex structures on symplectic manifolds. J. Symplectic Geom. **17**(5), 1251–1295 (2019)
89. Della Vedova, A., Gatti A.: Almost Kaehler geometry of adjoint orbits of semisimple Lie groups. arXiv:1811.06958 (2018)
90. Gatti, A.: Special almost-Kähler geometry of some homogeneous manifolds, PhD of Università degli Studi di Pavia, supervised by Dr. Alberto Della Vedova, December 2019
91. Biquard O.: Extended correspondence of Kostant-Sekiguchi-Vergne, preprint
92. Vergne, M.: Instantons et correspondance de Kostant-Sekiguchi, C. R. Acad. Sci. Paris Sr. I Math. 320, 901–906 (1995)
93. Bielawski, R.: Lie groups, Nahm's equations and hyper-Kähler manifolds, Algebraic groups. Proceedings of the summer school, Göttingen, June 27-July 13 (2005)
94. Kirwin, W.: Isotropic foliations of coadjoint orbits from the Iwasawa decomposition. Geom. Dedicata. **166**, 185–202 (2013)
95. Martínez Torres, D.: Semisimple coadjoint orbits and cotangent bundles. Bull. London Math. Soc. **48**(6), 977–984 (2016)
96. Bernatska, J., Holod, P.: Geometry and topology of coadjoint orbits of semisimple Lie groups. Proceedings of the 9th international conference on 'Geometry, Integrability and Quantization', June 8–13, 2007, Varna, Bulgarian Academy of Sciences, Sofia, 2008, 146–166
97. Shima H., Symmetric spaces with invariant locally Hessian structures. J. Math.Soc. Japan, pp. 581–589, 1977
98. Shima, H.: Homogeneous Hessian manifolds. Ann. Inst. Fourier, 91–128 (1980)
99. Shima H., Vanishing theorems for compact Hessian manifolds. Ann. Inst. Fourier, pp. 183–205, 1986

100. Shima, H.: Harmonicity of gradient mappings of level surfaces in a real affine space. Geometriae Dedicata, pp. 177–184 (1995)
101. Shima, H.: Hessian manifolds of constant Hessian sectional curvature. J. Math. Soc. Japan, 735–753 (1995)
102. Shima, H.: Homogeneous spaces with invariant projectively flat affine connections. Trans. Am. Math. Soc. 4713–4726 (1999)
103. Shima, H.: The Geometry of Hessian Structures. World Scientific (2007)
104. Shima, H.: Geometry of Hessian Structures, Springer Lecture Notes in Computer Science, Vol. 8085, F. Nielsen, Barbaresco, Frederic (Eds.), pp. 37–55 (2013) (planches: https://www.see.asso.fr/file/5104/download/25050); (vidéos GSI'13: https://www.youtube.com/watch?time_continue=139&v=6pyXxdIzDNQ, https://www.youtube.com/watch?time_continue=182&v=jG2tUjniOUs, https://www.youtube.com/watch?time_continue=6&v=I5kdMJvuNHA)
105. Chu, B.Y.: Symplectic homogeneous spaces. Trans. Am. Math. Soc. **197**, 14–159 (1974)
106. Sternberg, S.: Symplectic homogeneous spaces. Trans. Am. Math. Soc. **212**, 113–130 (1975)
107. Barbaresco, F.: Lie Group machine learning and gibbs density on poincaré unit disk from Souriau Lie groups thermodynamics and SU(1,1) coadjoint orbits. In: Nielsen, F., Barbaresco, F. (eds.) GSI 2019. LNCS, vol. 11712. Springer (2019)
108. Goze, M., Remm, E.: Coadjoint Orbits of Lie Algebras and Cartan Class, Symmetry, Integrability and Geometry: Methods and Applications, SIGMA 15 (2019)
109. Björn Villa, P.: Kählerian structures of coadjoint orbits of semisimple Lie groups and their orbihedra, Dissertation zur Erlangung des Doktorgrades der Naturwissenschaften an der Fakultät für Mathematik der Ruhr-Universität Bochum, Jan. (2015)
110. Barbaresco, F.: Jean-Louis Koszul and the Elementary Structures of Information Geometry, Geometric Structures of Information, pp. 333–392. Springer, Nov. (2018)
111. Koszul, J.L.: Introduction to Symplectic Geometry. Science Press, Beijing (1986) (in chinese); translated by SPRINGER, with F.Barbaresco, C.M. Marle and M. Boyom forewords, SPRINGER, 2019
112. Barbaresco, F., Gay-Balmaz, F.: Lie group cohomology and (Multi)symplectic integrators: new geometric tools for lie group machine learning based on Souriau geometric statistical mechanics. Entropy **22**, 498 (2020)
113. Barbaresco, F.: Lie Group Statistics and Lie Group Machine Learning based on Souriau Lie Groups Thermodynamics & Koszul-Souriau-Fisher Metric: New Entropy Definition as Generalized Casimir Invariant Function in Coadjoint Representation, MDPI Entropy (2020)
114. Barbaresco, F.: Radar Processing based on Matrix Lie Groups Geometry & Souriau Coadjoint Orbits Method, Preprint Academia (2020)
115. Barbaresco, F., Cellodoni, E., Gay-Balmaz, F., Bensoam, J.: Special Issue MDPI Entropy "Lie Group Machine Learning and Lie Group Structure Preserving Integrators". https://www.mdpi.com/journal/entropy/special_issues/Lie_group
116. Les Houches Summer Week, Joint Structures and Common Foundation of Statistical Physics, Information Geometry and Inference for Learning (SPIGL'20); 26th July to 31st July 2020. https://franknielsen.github.io/SPIG-LesHouches2020/
117. Tojo, K., Yoshino, T.: On a method to construct exponential families by representation theory. In: Nielsen, F., Barbaresco, F. (eds.) GSI 2019. LNCS, vol. 11712, Springer (2019)
118. Tojo, K., Yoshino, T.: A method to construct exponential families by representation theory. arXiv:1811.01394v3
119. Cishahayo, C., De Bièvre, S.: On the contraction of the discrete series of SU(1,1). Ann. Inst. Fourier **43**, 551–567 (1993). https://doi.org/10.5802/aif.1346
120. Cahen, B.: Contraction de SU(1,1) vers le Groupe de Heisenberg, pp. 19–43. Université de Metz, Fascicule XV, Travaux Mathématiques (2004)
121. Marle, C.-M., Projection stéréographique et moments, hal-02157930, version 1, Juin 2019
122. Adhumeau, T.: Julien Koszul – Correspondances, Les Cahiers BoËllmann-Gigout, 2005
123. Rothaus O. S., The Construction of Homogeneous Convex Cones, Annals of Mathematics, Ser.2, Vol.83, pp. 358-376, 1966
124. Koszul, J.L.: Formes hermitiennes canoniques des espaces homogènes complexes, Séminaire Bourbaki, Tome 3, Exposé no. 108, pp. 69–75 (1954–1956)

Chapter 6
Gauge Freedom of Entropies on q-Gaussian Measures

Hiroshi Matsuzoe and Asuka Takatsu

Abstract A q-Gaussian measure is a generalization of a Gaussian measure. This generalization is obtained by replacing the exponential function with the power function of exponent $1/(1-q)$ ($q \neq 1$). The limit case $q = 1$ recovers a Gaussian measure. On the set of all q-Gaussian densities over the real line with $1 \leq q < 3$, escort expectations determine information geometric structures such as an entropy and a relative entropy. The ordinary expectation of a random variable is the integral of the random variable with respect to its law. Escort expectations admit us to replace the law by any other measures. One of the most important escort expectations on the set of all q-Gaussian densities is the q-escort expectation since this escort expectation determines the Tsallis entropy and the Tsallis relative entropy. The phenomenon *gauge freedom of entropies* is that different escort expectations determine the same entropy, but different relative entropies. In this chapter, we first introduce a refinement of the q-logarithmic function. Then we demonstrate the phenomenon on an open set of all q-Gaussian densities over the real line by using the refined q-logarithmic functions. We write down the corresponding Riemannian metric.

Keywords Information geometry · Gauge freedom of entropies · Refined q-logarithmic function · q-Gaussian measure

H. Matsuzoe (✉)
Department of Computer Science, Nagoya Institute of Technology, Nagoya, Japan
e-mail: matsuzoe@nitech.ac.jp

A. Takatsu
Department of Mathematical Sciences, Tokyo Metropolitan University, Tokyo, Japan
e-mail: asuka@tmu.ac.jp

RIKEN Center for Advanced Intelligence Project (AIP), Tokyo, Japan

F. Nielsen (ed.), *Progress in Information Geometry*,
Signals and Communication Technology,
https://doi.org/10.1007/978-3-030-65459-7_6

6.1 q-Logarithmic Functions and Their Refinements

6.1.1 Definitions

For $q \in \mathbb{R}$, we set $\chi_q : (0, \infty) \to (0, \infty)$ by

$$\chi_q(s) := s^q.$$

We define a strictly increasing function $\ln_q : (0, \infty) \to \mathbb{R}$ by

$$\ln_q(t) := \int_1^t \frac{1}{\chi_q(s)} ds$$

and we denote by $\exp_q$ the inverse function of $\ln_q : (0, \infty) \to \ln_q(0, \infty)$. The functions $\ln_q$ and $\exp_q$ are called the *q-logarithmic function* and the *q-exponential function*, respectively. We observe that

$$\frac{d}{dt} \ln_q(t) = \frac{1}{\chi_q(t)} = t^{-q} \qquad \text{for } t \in (0, \infty),$$

$$\frac{d}{d\tau} \exp_q(\tau) = \chi_q(\exp_q(\tau)) = \exp_q(\tau)^q \qquad \text{for } \tau \in \ln_q(0, \infty).$$

It holds for $q \in \mathbb{R}$ that $\chi_q(1) = 1$ and $\ln_q(1) = 0$.

Remark 1 (1) For $q = 1$, we have that

$$\ln_1(t) = \log(t) \qquad \text{for } t \in (0, \infty),$$
$$\ln_1(0, \infty) = \mathbb{R},$$
$$\exp_1(\tau) = \exp(\tau) \qquad \text{for } \tau \in \mathbb{R}.$$

(2) For $q \neq 1$, we have that

$$\ln_q(t) = \frac{t^{1-q} - 1}{1 - q} \qquad \text{for } t \in (0, \infty),$$

$$\ln_q(0, \infty) = \begin{cases} \left(-\infty, \dfrac{1}{q-1}\right) & \text{if } q > 1, \\ \left(-\dfrac{1}{1-q}, \infty\right) & \text{if } q < 1, \end{cases}$$

$$\exp_q(\tau) = \{1 + (1-q)\tau\}^{\frac{1}{1-q}} \qquad \text{for } \tau \in \ln_q(0, \infty).$$

Taking account into the negativity of $\ln_q$ in $(0, 1)$, we introduce a refinement of the q-logarithmic function and the q-exponential function. For $q \in \mathbb{R}$ and $a \in \mathbb{R} \setminus \{0\}$, define two functions $\chi_{q,a} : (0, 1) \to (0, \infty)$ and $\ln_{q,a} : (0, 1) \to \mathbb{R}$ by

$$\chi_{q,a}(s) := \chi_q(s) \cdot (-\ln_q(s))^{1-a}, \qquad \ln_{q,a}(t) := -\frac{1}{a}\big(-\ln_q(t)\big)^a,$$

respectively. It turns out that

$$\begin{aligned} \frac{d}{ds}\chi_{q,a}(s) &= \chi_q'(s)(-\ln_q(s))^{1-a} - (1-a)(-\ln_q(s))^{-a} && \text{for } s \in (0,1), \\ \frac{d}{dt}\ln_{q,a}(t) &= \frac{1}{\chi_{q,a}(t)} > 0 && \text{for } t \in (0,1). \end{aligned} \tag{6.1}$$

Hence the function $\ln_{q,a} : (0, 1) \to \mathbb{R}$ is strictly increasing. We denote by $\exp_{q,a}$ the inverse function of $\ln_{q,a} : (0, 1) \to \ln_{q,a}(0, 1)$, which is given by

$$\exp_{q,a}(\tau) = \exp_q\left(-(-a\tau)^{\frac{1}{a}}\right) \qquad \text{for } \tau \in \ln_{q,a}(0,1). \tag{6.2}$$

The functions $\ln_{q,a}$ and $\exp_{q,a}$ are called the *a-refined q-logarithmic function* and the *a-refined q-exponential function*, respectively.

On one hand, it holds for $q \geq 1$ that

$$\ln_{q,a}(0,1) = \begin{cases} (-\infty, 0) & \text{if } a > 0, \\ (0, \infty) & \text{if } a < 0. \end{cases}$$

On the other hand, it holds for $q < 1$ that

$$\ln_{q,a}(0,1) = \begin{cases} \left(-\frac{1}{a}(1-q)^{-a}, 0\right) & \text{if } a > 0, \\ \left(-\frac{1}{a}(1-q)^{-a}, \infty\right) & \text{if } a < 0. \end{cases}$$

Remark 2 (1) The refinement of the ordinary logarithmic function, that is the case $q = 1$, was introduced by Ishige, Salani and the second named author [2], where they studied the preservation of concavity by the heat flow in Euclidean space.

(2) For a positive function $\chi : (0, \infty) \to (0, \infty)$ and $a \in \mathbb{R} \setminus \{0\}$, the χ-logarithmic function $\ln_\chi : (0, \infty) \to \mathbb{R}$ and its refinement $\ln_{\chi,a} : (0, 1) \to \mathbb{R}$ are defined in the same way as those of χ_q, respectively.

6.1.2 Properties

In this section, we give a condition for $\ln_{q,a}$ to be concave and compute the higher order derivatives of $\exp_{q,a}$, which will be used to define information geometric structures.

For $q \in \mathbb{R}$ and $a \in \mathbb{R} \setminus \{0\}$, define

$$t_{q,a} := \begin{cases} 0 & \text{if either } q > 0 \text{ or } q = 0 \text{ with } a - 1 > 0, \\ 1 & \text{if } q \leq 0 \text{ with } a - 1 \leq 0, \\ \dfrac{1}{\exp_q\left(\frac{1-a}{q}\right)} & \text{otherwise,} \end{cases}$$

$$T_{q,a} := \begin{cases} 0 & \text{if } q > 1 \text{ with } 1 - a \geq \dfrac{q}{q-1}, \\ 1 & \text{if } q \leq 0, \\ \dfrac{1}{\exp_q\left(\max\left\{0, \frac{1-a}{q}\right\}\right)} & \text{otherwise,} \end{cases}$$

and set $I_{q,a} := (t_{q,a}, T_{q,a}) \subset (0, 1)$. Note that $I_{q,a}$ is nonempty if and only if one of the following three conditions holds:

- $q > 1$ with $1 - a < \dfrac{q}{q-1}$;
- $0 < q \leq 1$;
- $q \leq 0$ with $a - 1 > 0$.

Proposition 1 *Fix $q \in \mathbb{R}$ and $a \in \mathbb{R} \setminus \{0\}$. For an interval $I \subset (0, 1)$, the strict concavity of $\ln_{q,a}$ in I is equivalent to the strict convexity of $\exp_{q,a}$ in $\ln_{q,a}(I)$. Moreover, if $I_{q,a} \neq \emptyset$, then $\ln_{q,a}$ is strictly concave in $I_{q,a}$.*

Proof Due to Eq. (6.1), $\ln_{q,a}$ is strictly increasing in $(0, 1)$ and so is $\exp_{q,a}$ in $\ln_{q,a}(0, 1)$. Fix an interval $I \subset (0, 1)$. For $t_i \in I$, $\tau_i \in \ln_{q,a}(I)$ $(i = 0, 1)$ with

$$\tau_i = \ln_{q,a}(t_i) \quad \text{or equivalently} \quad t_i = \exp_{q,a}(\tau_i)$$

and $\lambda \in (0, 1)$, it follows from the continuity of $\ln_{q,a}$ that

$$(1 - \lambda)t_0 + \lambda t_1 \in I, \quad (1 - \lambda)\tau_0 + \lambda \tau_1 \in \ln_{q,a}(I).$$

We observe from the monotonicity of $\ln_{q,a}$ and $\exp_{q,a}$ that

$$
\begin{aligned}
&\ln_{q,a}\left((1-\lambda)t_0+\lambda t_1\right) > (1-\lambda)\ln_{q,a}(t_0)+\lambda\ln_{q,a}(t_1) \\
\Leftrightarrow\ &\ln_{q,a}\left((1-\lambda)t_0+\lambda t_1\right) > (1-\lambda)\tau_0+\lambda\tau_1 \\
\Leftrightarrow\ &\exp_{q,a}\left(\ln_{q,a}\left((1-\lambda)t_0+\lambda t_1\right)\right) > \exp_{q,a}\left((1-\lambda)\tau_0+\lambda\tau_1\right) \\
\Leftrightarrow\ &(1-\lambda)t_0+\lambda t_1 > \exp_{q,a}\left((1-\lambda)\tau_0+\lambda\tau_1\right) \\
\Leftrightarrow\ &(1-\lambda)\exp_{q,a}(\tau_0)+\lambda\exp_{q,a}(\tau_1) > \exp_{q,a}\left((1-\lambda)\tau_0+\lambda\tau_1\right),
\end{aligned}
$$

where we used the fact that $\exp_{q,a}$ is the inverse function of $\ln_{q,a}$. This proves the first claim.

Assume $I_{q,a}\neq\emptyset$. A direct calculation provides that

$$
\begin{aligned}
\frac{d^2}{dt^2}\ln_{q,a}(t) &= \frac{d}{dt}\frac{1}{\chi_{q,a}(t)} = -\frac{1}{\chi_{q,a}(t)^2}\frac{d}{dt}\chi_{q,a}(t) \\
&= -\frac{(-\ln_q(t))^{-a}}{\chi_{q,a}(t)^2}\left\{\chi_q'(t)\left(-\ln_q(t)\right)-(1-a)\right\} \\
&= \frac{(-\ln_q(t))^{-a}}{\chi_{q,a}(t)^2}\left\{qt^{q-1}\ln_q(t)+(1-a)\right\}.
\end{aligned}
$$

Notice that $(-\ln_q(t))^{-a}/\chi_{q,a}(t)^2$ is positive in $t\in I_{q,a}$. In the case $q=0$, the condition $I_{0,a}\neq\emptyset$ leads to $a-1>0$, consequently

$$
\frac{d^2}{dt^2}\ln_{0,a}(t) = \frac{(-\ln_0(t))^{-a}}{\chi_{0,a}(t)^2}(1-a) < 0.
$$

Since the function given by

$$
t^{q-1}\ln_q(t) = -\ln_q\left(\frac{1}{t}\right) = \begin{cases} \log(t) & \text{if } q=1, \\ \dfrac{1-t^{q-1}}{1-q} & \text{if } q\neq 1, \end{cases}
$$

is strictly increasing in $t\in(0,1)$, on one hand, it holds for $q>0$ and $t\in I_{q,a}$ that

$$
\begin{aligned}
\frac{d^2}{dt^2}\ln_{q,a}(t) &= \frac{(-\ln_q(t))^{-a}}{\chi_{q,a}(t)^2}\left\{qt^{q-1}\ln_q(t)+(1-a)\right\} \\
&< \frac{(-\ln_q(t))^{-a}}{\chi_{q,a}(t)^2}\left\{-q\ln_q\left(\frac{1}{T_{q,a}}\right)+(1-a)\right\} \\
&= \frac{(-\ln_q(t))^{-a}}{\chi_{q,a}(t)^2}\left\{-q\cdot\max\left\{0,\frac{1-a}{q}\right\}+(1-a)\right\} \\
&= \frac{(-\ln_q(t))^{-a}}{\chi_{q,a}(t)^2}\left\{\min\{0,a-1\}+(1-a)\right\} \\
&\le 0.
\end{aligned}
$$

On the other hand, we see that

$$\begin{aligned}\frac{d^2}{dt^2}\ln_{q,a}(t) &= \frac{(-\ln_q(t))^{-a}}{\chi_{q,a}(t)^2}\left\{qt^{q-1}\ln_q(t)+(1-a)\right\}\\ &< \frac{(-\ln_q(t))^{-a}}{\chi_{q,a}(t)^2}\left\{-q\ln_q\left(\frac{1}{t_{q,a}}\right)+(1-a)\right\}\\ &= \frac{(-\ln_q(t))^{-a}}{\chi_{q,a}(t)^2}\left\{-q\cdot\frac{1-a}{q}+(1-a)\right\}\\ &= 0\end{aligned}$$

for $q < 0$ and $t \in I_{q,a}$. This completes the proof of the second claim. □

Lemma 1 *For $q \in \mathbb{R}$ and $a \in \mathbb{R}\setminus\{0\}$, there exists $\{b^n_j = b^n_j(q,a)\}_{n\in\mathbb{N},0\le j\le n-1}$ such that*

$$\frac{d^n}{d\tau^n}\exp_{q,a}(\tau) = \exp_{q,a}(\tau)^{(n-1)(q-1)+q}(-a\tau)^{\frac{n(1-a)}{a}}\sum_{j=0}^{n-1}b^n_j(q,a)\cdot(-a\tau)^{-\frac{j}{a}}$$

for $\tau \in \ln_{q,a}(0,1)$. Moreover, $\{b^n_j\}_{n\in\mathbb{N},0\le j\le n-1}$ satisfies

$$b^1_0 = 1,$$

$$b^{n+1}_j = \begin{cases}\{na(q-1)+1\}b^n_0 & \text{if } j=0,\\ \{(na+j)(q-1)+1\}b^n_j - \{n(1-a)-(j-1)\}b^n_{j-1} & \text{if } 1\le j\le n-1,\\ (na-1)b^n_{n-1} & \text{if } j=n.\end{cases}$$

Proof We observe that

$$\begin{aligned}\frac{d}{d\tau}\exp_{q,a}(\tau) &= \chi_{q,a}\left(\exp_{q,a}(\tau)\right)\\ &= \chi_q\left(\exp_{q,a}(\tau)\right)\cdot\left\{-\ln_q\left(\exp_{q,a}(\tau)\right)\right\}^{1-a}\\ &= \exp_{q,a}(\tau)^q\cdot(-a\tau)^{\frac{1-a}{a}},\end{aligned}$$

where we used Eq. (6.2). Thus the lemma holds for $n = 1$.

If the lemma holds for n, then we compute that

$$
\begin{aligned}
&\frac{d^{n+1}}{d\tau^{n+1}} \exp_{q,a}(\tau) \\
=&\frac{d}{d\tau}\left(\exp_{q,a}(\tau)^{(n-1)(q-1)+q} (-a\tau)^{\frac{n(1-a)}{a}} \sum_{j=0}^{n-1} b_j^n \cdot (-a\tau)^{-\frac{j}{a}} \right) \\
=&\left(\frac{d}{d\tau} \exp_{q,a}(\tau)^{(n-1)(q-1)+q}\right) \times (-a\tau)^{\frac{n(1-a)}{a}} \sum_{j=0}^{n-1} b_j^n \cdot (-a\tau)^{-\frac{j}{a}} \\
&+ \exp_{q,a}(\tau)^{(n-1)(q-1)+q} \times \frac{d}{d\tau}\left((-a\tau)^{\frac{n(1-a)}{a}} \sum_{j=0}^{n-1} b_j^n \cdot (-a\tau)^{-\frac{j}{a}} \right) \\
=& \{(n-1)(q-1)+q\} \exp_{q,a}(\tau)^{(n-1)(q-1)+q-1} \cdot \exp_{q,a}(\tau)^{q} (-a\tau)^{\frac{1-a}{a}} \\
&\times (-a\tau)^{\frac{n(1-a)}{a}} \sum_{j=0}^{n-1} b_j^n \cdot (-a\tau)^{-\frac{j}{a}} \\
&+ \exp_{q,a}(\tau)^{(n-1)(q-1)+q} \times \left\{ -a \sum_{j=0}^{n-1} \frac{n(1-a)-j}{a} b_j^n \cdot (-a\tau)^{\frac{n(1-a)-j}{a}-1} \right\} \\
=& \exp_{q,a}(\tau)^{n(q-1)+q} (-a\tau)^{\frac{(n+1)(1-a)}{a}} \\
&\times \left[\{(n-1)(q-1)+q\} \sum_{j=0}^{n-1} b_j^n (-a\tau)^{-\frac{j}{a}} \right. \\
&\left. \qquad - \exp_{q,a}(\tau)^{1-q} \sum_{j=0}^{n-1} \{n(1-a)-j\} b_j^n (-a\tau)^{-\frac{j+1}{a}} \right].
\end{aligned}
$$

We deduce from $\exp_{q,a}(\tau)^{1-q} = 1-(1-q)(-a\tau)^{\frac{1}{a}}$ that

$$
\begin{aligned}
&\exp_{q,a}(\tau)^{1-q} \sum_{j=0}^{n-1} \{n(1-a)-j\} b_j^n \cdot (-a\tau)^{-\frac{j+1}{a}} \\
=& \sum_{j=0}^{n-1} \{n(1-a)-j\} b_j^n \cdot (-a\tau)^{-\frac{j+1}{a}} - (1-q) \sum_{j=0}^{n-1} \{n(1-a)-j\} b_j^n \cdot (-a\tau)^{-\frac{j}{a}}.
\end{aligned}
$$

This completes the proof of the lemma. □

Remark 3 For $q \in \mathbb{R}$ and $a \in \mathbb{R} \setminus \{0\}$, we have that

$$
\begin{aligned}
b_0^1 &= 1, & b_0^2 &= a(q-1)+1, & b_0^3 &= \{2a(q-1)+1\}\{a(q-1)+1\},\\
& & b_1^2 &= a-1, & b_1^3 &= (a-1)\{(4a+1)(q-1)+3\},\\
& & & & b_2^3 &= (a-1)(2a-1).
\end{aligned}
$$

Corollary 1 *For $a \in \mathbb{R} \setminus \{0\}$ and $n \in \mathbb{N}$, then $b_0^n(1, a) = 1$.*

Proof It follows from Lemma 1 that

$$b_0^{n+1}(1, a) = \{na(1-1)+1\}b_0^n(1, a) = b_0^n(1, a) = \cdots = b_0^1(1, a) = 1. \qquad \square$$

Corollary 2 *Let $q \in \mathbb{R}$ and $n \in \mathbb{N}$. For $1 \leq j < n$, then $b_j^n(q, 1) = 0$.*

Proof This holds for $1 = j < n = 2$ by Remark 3. For $n \geq 2$, if $b_j^n(q, 1) = 0$ holds for $1 \leq j \leq n-1$, then Lemma 1 implies that $b_n^{n+1}(q, 1) = (na-1)b_{n-1}^n(q, 1) = 0$. For $2 \leq j \leq n-1$, we have that

$$b_j^{n+1}(q, 1) = \{(n+j)(q-1)+1\}b_j^n(q, 1) + (j-1)b_{j-1}^n(q, 1) = 0$$

by the assumption $b_k^n(q, 1) = 0$ for $1 \leq k \leq n-1$. For $j = 1$, we have that

$$b_1^{n+1}(q, 1) = \{(n+1)(q-1)+1\}b_1^n(q, 1) + (1-1)b_0^n(q, 1) = 0. \qquad \square$$

6.2 Escort Expectations

The ordinary expectation of a random variable is the integral of the random variable with respect to its law. An introduction to escort expectations admits us to replace the law by any other measures. The literature is very large and we just refer to the paper by Naudts [4].

For a probability space $(\Omega, \mathfrak{p})$ and a probability measure $\mathfrak{r}$ on Ω, the expectation of a random variable on Ω with respect to $\mathfrak{r}$ is called the *escort expectation* of the random variable with respect to $\mathfrak{r}$. We emphasize that the expectation does not depend on the probability measure $\mathfrak{p}$. Moreover, it is possible to replace the probability measure $\mathfrak{r}$ with an arbitrary measure on Ω.

Definition 1 For a measure ν on a measurable space Ω, the *escort expectation* of a function $f \in L^1(\nu)$ with respect to ν is defined by

$$\mathbb{E}_\nu[f] := \int_\Omega f(\omega)d\nu(\omega). \tag{6.3}$$

In this section, we fix a manifold $\mathcal{S}$ consisting of positive probability densities on a measure space $(\Omega, \mathfrak{m})$. Take $T \in (0, \infty]$ such that

$$T > \sup\{p(\omega) \mid p \in \mathcal{S},\ \omega \in \Omega\}$$

if the above supremum is finite, otherwise $T := \infty$.

Definition 2 Let $\ell : (0, T) \to \mathbb{R}$ be a differentiable function such that $\ell' > 0$ in $(0, T)$. For $p \in \mathcal{S}$, we define a measure $\nu_{\ell;p}$ on Ω as the absolutely continuous measure with respect to $\mathfrak{m}$ with Radon–Nikodym derivative

$$\frac{d\nu_{\ell;p}}{d\mathfrak{m}}(\omega) = \frac{1}{\ell'(p(\omega))}.$$

Note that ℓ is often assumed to be concave such as the logarithmic function. In the case $\ell = \log$, we have that

$$\frac{d\nu_{\ell;p}}{d\mathfrak{m}} = p.$$

Therefore, the escort expectation (6.3) is nothing but the ordinary expectation, that is

$$\mathbb{E}_{\nu_{\ell;p}}[f] = \int_\Omega f(\omega)d\nu_{\ell;p}(\omega) = \int_\Omega f(\omega)p(\omega)d\mathfrak{m}(\omega).$$

Definition 3 Fix a differentiable function $\ell : (0, T) \to \mathbb{R}$ such that $\ell' > 0$ in $(0, T)$ and assume that

$$\ell(r) = \ell \circ r \in L^1(\nu_{\ell;p}) \qquad \text{for } p, r \in \mathcal{S}. \tag{6.4}$$

(1) For $p, r \in \mathcal{S}$, the *ℓ-cross entropy* of p with respect to r is defined by

$$d_\ell(p, r) := -\mathbb{E}_{\nu_{\ell;p}}[\ell(r)].$$

(2) The *ℓ-entropy* of $p \in \mathcal{S}$ is defined by

$$\mathrm{Ent}_\ell(p) := d_\ell(p, p).$$

(3) For $p, r \in \mathcal{S}$, the *ℓ-relative entropy* of p with respect to r is defined by

$$D^{(\ell)}(p, r) := -d_\ell(p, p) + d_\ell(p, r).$$

A choice of differentiable functions $\ell : (0, T) \to \mathbb{R}$ such that $\ell' > 0$ in $(0, T)$ determines an entropy and a relative entropy on $\mathcal{S}$. The phenomenon *gauge freedom of entropies* is that different escort expectations determine the same entropy, but different relative entropies. This is motivated by gauge freedom of Riemannian metrics proposed by Zhang and Naudts [6] (see also [5]).

In the next section, we demonstrate *gauge freedom of entropies* on an open set of q-Gaussian densities over $\mathbb{R}$ for $1 \le q < 3$, which is a typical example of manifolds consisting of probability densities. To be precise, we show that different escort expectations determine the same entropy up to scalar multiple, but different relative entropies, where the entropy coincides with the Boltzmann–Shannon entropy if $q = 1$, and the Tsallis entropy otherwise.

6.3 Gauge Freedom of Entropies

6.3.1 q-Gaussian Measures

To define q-Gaussian measures, we extend $\exp_q$ to the whole of $\mathbb{R}$ by

$$\mathrm{Rexp}_q(\tau) := \max\{0, 1 + (1-q)\tau\}^{\frac{1}{1-q}} \qquad \text{for } \tau \in \mathbb{R},$$

where by convention $0^c := \infty$ for $c < 0$. We recall the following improper integral. Although it is known, we prove it for the sake of completeness.

Lemma 2 *For $q \in \mathbb{R}$ and $(\mu, \lambda) \in \mathbb{R} \times (0, \infty)$, the improper integral of the function*

$$x \mapsto \mathrm{Rexp}_q(-\lambda(x-\mu)^2)$$

on $\mathbb{R}$ converges if and only if $q < 3$. For $q < 3$,

$$\sqrt{3-q}\int_{\mathbb{R}} \mathrm{Rexp}_q(-x^2)dx = Z_q := \begin{cases} \sqrt{\dfrac{3-q}{q-1}}\,B\left(\dfrac{3-q}{2(q-1)}, \dfrac{1}{2}\right) & \text{if } q > 1, \\ \sqrt{2\pi} & \text{if } q = 1, \\ \sqrt{\dfrac{3-q}{1-q}}\,B\left(\dfrac{2-q}{1-q}, \dfrac{1}{2}\right) & \text{if } q < 1, \end{cases}$$

where $B(\cdot, \cdot)$ stands for the beta function.

Proof By the change of variables, it is enough to show the case $(\mu, \lambda) = (0, 1)$. We omit the proof for the case $q = 1$, which is well-known.

Assume $q > 1$. There exist $c, C, R > 0$ depending on q such that

$$cx^{\frac{2}{1-q}} \le \mathrm{Rexp}_q(-x^2) = \left\{1 - (1-q)x^2\right\}^{\frac{1}{1-q}} < Cx^{\frac{2}{1-q}}$$

for $|x| > R$. Since the improper integral of the function

$$x \mapsto x^{\frac{2}{1-q}}$$

on $[1, \infty)$ converges if and only if $2/(1-q) < -1$, that is $1 < q < 3$, so does the improper integral of the function $x \mapsto \mathrm{Rexp}_q(-x^2)$ on $\mathbb{R}$. We observe that

$$\begin{aligned}\int_{\mathbb{R}} \mathrm{Rexp}_q(-x^2)dx &= 2\int_0^{\infty} \left\{1-(1-q)x^2\right\}^{\frac{1}{1-q}} dx \\ &= \frac{1}{\sqrt{q-1}}\int_0^{\infty} (1+r)^{\frac{1}{1-q}} r^{-\frac{1}{2}} dr \\ &= \frac{1}{\sqrt{q-1}} B\left(\frac{3-q}{2(q-1)}, \frac{1}{2}\right),\end{aligned}$$

where we used that

$$B(t-s, s) = \int_0^{\infty} \frac{r^{s-1}}{(1+r)^t} dr \qquad \text{for } t > s > 0.$$

In the case $q < 1$, the support of the function $x \mapsto \mathrm{Rexp}_q(-x^2)$ on $\mathbb{R}$ is

$$\left[-\frac{1}{\sqrt{1-q}}, \frac{1}{\sqrt{1-q}}\right]$$

implying that

$$\begin{aligned}\int_{\mathbb{R}} \mathrm{Rexp}_q(-x^2)dx &= 2\int_0^{\frac{1}{\sqrt{(1-q)}}} \left[1-(1-q)x^2\right]^{\frac{1}{1-q}} dx \\ &= \frac{1}{\sqrt{1-q}}\int_0^{1} [1-r]^{\frac{1}{1-q}} r^{-\frac{1}{2}} dx \\ &= \frac{1}{\sqrt{1-q}} B\left(\frac{2-q}{1-q}, \frac{1}{2}\right).\end{aligned}$$

□

Definition 4 For $q < 3$ and $\xi = (\mu, \sigma) \in \mathbb{R} \times (0, \infty)$, the q-*Gaussian measure* with location parameter μ and scale parameter σ on $\mathbb{R}$ is an absolutely continuous probability measure with respect to the one-dimensional Lebesgue measure with Radon–Nikodym derivative

$$p_q(x; \xi) = p_q(x; \mu, \sigma) := \frac{1}{Z_q \sigma} \mathrm{Rexp}_q\left(-\frac{1}{3-q}\left(\frac{x-\mu}{\sigma}\right)^2\right).$$

We call $p_q(x; \xi) = p_q(x; \mu, \sigma)$ the q-*Gaussian density* with location parameter μ and scale parameter σ.

A q-Gaussian density corresponds to a *normal* (Gaussian) *distribution* for $q = 1$, and a *Student t-distribution* for $1 < q < 3$. In the both cases, the support of each q-Gaussian measure is the whole of $\mathbb{R}$ and

$$p_q(x;\xi) = p_q(x;\mu,\sigma) = \frac{1}{Z_q\sigma}\exp_q\left(-\frac{1}{3-q}\left(\frac{x-\mu}{\sigma}\right)^2\right).$$

6.3.2 *Sufficient Conditions for* (6.4)

In order to give a rigorous treatment of an escort expectation associated with the a-refined q-logarithmic function, we only deal with the case $1 \le q < 3$. Set

$$\Sigma_q := \left\{\sigma > 0 \,\middle|\, \frac{1}{Z_q\sigma} < 1\right\}, \qquad \mathcal{S}_q := \{p_q(\cdot;\xi) \mid \xi \in \mathbb{R}\times\Sigma_q\}.$$

It holds for $\xi = (\mu,\sigma) \in \mathbb{R}\times\Sigma_q$ and $x \in \mathbb{R}$ that

$$\ln_q\left(\frac{1}{Z_q\sigma}\right) < \ln_q(1) = 0, \qquad \ln_q\left(p_q(x;\xi)\right) \in (-\infty, 0).$$

Definition 5 For $1 \le q < 3$ and $\xi \in \mathbb{R}\times\Sigma_q$, define $\ell_q(\cdot;\xi) : \mathbb{R} \to (-\infty, 0)$ by

$$\ell_q(x;\xi) := \ln_q\left(p_q(x;\xi)\right),$$

which is called the *q-likelihood function* of $p_q(\cdot;\xi)$.

For $1 \le q < 3$, $a \in \mathbb{R}\setminus\{0\}$ and $\xi \in \mathbb{R}\times\Sigma_q$, we define a measure $\nu_{q,a;\xi}$ on $\mathbb{R}$ as the absolutely continuous measure with respect to the one-dimensional Lebesgue measure with Radon–Nikodym derivative

$$\frac{d\nu_{q,a;\xi}}{dx}(x) = \frac{1}{\ln'_{q,a}\left(p_q(x;\xi)\right)}.$$

Since the inverse function of $\ln_{q,a}$ is $\exp_{q,a}$, Lemma 1 in the case $n = 1$ leads to

$$\frac{d\nu_{q,a;\xi}}{dx}(x) = \exp'_{q,a}(\ln_{q,a}(p_q(x;\xi))) = \left(-\ell_q(x;\xi)\right)^{1-a}\chi_q(p_q(x;\xi)).$$

A direct computation leads to the relation that

$$\begin{aligned}\ell_q(x;\xi) &= \ln_q\left(\frac{1}{Z_q\sigma}\right) - \frac{1}{(Z_q\sigma)^{1-q}(3-q)}\left(\frac{x-\mu}{\sigma}\right)^2,\\ \ln_{q,a}(p_q(x;\xi)) &= -\frac{1}{a}\left(-\ell_q(x;\xi)\right)^a.\end{aligned} \tag{6.5}$$

Lemma 3 *Let* $1 \le q < 3$, $a \in \mathbb{R}\setminus\{0\}$ *and* $\xi \in \mathbb{R}\times\Sigma_q$. *Then for* $\lambda, \gamma \in \mathbb{R}$ *with* $\lambda > 0$, $(\lambda + x^2)^\gamma \in L^1(\nu_{q,a;\xi})$ *if and only if*

$$\textit{either } q = 1 \quad \textit{or} \quad q > 1 \textit{ with } \gamma < \frac{1}{2} + \frac{1}{q-1} + a - 1.$$

Proof Since the decay rate of $\nu_{1,a;\xi}$ is $o(\exp(-x^\varepsilon))$ as $x \to \infty$ for $\varepsilon < 2$, the lemma holds for $q = 1$.

Assume $q > 1$. By the change of variables, it is enough to show the case $\xi = (0, 2/Z_q)$. Here we have that $Z_q\sigma = 2$. There exist $c, C, R > 0$ depending on q such that

$$\begin{aligned}&cx^{2(1-a)+\frac{2q}{1-q}+2\gamma}\\ &< \left(-\ell_q(x;\xi)\right)^{1-a}\cdot\chi_q(p_q(x;\xi))\cdot(\lambda+x^2)^\gamma\\ &= \left\{-\ln_q\left(\frac{1}{2}\right) + \frac{1}{2^{1-q}(3-q)}\frac{Z_q^2x^2}{4}\right\}^{1-a}\cdot\frac{1}{2^q}\left(1+\frac{q-1}{3-q}\frac{Z_q^2x^2}{4}\right)^{\frac{q}{1-q}}\cdot(\lambda+x^2)^\gamma\\ &< Cx^{2(1-a)+\frac{2q}{1-q}+2\gamma}\end{aligned}$$

for $|x| > R$. This means that $(\lambda + x^2)^\gamma \in L^1(\nu_{q,a;\xi})$ if and only if

$$2(1-a) + \frac{2q}{1-q} + 2\gamma < -1 \Leftrightarrow \gamma < \frac{1}{2} + \frac{1}{q-1} + a - 1.$$

□

Lemma 3 in the case $\gamma = 0$ provides the condition for (q, a) such that $\nu_{q,a;\xi}$ has a finite mass.

Corollary 3 *Let* $1 \le q < 3$, $a \in \mathbb{R}\setminus\{0\}$ *and* $\xi \in \mathbb{R}\times\Sigma_q$. *Then* $1 \in L^1(\nu_{q,a;\xi})$ *if and only if*

$$\textit{either } q = 1 \quad \textit{or} \quad q > 1 \textit{ with } \frac{1}{2} - \frac{1}{q-1} < a.$$

Note that

$$\frac{1}{2} - \frac{1}{q-1} < 0 \qquad \text{for } 1 < q < 3.$$

Corollary 4 *Let* $1 \le q < 3$ *and* $a \in \mathbb{R}\setminus\{0\}$. *Then* $\ln_{q,a}(r) \in L^1(\nu_{q,a;\xi})$ *for* $\xi \in \mathbb{R}\times\Sigma_q$ *and* $r \in \mathcal{S}_q$.

Proof The corollary trivially holds for $q = 1$.

Assume $q > 1$. We observe from (6.5) that

$$\ln_{q,a}(p(x;\mu,\sigma)) = -\frac{1}{a}\left\{-\ln_q\left(\frac{1}{Z_q\sigma}\right) + \frac{1}{(Z_q\sigma)^{1-q}(3-q)}\left(\frac{x-\mu}{\sigma}\right)^2\right\}^a$$

for $(\mu, \sigma) \in \mathbb{R} \times \Sigma_q$. This with Lemma 3 yields that

$$\ln_{q,a}(r) \in L^1(\nu_{q,a;\xi}) \Leftrightarrow a < \frac{1}{2} + \frac{1}{q-1} + a - 1, \tag{6.6}$$

which holds for $q < 3$. □

Following Definition 3, we define an entropy and a relative entropy on $\mathcal{S}_q$. Recall the escort expectation of a function $f \in L^1(\nu_{q,a;\xi})$ with respect to $\nu_{q,a;\xi}$ is defined by

$$\mathbb{E}_{\nu_{q,a;\xi}}[f] = \int_{\mathbb{R}} f(x)d\nu_{q,a;\xi}(x) = \int_{\mathbb{R}} f(x)\exp'_{q,a}\left(\ln_{q,a}\left(p_q(x;\xi)\right)\right)dx.$$

Definition 6 Let $1 \le q < 3$ and $a \in \mathbb{R} \setminus \{0\}$. Take $\xi \in \mathbb{R} \times \Sigma_q$ and set $p = p_q(\cdot;\xi) \in \mathcal{S}_q$.

(1) The (q, a)-*cross entropy* of p with respect to $r \in \mathcal{S}_q$ is defined by

$$d_{q,a}(p, r) := -\mathbb{E}_{\nu_{q,a;\xi}}[\ln_{q,a}(r)].$$

(2) The (q, a)-*entropy* of p is defined by

$$\mathrm{Ent}_{q,a}(p) := d_{q,a}(p, p).$$

(3) The (q, a)-*relative entropy* of p with respect to $r \in \mathcal{S}_q$ is defined by

$$D^{(q,a)}(p, r) := -d_{q,a}(p, p) + d_{q,a}(p, r).$$

Remark 4 The domain of the $(q, 1)$-entropy can be extended from $\mathcal{S}_q$ to the whole of q-Gaussian densities. The $(q, 1)$-entropy coincides with the *Boltzmann–Shannon entropy* if $q = 1$, and the *Tsallis entropy* otherwise.

Theorem 1 *(Gauge freedom of entropies) Let* $1 \le q < 3$ *and* $a \in \mathbb{R} \setminus \{0\}$. *Then*

$$\mathrm{Ent}_{q,1} = a\mathrm{Ent}_{q,a}, \qquad D^{(q,1)} \neq \lambda D^{(q,a)} \quad \textit{for } a \neq 1 \textit{ and } \lambda \in \mathbb{R}.$$

Proof By the definition, we have that

$$d_{q,a}(p_q(\cdot;\xi_0), p_q(\cdot;\xi)) = \frac{1}{a}\int_{\mathbb{R}} \left(-\ell_q(x;\xi)\right)^a \nu_{q,a;\xi_0}(x)$$
$$= \frac{1}{a}\int_{\mathbb{R}} \left(-\ell_q(x;\xi)\right)^a \left(-\ell_q(x;\xi_0)\right)^{1-a} \chi_q(p_q(x;\xi_0))dx$$

for $\xi_0, \xi \in \mathbb{R}\times\Sigma_q$, which implies that

$$\mathrm{Ent}_{q,1}(p) = a\mathrm{Ent}_{q,a}(p) = -\int_{\mathbb{R}} \ln_q(p(x))p(x)^q dx$$
$$= \begin{cases} -\displaystyle\int_{\mathbb{R}} \frac{p(x)-p(x)^q}{1-q}dx & if\ q>1, \\ -\displaystyle\int_{\mathbb{R}} p(x)\log(p(x))dx & if\ q=1, \end{cases}$$

for $p \in \mathcal{S}_q$.

Recall that $\Sigma_q = \{\sigma > 0 \mid 1/(Z_q\sigma) < 1\}$. Since we observe that

$$\lim_{\sigma\to\infty} \frac{\left(-\ell_q(x;0,\sigma)\right)^a}{-\ln_q\left(\frac{1}{Z_q\sigma}\right)} = \begin{cases} \infty & \text{if } a>1, \\ 1 & \text{if } a=1, \\ 0 & \text{if } a<1, a\neq 0, \end{cases}$$

for $x \in \mathbb{R}$, we apply the dominated convergence theorem $a \leq 1$ and the monotone convergence theorem for $a > 1$ to have

$$\frac{\lambda D^{(q,a)}(p, p_q(\cdot;0;\sigma)) - D^{(q,1)}(p, p_q(\cdot;0;\sigma))}{-\ln_q\left(\frac{1}{Z_q\sigma}\right)}$$
$$= -\frac{\lambda d_{q,a}(p,p) - d_{q,1}(p,p)}{-\ln_q\left(\frac{1}{Z_q\sigma}\right)} + \frac{\lambda d_{q,a}(p, p_q(\cdot;0;\sigma)) - d_{(q,1)}(p, p_q(\cdot;0;\sigma))}{-\ln_q\left(\frac{1}{Z_q\sigma}\right)}$$
$$\xrightarrow{\sigma\to\infty} \begin{cases} \lambda\cdot\infty - M & \text{if } a>1, \\ (\lambda-1)M & \text{if } a=1, \\ -M & \text{if } a<1, a\neq 0, \end{cases}$$

for $p \in \mathcal{S}_q$ and $\lambda \in \mathbb{R}$, where we put $0\cdot\infty := 0$ and

$$M := \int_{\mathbb{R}} \chi_q(p(x))dx.$$

This constant M is obviously positive, and M is finite due to Lemma 5 in the next section. This ensures that $D^{(q,a)} \neq \lambda D^{(q,1)}$ for $a \neq 1$ and $\lambda \in \mathbb{R}$. □

The proof of Theorem 1 immediately gives the following corollary.

Corollary 5 *Let* $1 \le q < 3$ *and* $a \in \mathbb{R} \setminus \{0\}$. *Then*

$$d_{q,1} \neq \lambda d_{q,a} \quad \text{for } a \neq 1 \text{ and } \lambda \in \mathbb{R}.$$

6.4 Refined Riemannian Metrics

Throughout of this section, we fix $1 \le q < 3$ and $a \in \mathbb{R} \setminus \{0\}$ such that $I_{q,a} \neq \emptyset$, namely

$$\text{either } q = 1 \quad \text{or} \quad q > 1 \text{ with } 1 - a < \frac{q}{q-1}.$$

In this case, $t_{q,a} = 0$. Set

$$\Sigma_{q,a} := \left\{ \sigma \in \Sigma_q \mid \frac{1}{Z_q \sigma} < T_{q,a} \right\}, \qquad \mathcal{S}_{q,a} := \left\{ p_q(\cdot; \xi) \in \mathcal{S}_q \mid \xi \in \mathbb{R} \times \Sigma_{q,a} \right\}.$$

The manifold $\mathcal{S}_{q,a}$ admits information geometric structures.

6.4.1 Derivatives of (q, a)-Relative Entropy

The (q, a)-relative entropy is nondegenerate on $\mathcal{S}_{q,a} \times \mathcal{S}_{q,a}$.

Lemma 4 *For distinct* $p, r \in \mathcal{S}_{q,a}$, $D^{(q,a)}(p, r) > 0$.

Proof Proposition 1 yields that $\exp''_{q,a}(\ln_{q,a}(p(x))) > 0$ in $x \in \mathbb{R}$ for $p \in \mathcal{S}_{q,a}$. The strict convexity of $\exp_{q,a}$ leads to the inequality that

$$\begin{aligned}
r(x) &= \exp_{q,a}(\ln_{q,a}(r(x))) \\
&> \exp_{q,a}(\ln_{q,a}(p(x))) + \left\{ \ln_{q,a}(r(x)) - \ln_{q,a}(p(x)) \right\} \exp'_{q,a}(\ln_{q,a}(p(x))) \\
&= p(x) + \ln_{q,a}(r(x)) \exp'_{q,a}(\ln_{q,a}(p(x))) - \ln_{q,a}(p(x)) \exp'_{q,a}(\ln_{q,a}(p(x)))
\end{aligned}$$

for $x \in \mathbb{R}$ and $p, r \in \mathcal{S}_{q,a}$. Integrating this inequality on $\mathbb{R}$ gives

$$1 > 1 - d_{q,a}(p, r) + d_{q,a}(p, p) = 1 - D^{(q,a)}(p, r).$$

□

Let us define a function $\rho^{(q,a)}$ on $(x, \xi_1, \xi_2) \in \mathbb{R} \times (\mathbb{R} \times \Sigma_{q,a})^2$ by

$$\rho^{(q,a)}(x; \xi_1, \xi_2) := \left\{ \ln_{q,a}(p_q(x; \xi_1)) - \ln_{q,a}(p_q(x; \xi_2)) \right\} \exp'_{q,a}\left(\ln_{q,a}(p_q(x; \xi_1)) \right).$$

The function $x \mapsto \rho^{(q,a)}(x; \xi_1, \xi_2)$ is the integrand of $D^{(q,a)}(p_q(\cdot; \xi_1), p_q(\cdot; \xi_2))$ for $(\xi_1, \xi_2) \in (\mathbb{R} \times \Sigma_{q,a})^2$.

Given $\xi_i = (\mu_i, \sigma_i) \in \mathbb{R} \times \Sigma_{q,a}$, it turns out that

$$
\begin{aligned}
&\frac{\partial}{\partial s_1}\frac{\partial}{\partial s_2}\rho^{(q,a)}(x; \xi_1, \xi_2)\Big|_{(\xi,\xi)} \\
&= -\frac{\partial}{\partial s_2}\ln_{q,a}\left(p_q\left(x; \xi_2\right)\right) \cdot \frac{\partial}{\partial s_1}\exp'_{q,a}\left(\ln_{q,a}\left(p_q(x; \xi_1)\right)\right)\Big|_{(\xi,\xi)} \\
&= -\frac{\partial}{\partial s_2}\ln_{q,a}\left(p_q\left(x; \xi_2\right)\right) \cdot \frac{\partial}{\partial s_1}\ln_{q,a}\left(p_q(x; \xi_1\right) \cdot \exp''_{q,a}\left(\ln_{q,a}\left(p_q(x; \xi_1)\right)\right)\Big|_{(\xi,\xi)} \\
&= -\frac{\partial}{\partial s_2}\left\{-\frac{1}{a}\left(-\ell_q(x; \xi_2)\right)^a\right\} \cdot \frac{\partial}{\partial s_1}\left\{-\frac{1}{a}\left(-\ell_q(x; \xi_1)\right)^a\right\}\Big|_{(\xi,\xi)} \\
&\quad \times p_q(x; \xi)^{(2-1)(q-1)+q}\left(-\ell_q(x; \xi)\right)^{2(1-a)}\sum_{j=0}^{1} b_j^2\left(-\ell_q(x; \xi)\right)^{-j} \\
&= -\sum_{j=0}^{1} b_j^2\left(\frac{\partial}{\partial s_2}\ell_q(x; \xi_2) \cdot \frac{\partial}{\partial s_1}\ell_q(x; \xi_1)\Big|_{(\xi,\xi)} \cdot \left(-\ell_q(x, \xi)\right)^{-j} p_q(x; \xi)^{2q-1}\right)
\end{aligned}
$$

for $s_i \in \{\mu_i, \sigma_i\}$, where we used Lemma 1 in the case $n = 2$.

Let us generalize Lemma 3.

Lemma 5 *Fix $n \in \mathbb{N}$ and $\gamma \geq 0$. Then $\exp_q(-x^2)^{(n-1)(q-1)+q} \cdot x^{2\gamma} \in L^1(dx)$ if and only if*

$$
\textit{either } q = 1 \quad \textit{or} \quad q > 1 \textit{ with } \gamma < \frac{1}{2} + \frac{1}{q-1} + n - 1.
$$

Proof The lemma trivially holds for $q = 1$. Assume $q > 1$. There exist $c, C, R > 0$ depending on q such that

$$
\begin{aligned}
&c x^{2\frac{(n-1)(q-1)+q}{1-q}+2\gamma} \\
&< \exp_q(-x^2)^{(n-1)(q-1)+q} \cdot x^{2\gamma} = \left\{1 - (1-q)x^2\right\}^{\frac{(n-1)(q-1)+q}{1-q}} \cdot x^{2\gamma} \\
&< C x^{2\frac{(n-1)(q-1)+q}{1-q}+2\gamma}
\end{aligned}
$$

for $|x| > R$. This yields that $\exp_q(-x^2)^{(n-1)(q-1)+q} x^{2\gamma} \in L^1(dx)$ if and only if

$$
2\frac{(n-1)(q-1)+q}{1-q} + 2\gamma < -1 \Leftrightarrow \gamma < \frac{1}{2} + \frac{1}{q-1} + n - 1.
$$

□

Corollary 6 *For $n \in \mathbb{N}$, $0 \leq \gamma \leq n$, $j \in \mathbb{Z}_{\geq 0}$ and $\xi \in \mathbb{R} \times \Sigma_{q,a}$, then*

$$
p_q(x; \xi)^{(n-1)(q-1)+q} \cdot x^{2\gamma} \cdot \left(-\ell_q(x; \xi)\right)^{-j} \in L^1(dx).
$$

Proof Since we have that

$$n < \frac{1}{2} + \frac{1}{q-1} + n - 1 \qquad \text{for } 1 < q < 3,$$

we apply Lemme 5 together with the change of variables to have that

$$p_q(x;\xi)^{(n-1)(q-1)+q} \cdot x^{2\gamma} \in L^1(dx) \qquad \text{for } 0 \le \gamma \le n.$$

Moreover, the fact that

$$-\ell_q(x;\xi) \ge -\ln_q\left(\frac{1}{Z_q\sigma}\right) > 0$$

for $x \in \mathbb{R}$ completes the proof of the corollary. □

Combining the computation that

$$\begin{aligned} \frac{\partial}{\partial\mu}\ell_q(x;\mu,\sigma) &= \frac{2}{(3-q)} \cdot \frac{1}{(Z_q\sigma)^{1-q}\sigma}\frac{x-\mu}{\sigma}, \\ \frac{\partial}{\partial\sigma}\ell_q(x;\mu,\sigma) &= -\frac{1}{(Z_q\sigma)^{1-q}\sigma}\left\{1-\left(\frac{x-\mu}{\sigma}\right)^2\right\}, \end{aligned} \tag{6.7}$$

with Corollary 6 in the case $n = 2$, we conclude that

$$x \mapsto \frac{\partial}{\partial s_1}\frac{\partial}{\partial s_2}\rho^{(q,a)}(x;\xi_1,\xi_2)\Big|_{(\xi,\xi)}$$

is integrable on $\mathbb{R}$ for $\xi \in \mathbb{R} \times \Sigma_{q,a}$. Since the function $x \mapsto \rho^{(q,a)}(x;\xi_1,\xi_2)$ is integrable on $\mathbb{R}$ for $(\xi_1,\xi_2) \in (\mathbb{R} \times \Sigma_{q,a})^2$, the dominated convergence theorem implies that

$$\begin{aligned} &\frac{\partial}{\partial s_1}\frac{\partial}{\partial s_2}D^{(q,a)}(p_q(\cdot;\xi_1),p_q(\cdot;\xi_2)\Big|_{(\xi,\xi)} \\ &= -\int_{\mathbb{R}}\frac{\partial}{\partial s_2}\ln_{q,a}\left(p_q(x;\xi_2)\right)\cdot\frac{\partial}{\partial s_1}\ln_{q,a}\left(p_q(x;\xi_1)\cdot\exp''_{q,a}\left(\ln_{q,a}\left(p_q(x;\xi_1)\right)\right)\Big|_{(\xi,\xi)}dx \\ &= -\sum_{j=0}^{1} b_j^2\int_{\mathbb{R}}\left(\frac{\partial}{\partial s_2}\ell_q(x;\xi_2)\cdot\frac{\partial}{\partial s_1}\ell_q(x;\xi_1)\right)\Big|_{(\xi,\xi)}\cdot\left(-\ell_q(x,\xi)\right)^{-j}p_q(x;\xi)^{2q-1}dx \end{aligned}$$

for $s_i \in \{\mu_i,\sigma_i\}$. This quantity provides a Riemannian metric on $\mathcal{S}_{q,a}$.

Definition 7 For $s, t \in \{\mu,\sigma\}$, define a function $g_{st}^{(q,a)} : \mathbb{R} \times \Sigma_{q,a} \to \mathbb{R}$ by

$$g_{st}^{(q,a)}(\xi) := \int_{\mathbb{R}}\frac{\partial}{\partial s}\ln_{q,a}\left(p_q(x;\xi)\right)\cdot\frac{\partial}{\partial t}\ln_{q,a}\left(p_q(x;\xi)\right)\cdot\exp''_{q,a}\left(\ln_{q,a}\left(p_q(x;\xi)\right)\right)dx.$$

Theorem 2 *For $\xi \in \mathbb{R} \times \Sigma_{q,a}$ and $s, t \in \{\mu, \sigma\}$,*

$$g^{(q,a)}\left(\frac{\partial}{\partial s}, \frac{\partial}{\partial t}\right)(p_q(\cdot;\xi)) := g_{st}^{(q,a)}(\xi)$$

determines a Riemannian metric on $\mathcal{S}_{q,a}$.

Proof It is enough to show that

$$g_{\mu\mu}^{(q,a)}, g_{\sigma\sigma}^{(q,a)} > 0 \quad \text{and} \quad g_{\mu\sigma}^{(q,a)} = 0 \quad \text{on } \mathbb{R} \times \Sigma_{q,a}.$$

The positivity of $g_{\mu\mu}^{(q,a)}, g_{\sigma\sigma}^{(q,a)}$ follows from that of

$$\frac{\partial}{\partial s}\ln_{q,a}\left(p_q(x;\xi)\right) \cdot \frac{\partial}{\partial s}\ln_{q,a}\left(p_q(x;\xi)\right) \cdot \exp''_{q,a}\left(\ln_{q,a}\left(p_q(x;\xi)\right)\right) \qquad \text{for } s \in \{\mu, \sigma\}.$$

We derive $g_{\mu\sigma}^{(q,a)} = 0$ from the fact that

$$\frac{\partial}{\partial \mu}\ln_{q,a}\left(p_q(x;\xi)\right) \cdot \frac{\partial}{\partial \sigma}\ln_{q,a}\left(p_q(x;\xi)\right) \cdot \exp''_{q,a}\left(\ln_{q,a}\left(p_q(x;\xi)\right)\right)$$

is an odd function in $x \in \mathbb{R}$ with respect to $x = \mu$ according to (6.7). □

Remark 5 In information geometry, a certain pair of affine connections is important as well as a Riemann metric. On $\mathcal{S}_{q,a}$, this pair of affine connections is induced from the Riemannian metric $g^{(q,a)}$ together with the cubic tensor of the form

$$\begin{aligned}
&C^{(q,a)}\left(\frac{\partial}{\partial s}, \frac{\partial}{\partial t}, \frac{\partial}{\partial u}\right)(p_q(\cdot;\xi)) \\
&:= \int_{\mathbb{R}} \frac{\partial}{\partial s}\ln_{q,a}\left(p_q(x;\xi)\right) \cdot \frac{\partial}{\partial t}\ln_{q,a}\left(p_q(x;\xi)\right) \cdot \frac{\partial}{\partial u}\ln_{q,a}\left(p_q(x;\xi)\right) \\
&\qquad\qquad \times \exp'''_{q,a}\left(\ln_{q,a}\left(p_q(x;\xi)\right)\right) dx \\
&= \int_{\mathbb{R}} \frac{\partial}{\partial s}\left\{-\frac{1}{a}\left(-\ell_q(x;\xi)\right)^a\right\} \cdot \frac{\partial}{\partial t}\left\{-\frac{1}{a}\left(-\ell_q(x;\xi)\right)^a\right\} \cdot \frac{\partial}{\partial u}\left\{-\frac{1}{a}\left(-\ell_q(x;\xi)\right)^a\right\} \\
&\qquad \times p_q(x;\xi)^{(3-1)(q-1)+q}\left(-\ell_q(x;\xi)\right)^{3(1-a)}\sum_{j=0}^{2} b_j^2\left(-\ell_q(x;\xi)\right)^{-j} \\
&= \sum_{j=0}^{2} b_j^3 \int_{\mathbb{R}} \frac{\partial}{\partial s}\ell_q(x;\xi) \cdot \frac{\partial}{\partial t}\ell_q(x;\xi) \cdot \frac{\partial}{\partial u}\ell_q(x;\xi) \cdot \left(-\ell_q(x;\xi)\right)^{-j} p_q(x;\xi)^{3q-2} dx,
\end{aligned}$$

where $s, t, u \in \{\mu, \sigma\}$. This improper integral converges due to Corollary 6 in the case $n = 3$. For $a = 1$, the Riemannian metric $g^{(q,1)}$ and the cubic tensor $C^{(q,1)}$ correspond to the *Fisher metric* and the *Amari–Čencov tensor*, respectively. See [3] for further details.

6.4.2 Expression of the Refined Riemann Metrics

We compute the exact value of

$$
\begin{aligned}
g_{\mu\mu}^{(q,a)}(\xi) &= \frac{4}{(3-q)^2}\sum_{j=0}^{1}\frac{b_j^2}{(Z_q\sigma)^{2(1-q)}\sigma^2}\int_{\mathbb{R}}\left(\frac{x-\mu}{\sigma}\right)^2\frac{p_q(x;\xi)^{2q-1}}{\left(-\ell_q(x,\xi)\right)^j}dx\\
&= \frac{4}{(3-q)^2}\sum_{j=0}^{1}\frac{b_j^2}{(Z_q\sigma)^{2(1-q)}\sigma^2}\Phi(q,2,1,j;\xi),\\
g_{\sigma\sigma}^{(q,a)}(\xi) &= \sum_{j=0}^{1}\frac{b_j^2}{(Z_q\sigma)^{2(1-q)}\sigma^2}\int_{\mathbb{R}}\left\{1-\left(\frac{x-\mu}{\sigma}\right)^2\right\}^2\frac{p_q(x;\xi)^{2q-1}}{\left(-\ell_q(x,\xi)\right)^j}dx\\
&= \sum_{j=0}^{1}\frac{b_j^2}{(Z_q\sigma)^{2(1-q)}\sigma^2}\sum_{k=0}^{2}\binom{2}{k}(-1)^k\Phi(q,2,k,j;\xi),
\end{aligned}
\tag{6.8}
$$

for $\xi \in \mathbb{R} \times \Sigma_{q,a}$, where we set

$$
\Phi(q,n,k,j;\xi) := \int_{\mathbb{R}}\left(\frac{x-\mu}{\sigma}\right)^{2k}\frac{p_q(x;\xi)^{(n-1)(q-1)+q}}{\left(-\ell_q(x,\xi)\right)^j}dx.
$$

Lemma 6 *For $n \in \mathbb{N}$, $k \in \{0, 1, \dots, n\}$ and $\xi = (\mu, \sigma) \in \mathbb{R} \times \Sigma_{q,a}$, then*

$$
\Phi(q,n,k,0;\xi) = \begin{cases} \dfrac{\sigma}{(Z_q\sigma)^{(n-1)(q-1)+q}}\left(\dfrac{3-q}{q-1}\right)^{k+\frac{1}{2}} B\left(\dfrac{3-q}{2(q-1)}+n-k, \dfrac{1}{2}+k\right) & \text{if } q>1,\\ (2k-1)!! & \text{if } q=1, \end{cases}
$$

where by convention $(2\cdot 0-1)!! := 1$.

Proof We apply the change of variables with

$$
y = \frac{1}{2}\left(\frac{x-\mu}{\sigma}\right)^2 \text{ if } q=1, \quad \text{and} \quad y = \frac{q-1}{3-q}\left(\frac{x-\mu}{\sigma}\right)^2 \text{ otherwise.}
$$

For $q = 1$, we observe that

$$\begin{aligned}
\Phi(1,n,k,0;\xi) &= \int_{\mathbb{R}} p_1(x;\xi)^{(n-1)(1-1)+1}\left(\frac{x-\mu}{\sigma}\right)^{2k} dx\\
&= 2\int_0^\infty \frac{1}{\sqrt{2\pi}\sigma}\exp\left(-\frac{1}{2}\left(\frac{x-\mu}{\sigma}\right)^2\right)\left(\frac{x-\mu}{\sigma}\right)^{2k} dx\\
&= \frac{2^k}{\sqrt{\pi}}\int_0^\infty e^{-y}y^{k-\frac{1}{2}}dy\\
&= \frac{2^k}{\sqrt{\pi}}\Gamma\left(k+\frac{1}{2}\right) = \frac{2^k}{\sqrt{\pi}}\frac{(2k-1)!!}{2^k}\sqrt{\pi}\\
&= (2k-1)!!,
\end{aligned}$$

where $\Gamma(\cdot)$ stands for the Gamma function, that is

$$\Gamma(s) := \int_0^\infty e^{-x}x^{s-1}dx \qquad \text{for } s > 0.$$

For $q > 1$, it tuns out that

$$\begin{aligned}
&\Phi(q,n,k,0;\xi)\\
&= \int_{\mathbb{R}} p_q(x;\xi)^{(n-1)(q-1)+q}\left(\frac{x-\mu}{\sigma}\right)^{2k} dx\\
&= 2\int_0^\infty \frac{1}{(Z_q\sigma)^{(n-1)(q-1)+q}}\left[1+\frac{q-1}{3-q}\left(\frac{x-\mu}{\sigma}\right)^2\right]^{\frac{(n-1)(q-1)+q}{1-q}}\left(\frac{x-\mu}{\sigma}\right)^{2k} dx\\
&= \frac{\sigma}{(Z_q\sigma)^{(n-1)(q-1)+q}}\left(\frac{3-q}{q-1}\right)^{k+\frac{1}{2}}\int_0^\infty \frac{y^{k-\frac{1}{2}}}{(1+y)^{n-1+\frac{q}{q-1}}}dy\\
&= \frac{\sigma}{(Z_q\sigma)^{(n-1)(q-1)+q}}\left(\frac{3-q}{q-1}\right)^{k+\frac{1}{2}} B\left(\frac{3-q}{2(q-1)}+n-k,\frac{1}{2}+k\right).
\end{aligned}$$

□

Proposition 2 *For $a = 1$ and $\xi = (\mu,\sigma) \in \mathbb{R}\times\Sigma_{q,a}$, we have that*

$$g_{\mu\mu}^{(q,1)}(\xi) = \frac{1}{\sigma^2}, \qquad g_{\sigma\sigma}^{(q,1)}(\xi) = \frac{3-q}{\sigma^2}.$$

Proof It follows from Lemma 6 that

$$\Phi(1,2,0,0;\xi) = 1, \qquad \Phi(1,2,1,0;\xi) = 1, \qquad \Phi(1,2,2,0;\xi) = 3,$$

implying

$$g_{\mu\mu}^{(1,1)}(\xi) = b_0^2(1,1)\frac{1}{\sigma^2} = \frac{1}{\sigma^2}, \qquad g_{\sigma\sigma}^{(1,1)}(\xi) = b_0^2(1,1)\sum_{j=0}^{1}\frac{1}{\sigma^2}(1-2+3) = \frac{2}{\sigma^2}.$$

Assume $q > 1$. By the property that

$$B(s+1,t) = \frac{s}{s+t} B(s,t) \qquad \text{for } s,t > 0,$$

we have that

$$\begin{aligned}
\Phi(q,2,k,0;\xi) &= \frac{\sigma}{(Z_q\sigma)^{(2-1)(q-1)+q}} \left(\frac{3-q}{q-1}\right)^{k+\frac{1}{2}} B\left(\frac{3-q}{2(q-1)}+2-k, \frac{1}{2}+k\right) \\
&= \frac{\sigma}{(Z_q\sigma)^{(q-1)+q}} \left(\frac{3-q}{q-1}\right)^{k+\frac{1}{2}} \frac{f_2(k)}{(\frac{1}{q-1}+1)\cdot\frac{1}{q-1}} B\left(\frac{3-q}{2(q-1)}, \frac{1}{2}\right) \\
&= \frac{1}{(Z_q\sigma)^{2(q-1)}} \left(\frac{3-q}{q-1}\right)^{k} \frac{(q-1)^2 f_2(k)}{q},
\end{aligned}$$

where we set

$$\begin{aligned}
f_2(0) &:= \left(\frac{3-q}{2(q-1)}+1\right)\cdot\frac{3-q}{2(q-1)} = \frac{(q+1)(3-q)}{4(q-1)^2}, \\
f_2(1) &:= \frac{3-q}{2(q-1)}\cdot\frac{1}{2} = \frac{3-q}{4(q-1)}, \\
f_2(2) &:= \frac{3}{2}\cdot\frac{1}{2} = \frac{3}{4}.
\end{aligned}$$

This leads to that

$$\begin{aligned}
g_{\mu\mu}^{(q,1)}(\xi) &= \frac{4}{(3-q)^2} \frac{b_0^2(q,1)}{(Z_q\sigma)^{2(1-q)}\sigma^2} \Phi(q,2,1,0;\xi) = \frac{1}{\sigma^2}, \\
g_{\sigma\sigma}^{(q,1)}(\xi) &= \frac{b_0^2(q,1)}{(Z_q\sigma)^{2(1-q)}\sigma^2} \sum_{k=0}^{2} \binom{2}{k} (-1)^k \Phi(q,2,k,0;\xi) \\
&= \frac{1}{\sigma^2} \sum_{k=0}^{2} \binom{2}{k} \left\{(-1)^k \left(\frac{3-q}{q-1}\right)^k (q-1)^2 f_2(k)\right\} = \frac{3-q}{\sigma^2}.
\end{aligned}$$

□

Fix $n, j \in \mathbb{N}, k \in \{0, 1, \ldots, n\}$ and $\xi = (\mu, \sigma) \in \mathbb{R} \times \Sigma_{q,a}$. Let us compute $\Phi(q, n, k, j; \xi)$ with the use of the residue theorem. Note that

$$\Phi(q,n,k,j;\mu,\sigma) = \Phi(q,n,k,j;0,\sigma).$$

Define a complex valued function $\phi_{q,n,k,j;\sigma}$ on $\mathbb{C}$ by

$$\phi_{q,n,k,j;\sigma}(z) := \left(\frac{z}{\sigma}\right)^{2k} \frac{p_q(z;0,\sigma)^{(n-1)(q-1)+q}}{(-\ell_q(z;0,\sigma))^j}$$
$$= \left(\frac{z}{\sigma}\right)^{2k} p_q(z;0,\sigma)^{(n-1)(q-1)+q} \left\{\frac{z^2 + r(q,\sigma)^2}{(Z_q\sigma)^{1-q}(3-q)\sigma^2}\right\}^{-j},$$

where we set

$$r(q,\sigma) := \sqrt{-\ln_q\left(\frac{1}{Z_q\sigma}\right)\cdot (Z_q\sigma)^{1-q}(3-q)\sigma^2}.$$

The function $\phi_{q,n,k,j;\sigma}$ has poles of order j at $\pm \mathrm{i}r(q,\sigma)$. For $R > r(q,\sigma)$, let L_R and C_R be smooth curves in $\mathbb{C}$ defined by

$$L_R := \{z : [-R, R] \to \mathbb{C} \mid z(\theta) = \theta\}, \qquad C_R := \{z : [0,\pi] \to \mathbb{C} \mid z(\theta) = Re^{\mathrm{i}\theta}\},$$

respectively. The residue theorem yields that

$$\int_{L_R\cup C_R} \phi_{q,n,k,j;\sigma}(z)dz = 2\pi\mathrm{i}\cdot \mathrm{Res}(\phi_{q,n,k,j;\sigma}; \mathrm{i}r(q,\sigma)), \tag{6.9}$$

where $\mathrm{Res}(\phi_{q,n,k,j;\sigma}; \mathrm{i}r(q,\sigma))$ stands for the residue of $\phi_{q,n,k,j;\sigma}$ at $z = \mathrm{i}r(q,\sigma)$.

Lemma 7 *For $n, j \in \mathbb{N}$, $k \in \{0,1,\ldots,n\}$ and $(\mu,\sigma) \in \mathbb{R}\times\Sigma_{q,a}$, then*

$$\Phi(q,n,k,j;\mu,\sigma) = 2\pi\mathrm{i}\cdot \mathrm{Res}(\phi_{q,n,k,j;\sigma}, \mathrm{i}r(q,\sigma)).$$

Proof If we show that

$$\lim_{R\to\infty}\int_{C_R} \phi_{q,n,k,j;\sigma}(z)dz = 0,$$

then we have the desired result by letting $R \to \infty$ in (6.9).

Take $R > r(q,\sigma)$ large enough. We calculate that

$$\begin{aligned}
&\left|\int_{C_R} \phi_{q,n,k,j;\sigma}(z)dz\right| \\
&\le R\int_0^\pi \left|\phi_{q,n,k,j;\sigma}(Re^{\mathrm{i}\theta})\right| d\theta \\
&= R\int_0^\pi \left(\frac{R}{\sigma}\right)^{2k} \left|p_q(Re^{\mathrm{i}\theta};0,\sigma)\right|^{(n-1)(q-1)+q} \left|\frac{R^2e^{2\mathrm{i}\theta} + r(q,\sigma)^2}{(Z_q\sigma)^{1-q}(3-q)\sigma^2}\right|^{-j} d\theta \\
&\le CR^{2(k-j)+1}\int_0^\pi \left|\exp_q\left(-\frac{R^2e^{2\mathrm{i}\theta}}{(3-q)\sigma^2}\right)\right|^{(n-1)(q-1)+q} d\theta,
\end{aligned}$$

where the constant C depends on q and σ.

In the case $q = 1$, we have that

$$\left|\exp_1\left(-\frac{R^2e^{2i\theta}}{(3-1)\sigma^2}\right)\right|^{(n-1)(1-1)+1} = \exp\left(-\frac{R^2\cos 2\theta}{2\sigma^2}\right),$$

consequently

$$\left|\int_{C_R}\phi_{q,n,k,j;\sigma}(z)dz\right| \leq CR^{2(k-j)+1}\int_0^{\pi}\exp\left(-\frac{R^2\cos 2\theta}{2\sigma^2}\right)d\theta \xrightarrow{R\to\infty} 0.$$

In the case $q > 1$, we observe that

$$\left|\exp_q\left(-\frac{R^2e^{2i\theta}}{(3-q)\sigma^2}\right)\right|^{(n-1)(q-1)+q} = \left|1+\frac{q-1}{3-q}\frac{R^2e^{2i\theta}}{\sigma^2}\right|^{\frac{(n-1)(q-1)+q}{1-q}} \leq C'R^{-2n+\frac{2}{1-q}},$$

where the constant C' depends on q and σ. This yields that

$$\left|\int_{C_R}\phi_{q,n,k,j;\sigma}(z)dz\right| \leq C\cdot C'R^{2(k-j)+1-2n+\frac{2}{1-q}}\cdot\pi.$$

The right-hand side converges to 0 as $R\to\infty$ since we have

$$2(k-j)+1-2n+\frac{2}{1-q} \leq -1+\frac{2}{1-q} < 0$$

due to the assumption $k \leq n$ and $j \in \mathbb{N}$. □

Proposition 3 *For* $\xi = (\mu,\sigma) \in \mathbb{R}\times\Sigma_{q,a}$, *then*

$$g_{\mu\mu}^{(q,a)}(\xi) = \frac{b_0^2(q,a)}{b_0^2(q,1)\sigma^2} - \frac{4}{3-q}\frac{\pi b_1^2(q,a)}{(Z_q\sigma)^{1-q}\sigma^2}r(q,\sigma),$$

$$g_{\sigma\sigma}^{(q,a)}(\xi) = \frac{(3-q)b_0^2(q,a)}{b_0^2(q,1)\sigma^2} + \frac{\pi(3-q)b_1^2(q,1)}{(Z_q\sigma)^{1-q}r(q,\sigma)}\left\{1+\left(\frac{r(q,\sigma)}{\sigma}\right)^2\right\}^2.$$

Proof It follows from Lemma 7 that

$$\begin{aligned}
&\Phi(q,n,k,1;\xi)\\
&= 2\pi\mathrm{i}\cdot\mathrm{Res}(\phi_{q,n,k,1;\sigma}, \mathrm{i}r(q,\sigma))\\
&= 2\pi\mathrm{i}\lim_{z\to\mathrm{i}r(q,\sigma)}\left\{(z-\mathrm{i}r(q,\sigma))\cdot\phi_{q,n,k,j;\sigma}(z)\right\}\\
&= 2\pi\mathrm{i}\lim_{z\to\mathrm{i}r(q,\sigma)}\left(\frac{z}{\sigma}\right)^{2k} p_q(z;0,\sigma)^{(n-1)(q-1)+q}\left\{\frac{z+\mathrm{i}r(q,\sigma)}{(Z_q\sigma)^{1-q}(3-q)\sigma^2}\right\}^{-1}\\
&= 2\pi\mathrm{i}\cdot\left(\frac{\mathrm{i}r(q,\sigma)}{\sigma}\right)^{2k}\frac{(Z_q\sigma)^{1-q}(3-q)\sigma^2}{2\mathrm{i}r(q,\sigma)}\\
&= (-1)^k\frac{\pi(Z_q\sigma)^{1-q}(3-q)}{\sigma^{2(k-1)}}r(q,\sigma)^{2k-1},
\end{aligned}$$

where we used $p_q(\mathrm{i}r(q,\sigma);0,\sigma)=1$. This with Proposition 2 and (6.8) concludes the proof of the proposition. □

Remark 6 In the case $a=1$, the Riemannian manifold $(\mathcal{S}_{q,1}, g^{(q,1)})$ has a constant curvature $-1/(3-q)$. This means that all $(\mathcal{S}_{q,1}, g^{(q,1)})$ for $1\le q<3$ are homothetic to each other. However, Proposition 3 suggests that this homothety may fail for $a\neq 1$.

6.5 Concluding Remarks

In this chapter, we presented gauge freedom of entropies on the open set $\mathcal{S}_q$ of all q-Gaussian densities for $1\le q<3$. We showed that a constant multiple of each (q,a)-entropy coincides with the Boltzmann–Shannon entropy if $q=1$, and the Tsallis entropy otherwise. However, any constant multiple of the (q,a)-relative entropy differs from the $(q,1)$-relative entropy for $a\neq 1$. We remark that the $(q,1)$-relative entropy coincides with the Kullback–Leibler divergence if $q=1$, and the Tsallis relative entropy of the Csiszár type otherwise.

In information geometry, the Kullback–Leibler divergence projection from observed data to a statistical model attains the maximum likelihood estimator (see [1, Chapter 4]). The terminology "maximum" depends on a criterion. It is known that higher-order asymptotic theory of estimation and Bayesian statistics improve the maximum likelihood estimator in another criterion. Ishige, Salani and the second named author showed in [2, Theorem 3.2] that the concavity related to the case $(q,a)=(1,1/2)$ is the strongest concavity among all admissible concavities preserved by the heat flow in Euclidean space. We expect that the $(1,1/2)$-relative entropy improves the maximum likelihood estimator.

Acknowledgements The both authors were supported in part by JSPS Grant-in-Aid for Scientific Research (KAKENHI) 16KT0132. HM was supported in part by KAKENHI 19K03489. AT was supported in part by KAKENHI 19K03494, 19H01786.

Disclaimer: Views and opinions expressed are those of the authors and do not necessarily represent official positions of their respective companies.

References

1. Amari, S., Nagaoka, H.: Methods of information geometry, Translations of Mathematical Monographs, vol. 191. American Mathematical Society, Providence, RI; Oxford University Press, Oxford (2000), translated from the 1993 Japanese original by Daishi Harada
2. Ishige, K., Salani, P., Takatsu, A.: To logconcavity and beyond. Commun. Contemp. Math. **22**(2), 1950009, 17 (2020)
3. Matsuzoe, H.: A sequence of escort distributions and generalizations of expectations on q-exponential family. Entropy **19**(1), Paper No. 7, 13 (2017)
4. Naudts, J.: Estimators, escort probabilities, and ϕ-exponential families in statistical physics. JIPAM. J. Inequal. Pure Appl. Math. **5**(4), Article 102, 15 (2004)
5. Naudts, J., Zhang, J.: Rho-tau embedding and gauge freedom in information geometry. Inf. Geom. **1**(1), 79–115 (2018)
6. Zhang, J., Naudts, J.: Information geometry under monotone embedding. Part I: divergence functions. In: Geometric science of information, Lecture Notes in Comput. Sci., vol. 10589, pp. 205–214. Springer, Cham (2017)

Chapter 7
On Geodesic Triangles with Right Angles in a Dually Flat Space

Frank Nielsen

Abstract The dualistic structure of statistical manifolds in information geometry yields eight types of (possibly mixed type) geodesic triangles passing through three given points, the triangle vertices. The interior angles of geodesic triangles can sum up to π like in Euclidean/Mahalanobis flat geometry, or exhibit otherwise angle excesses or angle defects. In this work, we initiate the study of geodesic triangles in dually flat spaces, termed Bregman manifolds, where a generalized Pythagorean theorem holds. We consider non-self dual Bregman manifolds since Mahalanobis self-dual manifolds amount to Euclidean geometry. First, we show how to construct geodesic triangles with either one, two, or three interior right angles, whenever it is possible. Second, we report a construction of triples of points for which the dual Pythagorean theorems hold simultaneously at a point, yielding two dual pairs of dual-type geodesics with right angles at that point.

Keywords Dually flat space · Bregman divergence · Geodesic triangle · Right angle triangle · Pythagorean theorem · Angle excess/defect · Mahalanobis manifold · Itakura–Saito manifold · (Extended) Kullback–Leibler manifold · Multinoulli manifold

7.1 Introduction and Motivation

In Euclidean geometry, it is well-known that the sum of the three interior angles of any triangle *always* sum up to π (Fig. 7.1), and that Euclidean triangles can have *at most* one right angle ($\frac{\pi}{2}$ radians or 90°). These facts are not true anymore in hyperbolic geometry nor in spherical geometry [27], where the total sum of the interior angles of a triangle may vary [31]:

F. Nielsen (✉)
Sony Computer Science Laboratories, Inc., Tokyo, Japan
e-mail: Frank.Nielsen@acm.org
URL: https://FrankNielsen.github.io/

F. Nielsen (ed.), *Progress in Information Geometry*,
Signals and Communication Technology,
https://doi.org/10.1007/978-3-030-65459-7_7

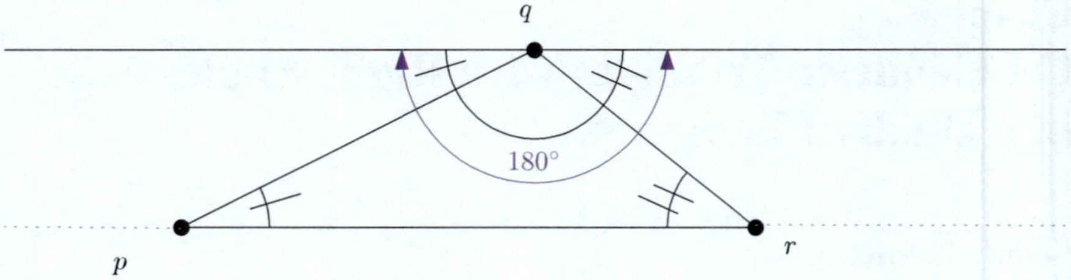

Fig. 7.1 Proof without words: The sum of the three interior angles of any triangle in Euclidean geometry always sum up to π (180°). Euclidean triangles dot not show angle defect nor angle excess: The Euclidean space is flat

- In hyperbolic geometry, hyperbolic triangles have always *angle defects*, meaning that the total sum of the three interior angles of any hyperbolic triangle is always strictly less than π. In the extreme case of hyperbolic ideal triangles, the total sums of their interior angles vanish since the interior angle at an ideal vertex is always 0. Figure 7.2 displays a right angle hyperbolic triangle and a hyperbolic ideal triangle.
- In spherical geometry, spherical triangles have always *angle excesses*, meaning that the sum of interior angles of any spherical triangle is always strictly greater than π, and is provably upper bounded by 3π radians or 540°. Moreover, there exist spherical triangles with one, two, or three right angles, as depicted in Fig. 7.2.

More generally, in Riemannian geometry [9], the angle excess or defect of a geodesic triangle reflects the total curvature enclosed by the geodesic triangle via the Gauss-Bonnet formula; The general Gauss-Bonnet formula states that the integral of the scalar curvature of a closed surface is 2π times the Euler characteristic χ of that surface (a topological characteristic). The hyperbolic geometry, Euclidean geometry and spherical geometry can be studied under the framework of Riemannian geometry, as a manifold of constant negative curvature, a flat manifold, and a manifold of constant positive curvature, respectively.

In this work, we consider the *dualistic structure* (M, g, ∇, ∇^*) of information geometry [3, 8] which can be derived from the statistical manifold structure of Lauritzen [11] (M, g, C): Namely, a simply connected smooth manifold M is equipped with two dual torsion-free affine connections ∇ and ∇^* such that these connections are coupled with the metric tensor g, meaning that their induced dual parallel transport preserves the metric [15]. We shall describe in details that dualistic structure in Sect. 7.2.

Any distinct pair of points p and q on the manifold M can either be joined by a primal ∇-geodesic arc γ_{pq} or by a dual ∇^*-geodesic arc γ^*_{pq}. Thus any triple of points (p, q, r) of M can be connected pairwise using one of $2 \times \binom{3}{2} = 6$ primal/dual

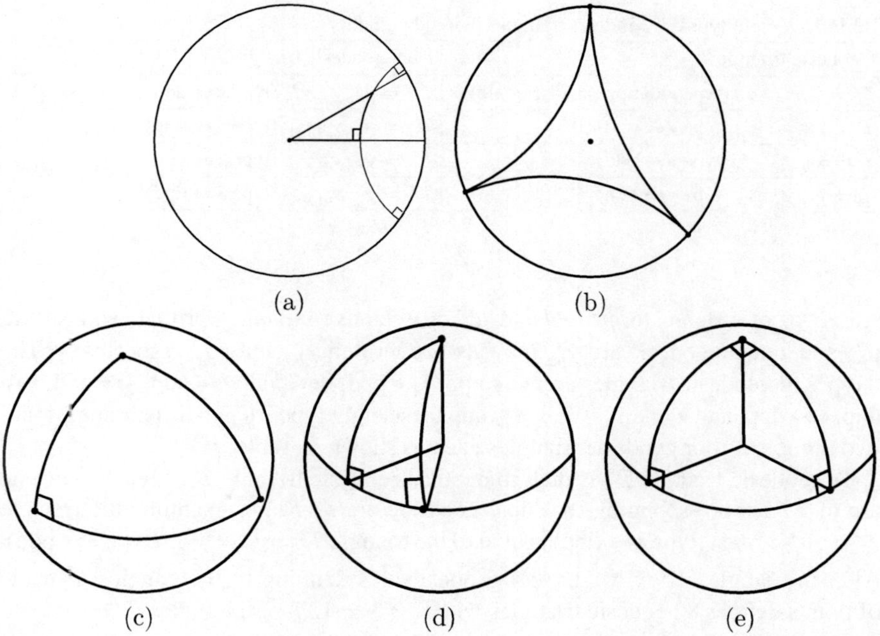

Fig. 7.2 In hyperbolic geometry, triangles have angle defects: Visualization in the Poincaré conformal disk model. **a** hyperbolic triangle with one right angle, **b** hyperbolic ideal triangle with all interior angles equal to zero (and area always π). In spherical geometry, triangles have angle excesses: Visualization of spherical triangles on the unit 3D sphere. **c** spherical triangle with one right angle, **d** spherical triangle with two right angles, and **e** spherical triangle with three right angles

geodesic arcs. Let us parameterize[1] these primal/dual geodesic arcs by $\gamma_{pq}(t)$ and $\gamma^*_{pq}(t)$ so that $\gamma_{pq}(0) = \gamma^*_{pq}(0) = p$ and $\gamma_{pq}(1) = \gamma^*_{pq}(1) = q$. Denote by $v_{pq} := \frac{d}{\mathrm{d}t}\gamma_{pq}(t)\big|_{t=0} = \dot{\gamma}_{pq}(0)$ and $v^*_{pq} :=:= \frac{d}{\mathrm{d}t}\gamma^*_{pq}(t)\big|_{t=0} = \dot{\gamma}_{pq}(0)$ the tangent vectors of the tangent plane T_p to the primal and dual geodesics, respectively. Two vectors $u, v \in T_p$ are orthogonal in T_p (denoted notationally by $u \perp_p v$) iff. $g_p(u, v) = 0$:

$$u \perp_p v \Leftrightarrow g_p(u, v) = 0. \tag{7.1}$$

More generally, two smooth curves $c_1(t)$ and $c_2(t)$ on the manifold are said orthogonal at a point $p = c_1(t_1) = c_2(t_2)$ iff. $g_p(\dot{c}_1(t_1), \dot{c}_2(t_2)) = 0$.

A geodesic triangle T passing through three points p, q, and r (i.e., the triangle vertices) is a triangle with edges linking two vertices defined either by primal or dual geodesic arcs. Thus the dualistic structure of information geometry yields $2^3 = 8$ *types* of geodesic triangles passing through any three distinct given points p, q and

[1]More precisely, a geodesic $\gamma^{\nabla}_{pq}(t)$ with respect to an affine connection ∇ satisfies $\nabla_{\dot{\gamma}_{pq}}\dot{\gamma}_{pq} = 0$. A ∇-geodesic is an autoparallel curve at it is invariant by affine reparameterization of t (i.e., $t' = at + b$).

Table 7.1 Eight types of geodesic triangles paired by duality

Geodesic triangle T	Dual geodesic triangle T^*
$T = \gamma_{pq}\gamma_{qr}\gamma_{rp}$ (type ●●● ppp : a ∇-triangle)	$T^* = \gamma^*_{pq}\gamma^*_{qr}\gamma^*_{rp}$ (type ●●● ddd: a ∇^*-triangle)
$T = \gamma_{pq}\gamma_{qr}\gamma^*_{rp}$ (type ●●● ppd)	$T^* = \gamma^*_{pq}\gamma^*_{qr}\gamma_{rp}$ (type ●●● ddp)
$T = \gamma_{pq}\gamma^*_{qr}\gamma^*_{rp}$ (type ●●● pdd)	$T^* = \gamma^*_{pq}\gamma_{qr}\gamma_{rp}$ (type ●●● dpp)
$T = \gamma_{pq}\gamma^*_{qr}\gamma_{rp}$ (type ●●● pdp)	$T^* = \gamma^*_{pq}\gamma_{qr}\gamma^*_{rp}$ (type ●●● dpd)

r. Let "p" or ● stands for *primal* and "d" or ● stands for *dual*. A primal geodesic arc γ_{pq} and a dual geodesic arc γ^*_{pq} may also be written $\gamma^{\bullet}_{pq}$ and $\gamma^{\bullet}_{pq}$, respectively. The 8 types of geodesic triangles are: ●●● ppp, ●●● ppd, ●●● pdd, ●●● pdp, ●●● ddd, ●●● ddp, ●●● dpp, and ●●● dpd. We can group these eight types of geodesic triangles into two groups of four geodesic triangles each as shown in Table 7.1.

A geodesic triangle T^* is dual to another geodesic triangle T iff. each geodesic arc of T^* is corresponding to a dual geodesic arc of T. For example, the triangle $T^* = \gamma^*_{pq}\gamma_{qr}\gamma_{rp}$ (type ●●● dpp) is dual to the triangle $T = \gamma_{pq}\gamma^*_{qr}\gamma^*_{rp}$ (type ●●● pdd). At each triangle vertex, we have 4 geodesic arcs defining $\binom{4}{2} = 6$ angles. A triple of points defines 8 geodesic triangles with $4 \times 3 = 12$ interior angles.[2] The interior angle $\alpha_p(c_{pq}, c_{pr})$ of a geodesic triangle $T = c_{pq}c_{qr}c_{rp}$ (where c_{ab} stands either for γ_{ab} or for γ^*_{ab}) at p is measured according to the metric tensor g as follows:

$$\alpha_p(c_{pq}, c_{pr}) = \arccos\left(\frac{g_p(\dot{c}_{pq}(0), \dot{c}_{pr}(0))}{\|\dot{c}_{pq}(0)\|_p \, \|\dot{c}_{pr}(0)\|_p}\right) = \alpha_p(c_{pr}, c_{pq}), \tag{7.2}$$

where $\|v\|_p = \sqrt{g_p(v, v)}$ measures the length of vector $v \in T_p$. The total sum $\alpha(T)$ of the three interior angles of a geodesic triangle $T = c_{pq}c_{qr}c_{rp}$ is defined by

$$\alpha(T) := \alpha_p(c_{pq}, c_{pr}) + \alpha_q(c_{qp}, c_{qr}) + \alpha_r(c_{rp}, c_{rq}). \tag{7.3}$$

We may specify a geodesic triangle as follows: $T_E(p, q, r)$ where $p, q, r \in M$ are the three triangle vertices (points on M), and $E \in (t_1 \in \{p, d\}, t_2 \in \{p, d\}, t_3 \in \{p, d\})$ is the type of primal/dual geodesic edges of the triangle so that $c^{t_i}_{ab} = \gamma_{ab}$ if $t_i = p$ and $c^{t_i}_{ab} = \gamma^*_{ab}$ if $t_i = d$. A geodesic triangle $T = T_E(p, q, r)$ with all primal geodesics arcs ($E = (p, p, p)$) is called a ∇-*triangle*, and its dual geodesic triangle $T^* = T_{E^*}(p, q, r)$ (with all dual geodesic arcs, i.e., $E^* = (d, d, d)$) is called a ∇^*-*triangle*.

By considering duality of geodesic triangles and permutations of the triple of points p, q and r, we may reduce the study of interior angles between a pair of geodesics at a vertex to three types of primal geodesic triangles (type ppp, pdp, and pdd) with their corresponding dual geodesic triangles (type ddd, dpd and dpp, respectively).

[2] We do consider at a triangle vertex only pairs of geodesics with interior angles linking the two other triangle vertices.

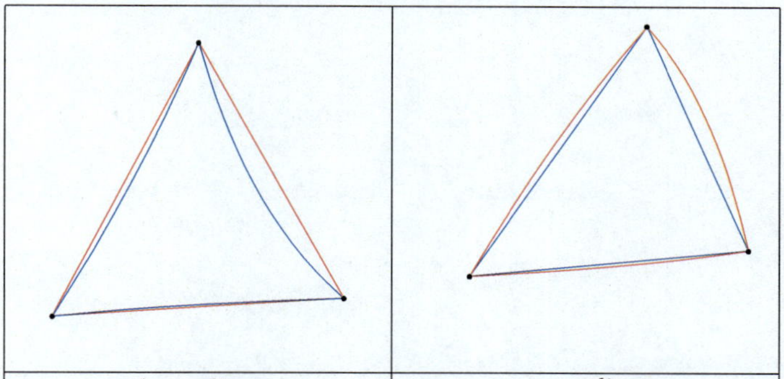

Fig. 7.3 The $2 \times 3 = 6$ potential geodesic arcs (edges) of a geodesic triangle T visualized both in the primal θ-coordinate system (left) and in the dual η-coordinate system (right). The vertices are $\theta(p) = (0.55, 0.575)$, $\theta(q) = (0.75, 0.95)$ and $\theta(r) = (0.95, 0.6)$ for the Bregman manifold defined by the 2D Burg negentropy: $F_{\mathrm{Burg}}(\theta^1, \theta^2) = -\log(\theta^1) - \log(\theta^2)$. Primal geodesic arcs are shown in red and are visualized as straight line segments in the θ-coordinate system. Dual geodesic arcs are shown in blue and are visualized as straight line segments in the η-coordinate system

A particular case of information-geometric manifolds are *dually flat spaces* [3] induced by a strictly convex and C^3 function F called the *potential function*. In a dually flat space M, there are two global dual affine coordinate systems θ and η (with $M = \{p : \theta(p) \in \mathrm{dom}(F)\}$), and a generalization of the Pythagoras' theorem holds [3]. The divergence $D_F(p : q)$ (potentially oriented distance, $D_F(p : q) \neq D_F(q : p)$) between two points p and q can be expressed using the Bregman divergence B_F on their primal coordinates:

$$D_F(p : q) = B_F(\theta(p) : \theta(q)) = F(\theta(p)) - F(\theta(q)) - (\theta(p) - \theta(q))^\top \nabla F(\theta(q)). \tag{7.4}$$

Whenever it is clear from context, we shall write D instead of D_F for conciseness.

Since the canonical divergence of a dually flat space (M, g, ∇, ∇^*) amounts to a Bregman divergence [3], we shall also call these dually flat spaces *Bregman manifolds* (M, F) in the remainder. When the Bregman generator is $F_{\mathrm{Euc}}(\theta) = \frac{1}{2}\theta^\top\theta$, we recover the Euclidean geometry (a self-dual Bregman manifold) where the *Euclidean divergence* is half of the squared of the Euclidean distance. Beware that the Euclidean distance is a metric distance but the Euclidean divergence is not.

Figure 7.3 visualizes in both the primal coordinate system θ and the dual coordinate system η the vertices of a triangle with all its primal and dual geodesic edges passing through its vertices: The considered Bregman manifold is the Itakura-Saito manifold induced by the Burg negentropy [20] yielding the Itakura-Saito divergence [5], a Bregman divergence. Figure 7.4 displays the 8 types of (possibly mixed-type) geodesic triangles that can be obtained from these 3 triangle vertices.

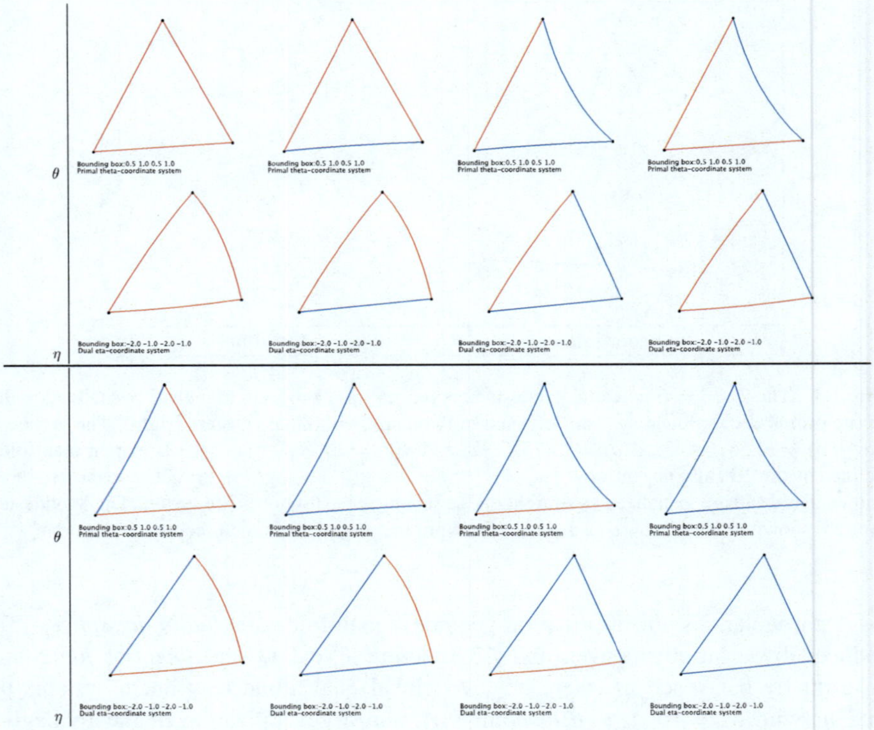

Fig. 7.4 The $2^3 = 8$ types of geodesic triangles visualized both in the primal θ-coordinate system and the dual η-coordinate system. Primal geodesic edges are shown in red and are straight in the θ-coordinate system. Dual geodesic edges are shown in blue and are straight in the η-coordinate system. Refer to Fig. 7.3 for point coordinates

In this paper, we raise and (partially) answer the following questions on a statistical manifold:

Q1. When and how can one build geodesic triangles with one, two and three right angles?

Q2. When and how can one build dual geodesic triangles T and T^* such that at a triangle vertex, we have two pairs of dual geodesics emanating from that vertex that are simultaneously orthogonal?

Q3. When is there a relationship or inequality between $\alpha(T)$ and $\alpha(T^*)$?

In this work, we are interested in studying these questions and unraveling some properties of geodesic triangles in the particular case of dually flat spaces, Bregman manifolds. In Bregman manifolds, the Pythagorean theorem allows one to check that a pair of dual geodesics is orthogonal at a given point by checking some divergence identity. In general, the Pythagorean theorem is useful to prove uniquess of projections in particular settings, interpret geometrically projections based on the divergence, and design derivative-free projection algorithms [1]. In information geometry, besides the dually flat spaces, Pythagorean theorems with corresponding divergence identities have also been reported for α-divergences on the probability simplex [3] and the logarithmic L_H^α-divergences [32] where H is an exponentially concave generator.

The paper is organized as follows: First, we quickly review the construction of Bregman manifolds in Sect. 7.2, thereby introducing familiar concepts and notations of information geometry [3], and give as examples of Bregman manifolds details for the Mahalanobis manifolds (Sect. 7.2.4.1), the extended Kullback–Leibler manifold (Sect. 7.2.4.2), the Itakura–Saito manifold (Sect. 7.2.4.3), and the multinoulli manifolds (Sect. 7.2.4.4). Then Sect. 7.3 shows, whenever it is possible, how to build geodesic ∇-triangles which have one, two or three right angles (thus necessarily exhibiting angle excesses for two and three right angle triangles). In Sect. 7.4, we prove that given two distinct points p and q, the locii of points r for which we have simultaneously the dual Pythagorean theorems holding at r are the intersection of an autoparallel ∇-submanifold with an autoparallel ∇^* submanifold (i.e., the intersection of a θ-flat with a η-flat [3]). We report explicitly the construction method for the 2D Itakura–Saito manifold and visualize several such triangles. Finally, we summarize and hint at further perspectives in Sect. 7.5.

7.2 Dually Flat Spaces: Bregman Manifolds

We first explain the mutually orthogonal primal basis and reciprocal basis in an inner product space in Sect. 7.2.1. Then we describe dual geodesics and their tangent vectors and the dual parallel transport in Sect. 7.2.2. In Sect. 7.2.3, we explain the Pythagorean theorem, and Sect. 7.2.4 provides some common examples of Bregman manifold: Mahalanobis self-dual manifolds, the extended Kullback–Leibler manifold and the Itakura–Saito manifold.

7.2.1 Preliminary: Inner Product Space and Reciprocal Basis

An *inner product space* is a vector space V equipped with a symmetric positive definite bilinear form $\langle\cdot,\cdot\rangle : V \times V \to \mathbb{R}$ called the inner product. The length of a vector $v \in V$ is given by its induced norm $\|v\| = \sqrt{\langle v, v\rangle}$, and the angle between any two vectors u and v is measured as $\alpha(u, v) = \arccos\left(\frac{\langle u,v\rangle}{\|u\|\|v\|}\right)$ (in radians). We consider finite D-dimensional inner product spaces where a vector v can be expressed in *any* basis $B = \{e_1, \ldots, e_D\}$ (a maximum set of linearly independent vectors) by its components $v_B = (v^1, \ldots, v^D)$: $v = \sum_{i=1}^D v^i e_i$. Vector v can also be expressed equivalently in an other basis $\hat{B} = \{\hat{e}_1, \ldots, \hat{e}_D\}$: $v = \sum_{i=1}^D \hat{v}^i \hat{e}_i$. Notice that in general the components v_B and $v_{\hat{B}}$ are different although they express the same geometric vector v according to their respective basis B and $\hat{B}$. We can express the basis vectors

$\hat{e}_i$ using basis B and the basis vectors e_i using basis $\hat{B}$ as $\hat{e}_i = \hat{A}_i^j e_j$ and $e_j = A_i^j \hat{e}_j$, respectively (using Einstein summation convention). We have $\hat{A}_i^j A_j^k = \delta_i^k$ where δ_i^j is the Krönecker symbol ($\delta_i^j = 1$ iff. $i = j$ and 0 otherwise), and the changes of basis reflects on components as $v^i = \hat{A}^i{}_j \hat{v}^j$ and $\hat{v}^i = A^i{}_j v_i$. When the basis B is orthonormal (i.e., $\langle e_i, e_j \rangle = \delta_{ij}$ where δ_{ij} is the Krönecker symbol: $\delta_{ij} = 1$ iff. $i = j$ and 0 otherwise) the vector components can be retrieved from the inner product as $v^i = \langle v, e_i \rangle = \langle \sum_{j=1}^D v^j e_j, e_i \rangle = \sum_{j=1}^D v^j \langle e_j, e_i \rangle$. This is no longer true for non-orthormal basis (e.g., an orthogonal but non-orthonormal basis or an oblique basis). Let us introduce the unique *reciprocal basis* $B^* = \{e^{*1}, \ldots, e^{*D}\}$ such that by construction, we have $\langle e_i, e^{*j} \rangle = \delta_i^j$. The vector v can be expressed in the reciprocal basis as $v = \sum_{i=1}^D v_i e^{*i}$. The vector components v^i (superscript notation) wrt. basis B are called the *contravariant components*, and $v^i = \langle v, e^{*i} \rangle$. The vector components v_i (subscript notation) wrt. basis B^* are called the *covariant components*, and $v_i = \langle v, e_i \rangle$. (We shall explain this contravariant/covariant component terminology at the end of this section.)

Let $G = [g_{ij} = \langle e_i, e_j \rangle]_{ij}$ and $G^* = [g^{*ij} = \langle e^{*i}, e^{*j} \rangle]_{ij}$ denote the $D \times D$ positive definite matrices, called dual metrics. These dual metric matrices are inverse of each other: $G^* = G^{-1}$. In textbooks, one often drops the superscript star '*' in the notation of the reciprocal basis and the dual riemannian metric, see [3]. Here, we keep them explicitly for easing the understanding, even if they load the notations.

We can convert the contravariant components v^i of a vector v to its covariant components v_i, and vice versa, using these metric matrices: $v_i = \sum_{i=1}^D g_{ij} v^j$ and $v^i = \sum_{i=1}^D g^{*ij} v_j$. Let $[u]_B$ denote the vector components of u in basis B arranged in a column vector. Then we rewrite the contravariant/covariant conversions as matrix-vector multiplications of linear algebra: $[v]_{B^*} = G \times [v]_B$ and $[v]_B = G^* \times [v]_{B^*}$. The inner product between two vectors can be written equivalently using algebra as

$$\langle u, v \rangle = \sum_{i=1}^D u_i v^i = [u]_{B^*}^\top \times [v]_B, \tag{7.5}$$

$$= \sum_{i=1}^D u^i v_i = [u]_B^\top \times [v]_{B^*}, \tag{7.6}$$

$$= \sum_{i=1}^D [u]_B^\top \times G^* \times [v]_B, \tag{7.7}$$

$$= \sum_{i=1}^D [u]_{B^*}^\top \times G \times [v]_{B^*}. \tag{7.8}$$

In differential geometry [3, 15], a smooth manifold M is equipped with a *metric tensor field* g that defines on each tangent plane T_p of $p \in M$ an inner product. The dual of a tangent plane T_p is the cotangent plane T_p^*, a vector space of linear functionals. In general, tensor fields define at each point of the manifold component-free geometric entities that can be interpreted as multilinear functionals over Cartesian products of dual covector and vector spaces. A vector v of T_p is a rank-1 tensor that can be expressed either using the covariant or contravariant components of a reciprocal basis. Thus one should not confuse the notion of "geometric vectors" that are tensors (coordinate-free objects independent of the choice of the basis) with the "column vectors" of vector components in a basis which are used to perform linear algebra calculations on (geometric) vectors. When the components of a tensor vector varies with the inverse transformation of the change of basis, we say that we have a contravariant (tensor) vector (a $(0, 1)$-tensor), and its components are called contravariant components. When the components of a tensor vector varies according to the transformation of the change of basis, we say that we have a covariant (tensor) vector (a covector or $(1, 0)$-tensor, i.e., an element of the dual vector space V^* of linear functionals or linear forms), and its components are called covariant components. The metric tensor g is a $(2, 0)$ covariant tensor [15].

7.2.2 Dual Geodesics and Their Tangent Vectors, and Dually Coupled Parallel Transport

Let $F(\theta)$ be a D-dimensional C^3 real-valued function defined on an *open* convex domain Θ, and denote by $F^*(\eta)$ its Legendre-Fenchel convex conjugate [3, 15]: $F^*(\eta) = \sup_{\theta\in\Theta} \theta^\top \eta - F(\theta)$. The dual potential functions F and F^* induce two torsion-free flat affine connections [15] ∇ and ∇^*, respectively. A Bregman manifold M is equipped with two global affine coordinate systems $\theta(\cdot)$ (the ∇-affine coordinate system) and $\eta(\cdot)$ (the ∇^*-affine coordinate system) such that it comes from Legendre-Fenchel transformation that $\eta(\theta) = \nabla F(\theta)$ and $\theta(\eta) = \nabla F^*(\eta)$. Let $\theta^i(p)$ and $\eta_i(p)$ denote the primal i-th θ-coordinate functions and the dual i-th η-coordinate functions of a point p, for $i \in \{1, \ldots, D\}$ so that $\theta(p) = (\theta^1(p), \ldots, \theta^D(p))$ and $\eta(p) = (\eta_1(p), \ldots, \eta_D(p))$. Notations are summarized in Appendix A. Any point $p \in M$ can be expressed equivalently either in the primal global θ-chart or the dual global η-chart. The dual geodesics[3] γ_{pq} and γ^*_{pq} passing through two given points $p, q \in M$ write simply using the dual coordinate systems as follows:

$$\gamma_{pq} = \{x_\lambda \in M \ : \ \theta(x_\lambda) = (1-\lambda)\theta(p) + \lambda\theta(q), \quad \lambda \in [0, 1]\}, \tag{7.9}$$

$$\gamma^*_{pq} = \{x_\lambda \in M \ : \ \eta(x_\lambda) = (1-\lambda)\eta(p) + \lambda\eta(q), \quad \lambda \in [0, 1]\}. \tag{7.10}$$

In general, a vector field $v(t)$ is parallel along a smooth curve $c(t)$ iff.

$$\forall i \in \{1, \ldots, D\}, \ \dot{v}^i + \sum_{j,k=1}^{D} \Gamma^i_{jk} \dot{x}^j v^k = 0. \tag{7.11}$$

[3]Given an affine connection ∇, the ∇-geodesic is an autoparallel curve [15].

Since the Christoffel symbols $\Gamma^i_{jk}(\theta) = 0$ (and $\Gamma^{*i}_{jk}(\eta) = 0$) in a dually flat space, we have $\dot{v}^i = 0$ and recover the equation of the primal geodesic (and $\dot{v}_i = 0$ for the equation of the dual geodesic, respectively).

The dual Riemannian metrics $[g_{ij}] = [g_{ji}]$ and $[g^{*ij}] = [g^{*ji}]$ are induced by the Hessians of the dual potential functions F and F^* (both symmetric positive-definite matrices), respectively. At any given point $p \in M$, we consider the natural basis $\{e_i = \partial_i = \frac{\partial}{\partial \theta^i}\}$ and the reciprocal basis $\{e^{*i} = \partial^i = \frac{\partial}{\partial \eta_i}\}_i$ of the tangent plane T_p so that $g(e_i, e^{*j}) = \delta_i^j$. That is, the basis vectors of the primal and reciprocal basis are *mutually orthogonal*. We have $[g_{ij}] = [g(e_i, e_j)]_{ij} = \nabla^2 F(\theta)$ and $[g^{*ij}] = [g^*(e^{*i}, e^{*j})]_{ij} = \nabla^2 F^*(\eta)$. We can check that the natural primal basis and the dual reciprocal basis are mutually orthogonal from Crouzeix identity:

$$\nabla^2 F(\theta) \nabla^2 F^*(\eta) = I, \tag{7.12}$$

where I denotes the $D \times D$ identity matrix. Note that Eq. 7.12 can be read using a single coordinate system as $\nabla^2 F(\theta) \nabla^2 F^*(\eta(\theta)) = \nabla^2 F(\theta) \nabla^2 F^*(\nabla F(\theta)) = I$ or $\nabla^2 F(\theta(\eta)) \nabla^2 F^*(\eta) = \nabla^2 F(\nabla F^*(\eta)) \nabla^2 F^*(\eta) = I$.

A tangent vector v_{pq} of the primal geodesic γ_{pq} at p is a vector of the tangent plane T_p with contravariant components $\theta(q) - \theta(p)$ (say, expressed in the primal basis $B_p = \{e_1, \ldots, e_D\}$): $v = \sum_{i=1}^D (\theta^i(q) - \theta^i(p)) e_i$ with $v^i = \theta^i(q) - \theta^i(p)$. Indeed, we have

$$\frac{d\gamma_{pq}(t)}{\mathrm{d}t} = \frac{d}{\mathrm{d}t}((1-t)\theta(p) + t\theta(q)) = \sum_i (\theta^i(q) - \theta^i(p)) \partial_i, \tag{7.13}$$

where $\partial_i = \frac{\partial}{\partial \theta^i}$. Similarly, a tangent vector v^*_{pq} at the dual geodesic γ^*_{pq} is a vector of the tangent plane T_p with covariant components $\eta(q) - \eta(p)$ (expressed in the reciprocal basis $B^*_p = \{e^*1, \ldots, e^{*D}\}$): $v = \sum_{i=1}^D (\eta_i i(q) - \eta_i(p)) e^{*i}$. Indeed, we check that

$$\frac{d\gamma^*_{pq}(t)}{\mathrm{d}t} = \frac{d}{\mathrm{d}t}((1-t)\eta(p) + t\eta(q)) = \sum_i (\eta_i(q) - \eta_i(p)) \partial^i, \tag{7.14}$$

where $\partial^i = \frac{\partial}{\partial \eta_i}$. In general, for separable Bregman generators, i.e., $F(\theta) = \sum_{i=1}^D F_i(\theta^i)$ where the F_i's are scalar strictly convex and C^3 functions, we can choose both the primal and reciprocal basis to be orthogonal but they are not necessarily orthonormal (except in the special case of the Euclidean geometry obtained by $F(\theta) = \sum_{i=1}^D \mathrm{sqr}(\theta^i)$ where $\mathrm{sqr}(x) := x^2$ denotes the square function).

The metric tensor field g defines a smooth scalar product $g(\cdot, \cdot)$ on the tangent bundle TM (informally, the union of all tangent planes) such that for any two vectors $u, v \in T_p$, we have $g_p(u, v) = u_i v^i = u^i v_i$. In each tangent plane, we thus have an inner product space. The contravariant and covariant components of a vector v can be retrieved using the inner product with the reciprocal basis and the primal basis, respectively: $v^i = g(v, e^{*i})$, and $v_i = g(v, e_i)$. We can also use vector-matrix multiplications of linear algebra to calculate the inner product: $g_p(u, v) = [u^i]^\top \times \nabla^2 F(\theta(p)) \times [v^i] = [u_i]^\top \times \nabla^2 F^*(\eta(p)) \times [v_i]$, where $[u^i]$ and $[u_i]$ denote the column-vector of the contravariant components of u and covariant components of u, respectively. Since $v^i = \nabla^2 F^*(\eta(p)) \times [v_i]$ and $v_i = \nabla^2 F(\theta(p)) \times [v^i]$, we therefore have the following equivalent rewritings of the inner product:

$$\begin{aligned} g_p(u, v) &= [u_i]^\top \times [v^i], &(7.15)\\ &= [u^i]^\top \times [v_j], &(7.16)\\ &= [u^i]^\top \times \nabla^2 F(\theta(p)) \times [v^i], &(7.17)\\ &= [u_i]^\top \times \nabla^2 F^*(\eta(p)) \times [v_i]. &(7.18) \end{aligned}$$

When the generator $F(\theta)$ is separable, i.e., $F(\theta) = \sum_{i=1}^D F_i(\theta^i)$ for univariate generators F_i's, the inner product at a tangent plane T_p writes equivalently as (using Einstein summation convention):

$$\begin{aligned} g_p(u, v) &= u^i F_i''(\theta^i(p)) v^i, &(7.19)\\ &= u_i F_i^{*\prime\prime}(\eta_i(p)) v_i, &(7.20)\\ &= u_i v^i, &(7.21)\\ &= u^i v_i. &(7.22) \end{aligned}$$

Given two vectors $u, v \in T_p$, we measure their lengths $\|u\|_p$ and $\|v\|_p$ and the interior angle $\alpha_p(u, v)$ between them as:

$$\begin{aligned} \|u\|_p &= \sqrt{g_p(u, u)} = \sqrt{u^i u_i}, &(7.23)\\ \|v\|_p &= \sqrt{g_p(v, v)} = \sqrt{v^i v_i}, &(7.24)\\ \alpha_p(u, v) &= \arccos\left(\frac{u^i v_i}{\|u\|_p \|v\|_p}\right) = \alpha_p(v, u). &(7.25) \end{aligned}$$

Notice that the Hessian metric g and g^* are not conformal (i.e., not a positive scalar function of the Euclidean metric), and that we cannot "read" the angles directly in the θ- or η-coordinate systems. In other words, the Euclidean angles displayed in the θ- or η-coordinate systems do not correspond to the intrinsic angles of the underlying Bregman geometry. Figure 7.5 shows a visualization of the metric tensor field for the Itakura–Saito manifold (with $[g_{ij}](\theta) = \nabla^2 F(\theta) = \mathrm{diag}(\frac{1}{\mathrm{sqr}(\theta^1)}, \frac{1}{\mathrm{sqr}(\theta^2)})$ for 2D $\theta = (\theta^1, \theta^2)$). Dually flat spaces are a particular case of more general (local) Hessian structures studied in [29].

In information geometry [3], the dually flat torsion-free affine connections[4] ∇ and ∇^* are coupled the metric tensor g. An affine connection ∇ is flat iff. there exists

[4]The notion of dual connections of information geometry is more general than the notion of conjugate connections of affine differential geometry [10] which stems from dual affine immersions.

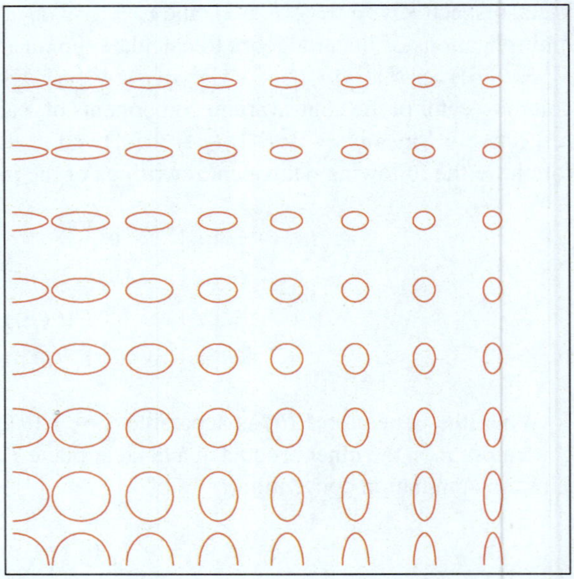

Fig. 7.5 Visualization of the metric tensor field $g(\theta) = \mathrm{diag}\left(\frac{1}{\mathrm{sqr}(\theta^1)}, \frac{1}{\mathrm{sqr}(\theta^2)}\right)$ for the 2D Itakura–Saito manifold. The metric tensor field g is not conformal. This can be seen from ellipses depicting the metric tensors (tiny Bregman circles) at regularly sampled grid positions which are not scaled (Euclidean) circles

a coordinate system such that its Christoffel symbols Γ expressed in a coordinate system vanishes. For Bregman manifolds, we have both $\Gamma_{ijk}(\theta) = 0$ and $\Gamma^{*ijk}(\eta) = 0$, see [3], so the connections ∇ and ∇^* are both flat. Notice that a cylinder is ∇-flat for the natural Euclidean connection but geodesic triangles on the cylinder have interior angles not summing up to π. That is, let $\mathcal{X}(M)$ denotes the space of smooth vector fields (the cross sections of the tangent bundles TM). We say that the two torsion-free affine connections ∇ and ∇^* are dual when it holds that

$$\forall X, Y, Z \in \mathcal{X}(M), \quad Xg(Y, Z) = g\left(\nabla_X Y, Z\right) + g\left(Y, \nabla^*_X Z\right). \tag{7.26}$$

An important consequence of the coupling of the dual connections to the metric is that the dual parallel transport preserves the metric.

Let us choose by convention to fix the primal basis of the tangent vectors at T_p for any $p \in M$ to be the e_i's, the one-hot vectors in the θ-coordinate system, i.e., $e_i = (\underbrace{0, \ldots, 0}_{i-1}, 1, 0, \ldots, 0)$ (the standard basis of $\mathbb{R}^D$). This is the canonical basis.[5]

Since the ∇-connection is flat, the primal parallel transport $\prod_{p,q}(v)$ of a vector v of T_p to a corresponding vector T_q is *independent* of the chosen smooth curve, and we have for $v = \sum_i v^i \, e_i|_p \in T_p$

$$\prod_{p,q}(v) = v^i \; e_i|_q \in T_q. \tag{7.27}$$

That is, the contravariant components of v do not change with primal parallel transport assuming that the primal basis is fixed for all tangent planes, i.e., $B_q = B_p, \forall p, q \in M$ (and $e_i|_p = e_i|_q$).

Similarly, since the ∇^*-connection is flat, the dual parallel transport $\prod^*_{p,q}(v)$ of a vector v of T_p to a corresponding vector of T_q is independent of the chosen smooth curve, and we have

$$\overset{*}{\prod_{p,q}}(v) = v_i \; e^{*i}\Big|_q. \tag{7.28}$$

However, because $B_q^* = \{e^{*i}|_q\}$ is the reciprocal basis of the fixed primal basis $B_q = \{e_i|_q\} = e_i$, the basis B_q^* varies accordingly to the metric tensor g (and $B_q^* \neq B_p^*$). Indeed, in general, we cannot fix both the primal and reciprocal basis in all tangent planes because they relate to each other by construction by the metric tensor. This is only possible when $G(\theta(p)) = G^*(\eta(p)) = I$, the $D \times D$ identity matrix, corresponding to the special case of Euclidean or Mahalanobis geometry.

Let us now check the metric compatibility of the dual parallel transport of two vectors $u, v \in T_p$ to T_q:

$$g_q\left(\prod_{p,q}(u), \overset{*}{\prod_{p,q}}(v)\right) = g_p(u, v). \tag{7.29}$$

Proof We have

$$g_q\left(\prod_{p,q}(u) = u^i \; e_i|_q \, , \overset{*}{\prod_{p,q}}(v) = v_i \; e^{*i}\Big|_q\right) = u^i v_i g_q\left(e_i|_q \, , e^{*i}\Big|_q\right), \tag{7.30}$$

$$= u^i v_i, \tag{7.31}$$

[5]In differential geometry, the tangent plane at a point p is the space of all linear derivations that satisfies the Leibniz's rule. A basis $\{t_i\}_i$ of T_p is such that $t_i(f) = \frac{\partial f}{\partial \theta^i}$.

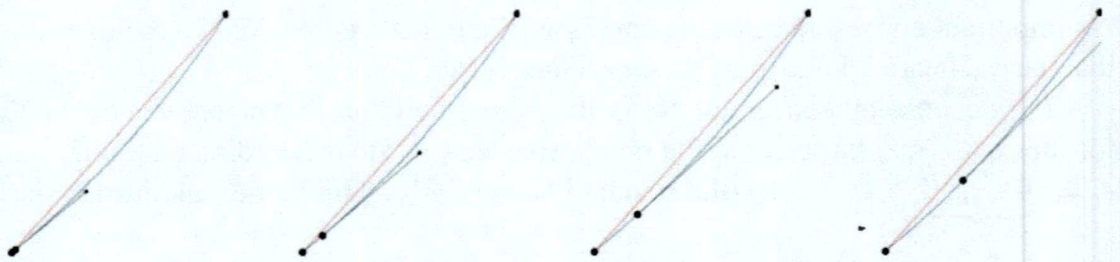

Fig. 7.6 Dual parallel transport of tangent vector $\dot{\gamma}^*_{pq}(0)$ at T_p along the ∇^*-geodesic: Since the dual geodesic $\gamma^*_{pq}(t)$ is a ∇^*-autoparallel curve, the tangent vector at $\dot{\gamma}^*_{pq}(0)$ is transported into tangent vectors along that dual geodesic

$$= u^i v_i g_p\left(e_i|_p, e^{*i}\Big|_p \right), \quad (7.32)$$

$$= g_p(u, v). \quad (7.33)$$

□

Figure 7.6 displays the parallel transport of the tangent vector $\dot{\gamma}^*_{pq}(0)$ from T_p to $T_{\gamma^*_{pq}(t)}$ for several time steps t. Although the dual parallel transport preserves the metric, the length of a vector transported by either the primal or the dual parallel transport varies.

Historically, the dual parallel transport coupled to the metric tensor was studied independently by Norden [25] and Sen [28] in the 1930s and 1940s, respectively.

7.2.3 Dual Pythagorean Theorems

The divergence $D_F(p : q)$ from a point p to a point q can be expressed using dual Bregman divergences as follows:

$$D_F(p : q) = B_F(\theta(p) : \theta(q)) = D^*(q : p) = B_{F^*}(\eta(q) : \eta(p)), \quad (7.34)$$

where $D^*_F(p : q) := D_F(q : p)$ denotes the reverse divergence, and

$$B_F(\theta_1 : \theta_2) = F(\theta_1) - F(\theta_2) - (\theta_1 - \theta_2)^\top \nabla F(\theta_2), \quad (7.35)$$

is the Bregman divergence associated to generator F. Bregman divergences generalize the squared Euclidean distance with the relative entropy, and thus offer a nice framework to unify or develop generic algorithms [4]. Bregman divergences proved useful in machine learning [5] and topological data analysis [7] among others.

Now, consider two smooth curves $c_1(t)$ and $c_2(t)$ intersecting at a point $p = c_1(0) = c_2(0)$. Denote by $\dot{c}_1(0) = \frac{d}{dt}c_1(t)\big|_{t=0}$ and $\dot{c}_2(0) = \frac{d}{dt}c_2(t)\big|_{t=0}$ the tangent vectors to the curves at point p, belonging to the tangent plane T_p. Curve c_1 is said *orthogonal* to curve c_2 at p, i.e., $c_1(t) \perp_p c_2(t)$, iff. $g_p(\dot{c}_1(0), \dot{c}_2(0)) = 0$.

In a Bregman manifold, the remarkable following generalized Pythagorean theorem holds:

Theorem 1 (Generalized Pythagorean theorem) *When a primal geodesic γ_{pq} is orthogonal to a dual geodesic γ^*_{qr} at point q (i.e., $\gamma_{pq} \perp_q \gamma^*_{qr}$), we have $(\theta(p) - \theta(q))^\top(\eta(r) - \eta(q)) = 0$, and the following Pythagorean divergence identity holds:*

$$D_F(p : q) + D_F(q : r) = D_F(p : r). \tag{7.36}$$

Proof The proof proceeds in two steps:

- First, Let us show that the test $(\theta(p) - \theta(q))^\top(\eta(r) - \eta(q)) = 0$ can be rewritten as an inner product showing that

$$(\theta(p) - \theta(q))^\top(\eta(r) - \eta(q)) = g_q(v_{qp}, v^*_{qr}) = g_q(\dot\gamma_{qp}(0), \dot\gamma^*_{qr}(0)), \tag{7.37}$$

where v_{qp} denotes the tangent vector at q of the primal geodesic γ_{pq}, and v^*_{qr} is the tangent vector at q of the dual geodesic γ^*_{qr}. Indeed, the inner product is $g(v_{qp}, v^*_{qr}) = (v_{qp})^i(v^*_{qr})_i$ and we have

$$(\theta(p) - \theta(q))^\top(\eta(r) - \eta(q)) = [\theta_{pq}]^\top \times \nabla^2 F(\theta(q))(\nabla^2 F(\theta(q)))^{-1} \times [\eta^*_{rq}], \tag{7.38}$$

where $\theta_{pq} := \theta(p) - \theta(q)$ (contravariant components of v_{qp}) and $\eta^*_{rq} := \eta(r) - \eta(q)$ (covariant components of v^*_{qr}). That is, $[(v_{qp})^i] = \theta_{pq}$ and $[v^*_{qp}]_i = \eta_{pq}$. Using the Crouzeix identity [6] of Eq. 7.12, we get

$$(\theta(p) - \theta(q))^\top(\eta(r) - \eta(q)) = [\theta_{pq}]^\top \nabla^2 F(\theta(q))(\nabla^2 F^*(\eta(q)))[\eta^*_{rq}]. \tag{7.39}$$

The term $(\nabla^2 F^*(\eta(q))) \times [\eta^*_{rq}]$ gives the contravariant components θ^*_{rq} of the vector v^*_{qr}. Thus we have checked that

$$(\theta(p) - \theta(q))^\top(\eta(r) - \eta(q)) = [\theta_{pq}]^\top \times \nabla^2 F(\theta(q)) \times [\theta^*_{rq}] = g_q(v_{qp}, v^*_{qr}). \tag{7.40}$$

That is,

$$(\theta(p) - \theta(q))^\top(\eta(r) - \eta(q)) = 0 \Leftrightarrow \dot\gamma_{pq}(0) \perp_q \dot\gamma^*_{qr}(0). \tag{7.41}$$

- Now, let us prove the Pythagorean identity when $\gamma_{pq} \perp_q \gamma^*_{qr}$. We use the Bregman 3-parameter identity [4] which generalizes the Euclidean law of cosines:

Property 1 (*Bregman* 3*-parameter identity*)

$$B_F(\theta_1 : \theta_2) = B_F(\theta_1 : \theta_3) + B_F(\theta_3 : \theta_2) - (\theta_1 - \theta_3)^\top(\nabla F(\theta_2) - \nabla F(\theta_3)) \geq 0 \tag{7.42}$$

Instantiating this identity with $\theta_1 = \theta(p)$, $\theta_2 = \theta(r)$ and $\theta_3 = \theta(q)$, and plugging the fact that $(\theta(p) - \theta(q))^\top(\nabla F(\theta(r)) - \nabla F(\theta(q))) = 0$, we get

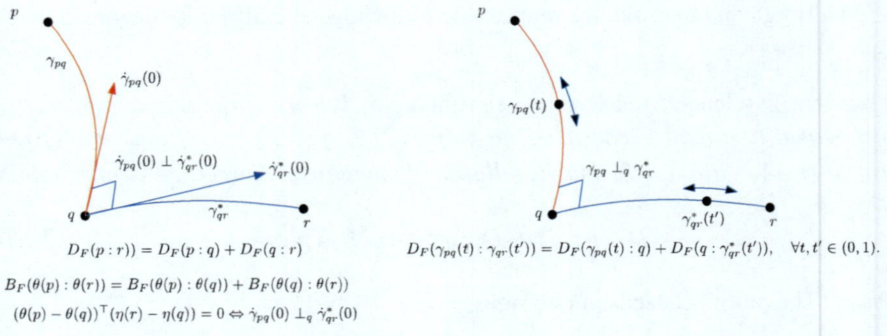

Fig. 7.7 The generalized Pythagorean theorem in a dually flat space. Left: ordinary statement for dual-type geodesics γ_{qp} and γ^*_{qr} orthogonal at a point q. Right: Extended statement where the points can move along the dual geodesics $\gamma_{pq}(t)$ and $\gamma qr^*(t')$

$$B_F(\theta(p) : \theta(r)) = B_F(\theta(p) : \theta(q)) + B_F(\theta(q) : \theta(r)), \tag{7.43}$$
$$D_F(p : r) = D_F(p : q) + D_F(q : r). \tag{7.44}$$

Notice that the Bregman 3-parameter identity can be proved by checking that the left-hand side equals the right-hand side of Eq. 7.42.

Another direct proof consists in writing:

$$B_F(\theta_1 : \theta_3) = F(\theta_1) - F(\theta_3) - (\theta_1 - \theta_3)^\top \nabla F(\theta_3), \tag{7.45}$$
$$B_F(\theta_3 : \theta_2) = F(\theta_3) - F(\theta_2) - (\theta_3 - \theta_2)^\top \nabla F(\theta_2), \tag{7.46}$$
$$B_F(\theta_1 : \theta_2) = F(\theta_1) - F(\theta_2) - (\theta_1 - \theta_2)^\top \nabla F(\theta_2). \tag{7.47}$$

Adding Eq. 7.46 with Eq. 7.47 and subtracting Eq. 7.47 from them, we get:

$$B_F(\theta_1 : \theta_3) + B_F(\theta_3 : \theta_2) - B_F(\theta_1 : \theta_2) = (\theta_1 - \theta_3)^\top(\nabla F(\theta_2) - \nabla F(\theta_3)), \tag{7.48}$$

from which it follows that

$$B_F(\theta_1 : \theta_2) = B_F(\theta_1 : \theta_3) + B_F(\theta_3 : \theta_2) - (\theta_1 - \theta_3)^\top(\nabla F(\theta_2) - \nabla F(\theta_3)). \tag{7.49}$$

More generally, we have

$$D_F(\gamma_{pq}(t) : q) + D_F(q : \gamma_{qr}(t')) = D_F(\gamma_{pq}(t) : \gamma_{qr}(t')), \quad \forall t, t' \in (0, 1). \tag{7.50}$$

That is, once we have a triple of points (p, q, r) for which the generalized Pythagorean theorem holds, we can build an infinite number of such triple of points: $(\gamma_{pq}(t), q, \gamma_{qr}(t'))$ for $t, t' \in (0, 1)$. Figure 7.7 illustrates this view of the generalized Pythagorean theorem.

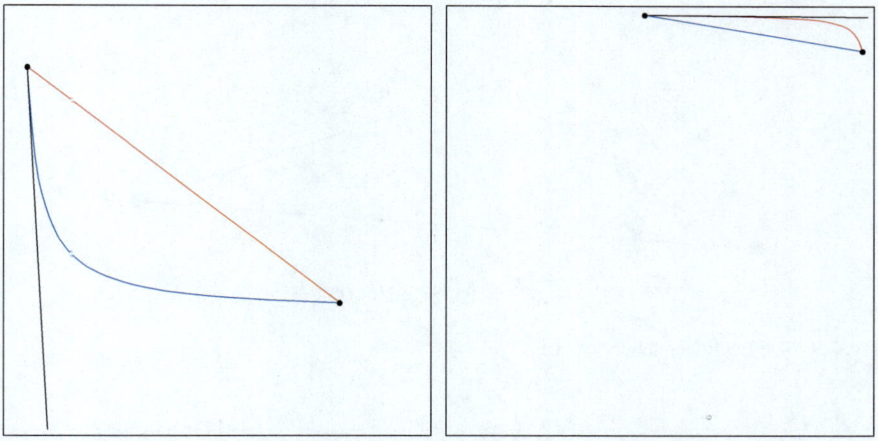

Fig. 7.8 Two points p and q and their primal and dual geodesics visualized using the θ-coordinates (left) and the η-coordinates (right). The tangent vector $v^*_{pq} \in T_p$ to γ^*_{pq} is visualized in the θ-coordinates (left, black line segment), and the tangent vector $v_{pq} \in T_p$ to γ_{pq} is visualized in the η-coordinates (right, black line segment). Here, we considered the Itakura–Saito manifold with $\theta(p) = (0.17823054175936948, 1.746348492830485)$ and $\theta(q) = (1.6105239241969733, 0.6712015045558234)$

Euclidean geometry with its ordinary Pythagoras' theorem is recovered as the special case of a self-dual Bregman manifold (when the dual potential functions coincide) induced by the Bregman generator $F_{\mathrm{Euc}}(\theta) = \frac{1}{2}\theta^\top\theta$.

Notice that the tangent vector $v^*_{pq} = \dot{\gamma}^*_{pq}(0)$ of the dual geodesic $\gamma^*_{pq}(t)$ is written using the covariant components in the reciprocal basis of T_p as $\eta(q) - \eta(p)$. This tangent vector can be written equivalently using the contravariant coordinates as $\nabla^2 F^*(\eta(p)) \times [\eta(q) - \eta(p)]$. That is, $[v^*_{pq\,i}] = \eta(q) - \eta(p)$ and $[{v^*_{pq}}^i] = \nabla^2 F^*(\eta(p)) \times [\eta(q) - \eta(p)]$. Similarly, the tangent vector $v_{pq} = \dot{\gamma}_{pq}(0)$ of the primal geodesic $\gamma_{pq}(t)$ is written using the contravariant components in the basis of T_p as $\theta(q) - \theta(p)$. The covariant coordinates of this vector is $\nabla^2 F(\theta(p)) \times [\theta(q) - \theta(p)]$: $[{v_{pq}}^i] = \theta(q) - \theta(p)$ and $[v_{pq_i}] = \nabla^2 F(\theta(p)) \times [\theta(q) - \theta(p)]$. Figure 7.8 displays an example of two primal and dual geodesic arcs linking p to q with the tangent vectors $\gamma^*_{pq}(0)$ at p of the dual geodesic γ_{pq}.

When we exchange the role played by points p and r in the above Pythagorean theorem, we obtain the following *dual* Pythagorean theorem:

Theorem 2 (Dual Pythagorean theorem) *When a dual geodesic γ^*_{pq} is orthogonal to a primal geodesic γ_{qr} at point q ($\gamma^*_{pq} \perp_q \gamma_{qr}$),*

we have $(\eta(p) - \eta(q))^\top(\theta(r) - \theta(q)) = 0$ and the following divergence identity:

$$D^*_F(p : q) + D^*_F(q : r) = D^*_F(p : r), \tag{7.51}$$

or equivalently

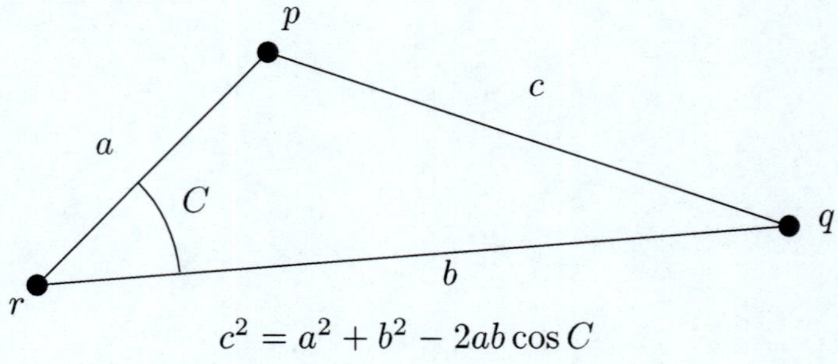

Fig. 7.9 The Euclidean law of cosines

$$D_F(r:q) + D_F(q:p) = D_F(r:p). \tag{7.52}$$

Notice that we can write the Bregman generalized law of cosines of Eq. 7.42 geometrically (i.e., without relying on any prescribed coordinate system) as follows:

$$D_F(p:q) = D_F(p:r) + D_F(r:q) - g_r(\dot{\gamma}_{rp}(0), \dot{\gamma}^*_{rq}(0)) \geq 0, \tag{7.53}$$

$$= D_F(p:r) + D_F(r:q) - \|\dot{\gamma}_{rp}(0)\|_r \, \|\dot{\gamma}^*_{rq}(0)\|_r \cos\left(\alpha_r(\dot{\gamma}_{rp}, \dot{\gamma}^*_{rq})\right). \tag{7.54}$$

It follows that

- when $\alpha_r(\dot{\gamma}_{rp}, \dot{\gamma}^*_{rq}) \in \left[0, \frac{\pi}{2}\right)$ (acute angle), we have $D_F(p:q) < D_F(p:r) + D_F(r:q)$,
- when $\alpha_r(\dot{\gamma}_{rp}, \dot{\gamma}^*_{rq}) = \frac{\pi}{2}$ (right angle), we have $D_F(p:q) = D_F(p:r) + D_F(r:q)$, and
- when $\alpha_r(\dot{\gamma}_{rp}, \dot{\gamma}^*_{rq}) \in \left(\frac{\pi}{2}, \pi\right)$ (obtuse angle), we have $D_F(p:q) > D_F(p:r) + D_F(r:q)$.

Notice that when $F(\theta) = F_{\mathrm{Euc}}(\theta) = \frac{1}{2}\theta^\top\theta$ (Euclidean geometry) and $\theta(r) = 0$, we get

$$\frac{1}{2}\|p-q\|^2 = \frac{1}{2}\|p\|^2 + \frac{1}{2}\|q\|^2 - \|p\|\|q\| \cos\alpha_r(p,q). \tag{7.55}$$

Multiplying by two both sides of Eq. 7.55, we recover the usual law of cosines of Euclidean geometry illustrated in Fig. 7.9 with $a = \|p\|, b = \|q\|, c = \|p-q\|$, and $C = \alpha_r([rp],[rq])$:

$$c^2 = a^2 + b^2 - 2ab\cos C. \tag{7.56}$$

We also have the following identity illustrated in Fig. 7.10:

$$B_F(\theta_1:\theta) + B_F(\theta_2:\theta) = B_F\left(\theta_1 : \frac{\theta_1+\theta_2}{2}\right) + B_F\left(\theta_2 : \frac{\theta_1+\theta_2}{2}\right) + 2B_F\left(\frac{\theta_1+\theta_2}{2} : \theta\right). \tag{7.57}$$

$$B_F(\theta_1 : \theta) + B_F(\theta_2 : \theta) = 2\mathrm{JB}_F(\theta_1, \theta_2) + 2B_F\left(\frac{\theta_1 + \theta_2}{2} : \theta\right), \tag{7.58}$$

$$B_F(\theta_1 : \theta) + B_F(\theta_2 : \theta) = 2J_F(\theta_1, \theta_2) + 2B_F\left(\frac{\theta_1 + \theta_2}{2} : \theta\right), \tag{7.59}$$

where JB_F and J_F denote the *Jensen-Bregman divergence* [21] and the *Jensen divergence* [16], respectively:

$$\mathrm{JB}_F(\theta_1, \theta_2) := \frac{1}{2}\left(B_F\left(\theta_1 : \frac{\theta_1 + \theta_2}{2}\right) + B_F\left(\theta_2 : \frac{\theta_1 + \theta_2}{2}\right)\right), \tag{7.60}$$

$$= \frac{F(\theta_1) + F(\theta_2)}{2} - F\left(\frac{\theta_1 + \theta_2}{2}\right) =: J_F(\theta_1, \theta_2). \tag{7.61}$$

The family of categorical distributions forms a mixture family [22] in information geometry for the Shannon negentropy generator F_{Shannon}, and we have $B_{F_{\text{Shannon}}}(\theta_1 : \theta_2) = \mathrm{KL}(p_{\theta_1} : p_{\theta_2})$. It follows that for any three categorical distributions p, q, and r, we have

$$\mathrm{KL}(p : r) + \mathrm{KL}(q : r) = 2\mathrm{JS}(p, q) + 2\mathrm{KL}\left(\frac{p + q}{2} : r\right), \tag{7.62}$$

since $\frac{p+q}{2}$ is a categorical distribution. In general, the identity does not hold for members of an exponential family (i.e., Gaussian family) since the mixture density $\frac{p+q}{2}$ does not belong to the exponential family. The categorical family is a very special case of family that is *both* an exponential family and a mixture family [3].

Since there exists a bijection between regular exponential families and regular Bregman divergences [4], we can state the parallelogram-type identity of Eq. 7.57 for parametric densities $\{p_\theta\}$ belonging to the same exponential family [18] (with cumulant function F) with respect to both the Kullback–Leibler divergence and the *Bhattacharyya divergence* [16]

$$\mathrm{Bhat}(p, q) := -\log \int \sqrt{p(x)q(x)}\mathrm{d}\mu(x), \tag{7.63}$$

as:

$$\mathrm{KL}(p_\theta : p_{\theta_1}) + \mathrm{KL}(p_\theta : p_{\theta_2}) = 2\mathrm{Bhat}(p_{\theta_1}, p_{\theta_2}) + 2\mathrm{KL}\left(p_{\frac{\theta_1+\theta_2}{2}} : p_\theta\right), \tag{7.64}$$

since $\mathrm{KL}(p_\theta : p_{\theta'}) = B_F(\theta' : \theta)$ and $\mathrm{Bhat}(p_\theta : p_{\theta'}) = J_F(\theta, \theta')$ for densities p_θ and $p_{\theta'}$ belonging to the same exponential family. For example, the identity of Eq. 7.64 applies to any two densities of the Gaussian family.

The 3-parameter property of Bregman divergences is a particular instance of the following 4-*parameter property* [26] (a quadrilateral relation):

Property 2 (*Bregman 4-parameter property*) For any four points p_1, p_2, q_1, q_2, we have the following identity:

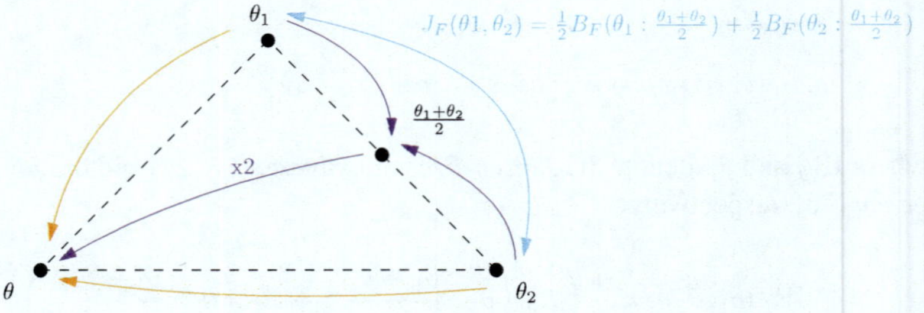

Fig. 7.10 A parallelogram-type identity for Bregman divergences

$$B_F(\theta(p_1) : \theta(q_1)) + B_F(\theta(p_2) : \theta(q_2)) \; -B_F(\theta(p_1) : \theta(q_2)) - B_F(\theta(p_2) : \theta(q_1)) \\ -(\theta(p_2) - \theta(p_1))^\top(\eta(q_1) - \eta(q_2)) = 0. \quad (7.65)$$

Indeed, let $q_1 = p_1$. Then we recove:r

$$B_F(\theta(p_2) : \theta(q_2)) = B_F(\theta(p_1) : \theta(q_2)) + B_F(\theta(p_2) : \theta(q_1)) - (\theta(p_2) - \theta(p_1))^\top(\eta(q_2) - \eta(p_1)), \quad (7.66)$$
$$D_F(p_2 : q_2) = D_F(p_2 : p_1) + D_F(p_1 : q_2) - g_{p_1}(\dot{\gamma}_{p_1 p_2}(0), \dot{\gamma}^*_{p_1 q_2}(0)). \quad (7.67)$$

Similarly, let $q_2 = p_2$. Then we get:

$$B_F(\theta(p_1) : \theta(q_1)) = B_F(\theta(p_1) : \theta(p_2)) + B_F(\theta(p_2) : \theta(q_1)) - (\theta(p_1) - \theta(p_2))^\top(\eta(q_1) - \eta(p_2)), \quad (7.68)$$
$$D_F(p_1 : q_1) = D_F(p_1 : p_2) + D_F(p_2 : q_1) - g_{p_2}(\dot{\gamma}_{p_2 p_1}(0), \dot{\gamma}^*_{p_2 q_1}(0)). \quad (7.69)$$

To derive the 4-parameter identity, we use twice the 3-parameter identity as follows:

$$B_F(\theta_1 : \theta_3) = B_F(\theta_1 : \theta_2) + B_F(\theta_2 : \theta_3) - (\theta_3 - \theta_2)^\top(\nabla F(\theta_1) - \nabla F(\theta_2)), \quad (7.70)$$
$$B_F(\theta_4 : \theta_3) = B_F(\theta_4 : \theta_2) + B_F(\theta_2 : \theta_3) - (\theta_3 - \theta_2)^\top(\nabla F(\theta_4) - \nabla F(\theta_2)). \quad (7.71)$$

Indeed, subtracting Eq. 7.71 from Eq. 7.70, we get

$$B_F(\theta_1 : \theta_3) - B_F(\theta_4 : \theta_3) = B_F(\theta_1 : \theta_2) - B_F(\theta_4 : \theta_2) - (\theta_3 - \theta_2)^\top(\nabla F(\theta_1) - \nabla F(\theta_4)). \quad (7.72)$$

We can interpret geometrically the 4-parameter identity as follows:

$$D_F(p_1 : p_3) - D_F(p_4 : p_3) = D_F(p_1 : p_2) - D_F(p_4 : p_2) - g_{p_2}(\dot{\gamma}_{p_2 p_3}(0), \dot{\gamma}^*_{p_2 p_1}(0)) \\ + g_{p_2}(\dot{\gamma}_{p_2 p_3}(0), , \dot{\gamma}^*_{p_2 p_4}(0)). \quad (7.73)$$

Note that the Bregman 4-parameter identity is said parallelogram-type, because if we choose $F_{\mathrm{Euc}}(\theta) = \frac{1}{2}\theta^\top\theta = \frac{1}{2}\|\theta\|_2^2$, then we have $B_{F_{\mathrm{Euc}}}(\theta : \theta') = \frac{1}{2}\|\theta - \theta'\|^2$, and the 4-parameter identity becomes for $\theta = 0$:

$$\frac{1}{2}\|\theta_1\|^2 + \frac{1}{2}\|\theta_2\|^2 = \left\|\frac{\theta_1 - \theta_2}{2}\right\|^2 + \left\|\frac{\theta_1 + \theta_2}{2}\right\|^2, \tag{7.74}$$

$$2\|\theta_1\|^2 + 2\|\theta_2\|^2 = \|\theta_1 - \theta_2\|^2 + \|\theta_1 + \theta_2\|^2, \tag{7.75}$$

the usual parallelogram identity for the L_2-normed space.

Finally, let us mention some special submanifolds in dually flat spaces: A k-dimensional submanifold $S \subset M$ of a dually flat manifold M is said ∇-autoparallel [3] (∇^*-autoparallel) iff. the sets of points of S can be described by a k-dimensional affine flat in the θ-coordinate system (η-coordinate system, respectively). Primal γ_{pq} geodesics are 1D ∇-autoparallel submanifolds and dual γ_{pq}^* geodesics are 1D ∇^*-autoparallel submanifolds. Notice that the intersection of θ-flats is a θ-flat, but the intersection of a θ-flat with a η-flat is in general neither a θ-flat nor a η-flat.

In a Bregman manifold, we can also express the divergence between two points p and q using the *Fenchel-Young divergence* A_F (also called the canonical divergence of dually flat spaces):

$$D_F(p : q) = A_F(\theta(p) : \eta(q)) := F((\theta(p)) + F^*(\eta(q)) - \theta(p)^\top \eta(q). \tag{7.76}$$

Thus we have

$$\begin{aligned} D_F(p : q) = B_F(\theta(p) : \theta(q)) &= A_F(\theta(p) : \eta(q)) = A_{F^*}(\eta(q) : \theta(p)) \\ &= B_{F*}(\eta(q) : \eta(p)) = D_{F^*}(q : p), \end{aligned} \tag{7.77}$$

and $D_{F^*}(q : p) = D_F{}^*(q : p) = D_F(p : q)$ (where $D_F{}^*$ is the reverse divergence).

7.2.4 Some Examples of Bregman Manifolds

We describe concisely the Mahalanobis manifolds in Sect. 7.2.4.1, the extended Kullback–Leibler manifold in Sect. 7.2.4.2, the Itakura–Saito manifold in Sect. 7.2.4.3, and the multinoulli manifolds (Sect. 7.2.4.4).

7.2.4.1 The Mahalanobis Manifolds

Consider the case where the Bregman generator is defined by

$$F_Q(\theta) = \frac{1}{2}\theta^\top Q\theta, \tag{7.78}$$

for a prescribed symmetric positive-definite $D \times D$ matrix $Q \succ 0$. The Legendre-Fenchel convex conjugate [13, 14] is

$$F^*(\eta) = \frac{1}{2}\eta^\top Q^{-1}\eta = F_{Q^{-1}}(\eta), \tag{7.79}$$

and it follows that

$$\eta(\theta) = \nabla F_Q(\theta) = Q\theta, \quad \theta(\eta) = \nabla F_Q^*(\eta) = Q^{-1}\eta. \tag{7.80}$$

The dual Riemannian metrics are

$$\left[g_{ij}\right] = \nabla^2 F_Q(\theta) = Q, \left[g^{*ij}\right] \tag{7.81}$$

$$= \nabla^2 F_Q^*(\eta) = Q^{-1}. \tag{7.82}$$

The dual geodesics γ_{pq} and γ_{qp}^* in a Mahalanobis manifold coincide. The dual Bregman divergences are squared Mahalanobis distances [12, 23]:

$$B_{F_Q}(\theta_1 : \theta_2) = \frac{1}{2}(\theta_1 - \theta_2)^\top Q(\theta_1 - \theta_2), \tag{7.83}$$

and

$$B_{F_Q^*}(\eta_1 : \eta_2) = \frac{1}{2}(\eta_1 - \eta_2)^\top Q^{-1}(\eta_1 - \eta_2). \tag{7.84}$$

We check that

$$B_{F_Q^*}(\eta_2 : \eta_1) = \frac{1}{2}(Q(\theta_2 - \theta_1))^\top Q^{-1}Q(\theta_2 - \theta_1) = B_{F_Q}(\theta_1 : \theta_2), \tag{7.85}$$

since $Q^\top = Q$. The squared Mahalanobis Bregman divergences are provably the only symmetric Bregman divergences [5]. In particular, when $Q = I$, the identity matrix, the dual Bregman divergences B_F and B_F^* coincide with half of the squared Euclidean distance $D_E(\theta_1, \theta_2) = \frac{1}{2}(\theta_2 - \theta_1)^\top(\theta_2 - \theta_1)$. By using the Cholesky decomposition of Q, i.e., $Q = LL^\top$ where L is a lower triangular matrix with positive diagonal entries, we have $B_{F_Q}(\theta_1 : \theta_2) = B_{F_I}(L^\top\theta_1 : L^\top\theta_2) = \|L^\top(\theta_1 - \theta_2)\|^2$, where I denotes the identity matrix. Notice that the squared Mahalanobis divergences B_{F_Q} are non-separable Bregman divergences whenever Q has non-zero off-diagonal elements.

7.2.4.2 The Kullback–Leibler Manifold

The extended Shannon negative entropy [16]

$$F_{\mathrm{KL}}(\theta) = \sum_{i=1}^{D} \theta^i \log \theta^i - \theta^i \tag{7.86}$$

is a strictly convex and C^3 function on $\Theta = \mathbb{R}^D_{++}$, i.e., a separable Bregman generator. Here, we consider the positive orthant domain instead of the probability simplex hence the name extended Shannon negentropy. We have the following conversion formula between the primal and dual coordinates:

$$\eta(\theta) = \nabla F_{\mathrm{KL}}(\theta) = [\log \theta^i], \quad \theta(\eta) = \nabla F^*_{\mathrm{KL}}(\eta) = [\exp \eta^i]. \tag{7.87}$$

The Legendre convex conjugate is

$$F^*_{\mathrm{KL}}(\eta) = \theta(\eta)^\top \eta - F(\theta(\eta)) = \sum_{i=1}^{D} \exp(\eta^i). \tag{7.88}$$

The dual Riemannian metrics are

$$[g_{ij}] = \nabla^2 F_{\mathrm{KL}}(\theta) = \mathrm{diag}\left(\frac{1}{\theta^1}, \ldots, \frac{1}{\theta^D}\right), \quad [g^{*ij}] = \nabla^2 F^*_{\mathrm{KL}}(\eta) = \mathrm{diag}\left(\exp(\eta^1), \ldots, \exp(\eta^D)\right). \tag{7.89}$$

The dual Bregman divergences are

$$B_{F_{\mathrm{KL}}}(\theta_1 : \theta_2) = \sum_{i=1}^{D} \theta_1^i \log \frac{\theta_1^i}{\theta_2^i} + \theta_2^i - \theta_1^i, \tag{7.90}$$

$$B_{F^*_{\mathrm{KL}}}(\eta_1 : \eta_2) = \sum_{i=1}^{D} \exp(\eta_1^i) - \exp(\eta_2^i) - (\eta_1^i - \eta_2^i)\exp(\eta_2^i) = B_{F_{\mathrm{KL}}}(\theta(\eta_2) : \theta(\eta_1)). \tag{7.91}$$

7.2.4.3 Itakura–Saito Manifold

The D-dimensional Burg information (i.e., Burg negative entropy [16, 20]) is defined by the following separable convex generator:

$$F_{\mathrm{IS}}(\theta) = -\sum_{i=1}^{D} \log(\theta^i). \tag{7.92}$$

We have

$$\eta(\theta) = \nabla F(\theta) = \left[-\frac{1}{\theta^i}\right]_i, \tag{7.93}$$

and the Bregman divergence is called the Itakura–Saito divergence [5]:

$$B_{F_{\mathrm{IS}}}(\theta_1 : \theta_2) = \sum_{i=1}^{D} \frac{\theta_1^i}{\theta_2^i} - \log \frac{\theta_1^i}{\theta_2^i} - 1. \tag{7.94}$$

We convex η-coordinates to θ-coordinates as follows:

$$\theta(\eta) = \left[-\frac{1}{\eta^i}\right]_i, \tag{7.95}$$

and the dual Bregman generator is

$$F_{\mathrm{IS}}^*(\eta) = \eta^\top \theta(\eta) - F_{\mathrm{IS}}(\theta(\eta)) = -D + \sum_{i=1}^{D} \log\left(-\frac{1}{\eta^i}\right) = -D - \sum_{i=1}^{D} \log(-\eta_i), \tag{7.96}$$

where $\eta \in \mathbb{R}_{--}^D$. The dual Bregman divergence is

$$B_{F_{\mathrm{IS}}^*}(\eta_1 : \eta_2) = \sum_{i=1}^{D} \frac{\eta_1^i}{\eta_2^i} - \log \frac{\eta_1^i}{\eta_2^i} - 1. \tag{7.97}$$

We check that

$$B_{F_{\mathrm{IS}}^*}(\eta_2 : \eta_1) = B_{F_{\mathrm{IS}}}(\theta_1 : \theta_2). \tag{7.98}$$

The dual Riemannian metric tensors are

$$[g_{ij}] = \nabla^2 F_{\mathrm{IS}}(\theta) = \mathrm{diag}\left(\frac{1}{\mathrm{sqr}(\theta^1)}, \ldots, \frac{1}{\mathrm{sqr}(\theta^D)}\right), \quad [g^{*ij}] = \nabla^2 F_{\mathrm{IS}}^*(\eta)$$
$$= \mathrm{diag}\left(\frac{1}{\mathrm{sqr}(\eta^1)}, \ldots, \frac{1}{\mathrm{sqr}(\eta^D)}\right). \tag{7.99}$$

7.2.4.4 The Multinoulli Manifolds

We can build a dually flat space from any strictly convex and C^3 convex function F [15] which also defines a Bregman generator. In particular, we can use the *log-normalizer* (also called cumulant function or log-partition function) of a regular exponential family [4] as such a Bregman generator [19]. Let us consider the multinoulli family (i.e., the multinomial family for a single trial also called the family of categorical distributions). The probability of a multinoulli distribution with d categories $c_1, \ldots, c_d$ such that $\Pr(x = c_i) = \lambda_i$ is:

$$\Pr(x) = \lambda_1^{x_1} \times \cdots \times \lambda_d^{x_d}, \tag{7.100}$$

with $x_i \in \{0, 1\}$ and $\sum_{i=1}^d x_i = 1$. Let us write the multinoulli probability mass function in the canonical form of an exponential family as

$$\lambda_1^{x_1} \times \cdots \times \lambda_d^{x_d} = \exp\left(\sum_{i=1}^{d} x_i \log \lambda_i\right), \tag{7.101}$$

$$= \exp\left(\sum_{i=1}^{d-1} x_i \log \lambda_i + \underbrace{(1 - \sum_{i=1}^{d-1} x_i)}_{x_d} \log \lambda_d\right), \quad (7.102)$$

$$= \exp\left(\sum_{i=1}^{d-1} x_i \theta_i - F(\theta)\right), \quad (7.103)$$

with the natural parameter $\theta_i = \log \frac{\lambda_i}{\lambda_d}$ for $i \in \{1, \ldots, d-1\}$. The multinoulli distribution is an exponential family of order $D = d - 1$. The natural parameter space Θ is $\mathbb{R}^D = \mathbb{R}^{d-1}$. We can convert back the natural parameter θ to the original parameters λ as follows:

$$\lambda(\theta) = \begin{cases} \lambda_i = \frac{\exp(\theta_i)}{1+\sum_{i=1}^{D} \exp(\theta_i)}, & \text{for} i \in \{1, \ldots, D\} \\ \lambda_d = \frac{1}{1+\sum_{i=1}^{D} \exp(\theta_i)}. \end{cases} \quad (7.104)$$

The log-normalizer is

$$F_{\mathrm{Mult}}(\theta) = -\log \lambda_d = \log\left(1 + \sum_{i=1}^{D} \exp(\theta_i)\right). \quad (7.105)$$

The gradient of the log-normalizer is

$$\eta(\theta) = \nabla F_{\mathrm{Mult}}(\theta) = \left[\frac{\exp(\theta_i)}{1 + \sum_{i=1}^{D} \exp(\theta_i)}\right]_i, \quad (7.106)$$

and the reciprocal gradient is

$$\theta(\eta) = \nabla G_{\mathrm{Mult}}(\eta) = \left[\log \frac{\eta_i}{1 - \sum_{i=1}^{D} \eta_i}\right]_i. \quad (7.107)$$

The convex conjugate of F is

$$G_{\mathrm{Mult}}(\eta) = \eta^\top \theta(\eta) - F(\theta(\eta)) = \sum_i \eta_i \log \eta_i + \left(1 - \sum_i \eta_i\right) \log\left(1 - \sum_i \eta_i\right). \quad (7.108)$$

It follows that the Riemannian metric and dual Riemannian metric are Hessians of the potential functions $F_{\mathrm{Mult}}(\theta)$ and $G_{\mathrm{Mult}}(\eta)$, respectively:

$$[\nabla^2 F_{\mathrm{Mult}}(\theta)]_{ij} = \begin{cases} -\frac{\exp(\theta_i+\theta_j)}{(1+\sum_i \exp(\theta_i))^2} & \text{if} i \neq j, \\ \frac{\exp(\theta_i)}{(1+\sum_i \exp(\theta_i))} - \frac{\exp(2\theta_i)}{(1+\sum_i \exp(\theta_i))^2} & \text{if} i = j \end{cases} \quad (7.109)$$

$$\nabla^2 G_{\text{Mult}}(\eta) = \frac{1}{1-\sum_i \eta_i} 1_M + \text{diag}\left(\frac{1}{\eta_1}, \ldots, \frac{1}{\eta_D}\right) = \begin{cases} \frac{1}{1-\sum_i \eta_i} & \text{if} i \neq j, \\ \frac{1}{1-\sum_i \eta_i} + \frac{1}{\eta_i} & \text{if} i = j \end{cases}, \tag{7.110}$$

where 1_M denotes the $D \times D$-matrix with all entries equal to 1, and $\text{diag}(x_1, \ldots, x_D)$ the diagonal matrix with diagonal elements $x_1, \ldots, x_D$.

The dual Bregman divergences are

$$B_{F_{\text{Mult}}}(\theta_1 : \theta_2) = \text{KL}^*(p_{\lambda(\theta_1)} : p_{\lambda(\theta_2)}) = \text{KL}(p_{\lambda(\theta_2)} : p_{\lambda(\theta_1)}), \tag{7.111}$$

$$B_{G_{\text{Mult}}}(\eta_1 : \eta_2) = \text{KL}(p_{\lambda(\eta_1)} : p_{\lambda(\eta_2)}), \tag{7.112}$$

where KL is the discrete Kullback–Leibler divergence and KL^* the reverse Kullback–Leibler divergence. We convert the expectation parameter η to the original parameter λ as follows:

$$\lambda(\eta) = \begin{cases} \lambda_i = \eta_i, & \text{for} i \in \{1, \ldots, D\} \\ \lambda_d = 1 - \sum_{i=1}^D \eta_i. \end{cases} \tag{7.113}$$

In general, the Bregman divergence induced by log-normalizer of an exponential family amounts to a reverse Kullback–Leibler divergence [19]. Notice that the non-separable log-normalizer of a multinomial family yields a nice example of a non-separable Bregman divergence. The multinoulli manifold is called the *Bernoulli manifold* when $D = 1$ (i.e., $d = 2$), and the trinoulli manifold when $D = 2$ (i.e., $d = 3$).

7.3 Geodesic ∇-Triangles with One, Two, or Three Interior Right Angles

7.3.1 Geodesic ∇-Triangles with One Interior Right Angle

To build a geodesic ∇-triangle with a right angle, fix two points p and q (i.e., the first two triangle vertices), and consider the location of the third triangle vertex point r such that $\gamma_{qr} \perp_q \gamma_{qp}$ (i.e., $\dot{\gamma}_{qr}(0) \perp_q \dot{\gamma}_{qp}(0)$). We end up with the following linear equation which defines a θ-flat H_q^θ:

$$H_q^\theta : \theta_r^\top \nabla^2 F(\theta_q)\theta_{pq} = \theta_q^\top \nabla^2 F(\theta_q)\theta_{pq}. \tag{7.114}$$

By restricting the solution $r \in H_q^\theta$ to the manifold M, we get:

Proposition 1 *The locii of points r of a ∇-triangle that form a right angle at q is $H_q^\theta \cap M$.*

Figure 7.11 displays that a ∇-triangle with one right angle $\alpha_q(r, p) = \frac{\pi}{2}$ in the Itakura–Saito manifold: $\theta(p) = (1.2885253880864789, 3.4136709176658546)$,

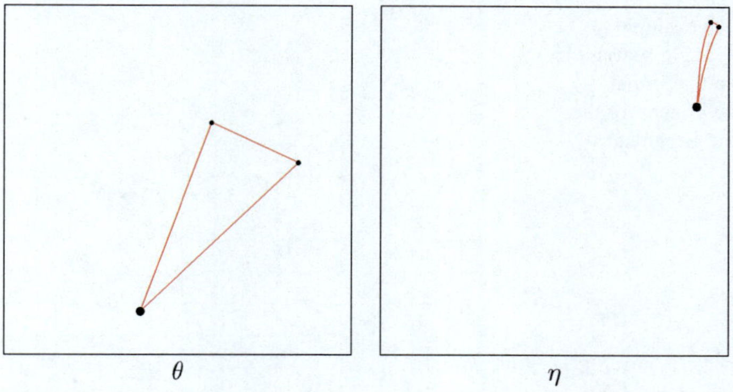

Fig. 7.11 An example of a geodesic ∇-triangle with a right angle at q: (thicker vertex is r)

$\theta(q) = (4.9336774965526065, 1.656631440605195)$,
$\theta(r) = (3.5399193730133236, 4.6263857851449846)$.
The interior angles of the ∇-triangle are $\alpha_p(q,r) = 1.8276508176456936$,
$\alpha_q(p,r) = 1.5707963267948966$
and $\alpha_r(p,q) = 1.1542328404967954$.
That total sum of the interior angles are 4.552679984937385 radians (equivalent to about 260.8°).

7.3.2 *Geodesic ∇-Triangles with Two Interior Right Angles*

We now report the construction of two right angle ∇-triangles. That is, geodesic triangles with all primal geodesic edges (i.e., ∇-triangle), with both the right angles $\alpha_p(q,r) = 90^o$ and $\alpha_q(p,r) = 90^o$. Figure 7.12 displays such a double right angle ∇-triangle for the Burg negentropy generator $F_{\mathrm{IS}}(\theta)$ yielding the Itakura–Saito divergence. Observe that because the metric tensor field g is not a scalar function of the Euclidean metric tensor g_{Euc}, the Itakura–Saito Bregman geometry is not conformal (see Fig. 7.5).

Now, fix two points p and q, and let us seek for the third point r of the ∇-triangle $\gamma_{pq}\gamma_{qr}\gamma_{rp}$ such that it holds both *simultaneously* that (i) $\gamma_{qr} \perp_q \gamma_{qp}$ (i.e., $\dot\gamma_{qr}(0) \perp_q \dot\gamma_{qp}(0)$) and (ii) $\gamma_{pq} \perp_p \gamma_{pr}$ (i.e., $\dot\gamma_{pq}(0) \perp_p \dot\gamma_{pr}(0)$). We end up with the following system of equations:

$$\begin{cases} \theta_r^\top \nabla^2 F(\theta_q)\theta_{pq} = \theta_q^\top \nabla^2 F(\theta_q)\theta_{pq}, \\ \theta_r^\top \nabla^2 F(\theta_p)\theta_{pq} = \theta_p^\top \nabla^2 F(\theta_p)\theta_{pq}. \end{cases} \tag{7.115}$$

It is a linear system $A\theta = b$ with

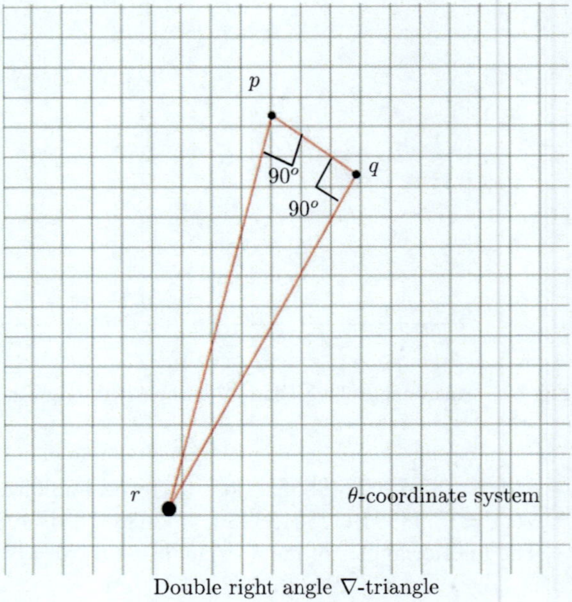

Fig. 7.12 An example of double right angle ∇-triangle visualized in the primal θ-coordinate system for the Itakura–Saito Bregman manifold

$$A = [a_{ij}] = \begin{bmatrix} \nabla^2 F(\theta_q)\theta_{pq} \\ \nabla^2 F(\theta_p)\theta_{pq} \end{bmatrix}, \quad b = [b_i] = \begin{bmatrix} \theta_q^\top \nabla^2 F(\theta_q)\theta_{pq} \\ \theta_p^\top \nabla^2 F(\theta_p)\theta_{pq} \end{bmatrix}. \tag{7.116}$$

When $\nabla^2 F(\theta) = Q \succ 0$ for a fixed positive-definite matrix and $p \neq q$, the system does not admit any solution (i.e., case of squared Mahalanobis distances which generalize the squared Euclidean distance and can have at most one right angle).

Otherwise, this linear system solves *uniquely* for asymmetric Bregman divergences [5] using Cramer's rule as

$$\theta_r^1 = \frac{\begin{vmatrix} b_1 & a_{12} \\ b_2 & a_{22} \end{vmatrix}}{|A|}, \quad \theta_r^2 = \frac{\begin{vmatrix} a_{11} & b_1 \\ a_{21} & b_2 \end{vmatrix}}{|A|}, \tag{7.117}$$

where $|\cdot|$ denotes the matrix determinant.

Similarly, we can build ∇^*-triangles with two right angles by exchanging $F \Leftrightarrow F^*$. Notice that having two right angles in a non-degenerate triangle makes it necessarily having an angle excess.

Figure 7.13 displays the following two double right angle ∇-triangles (up to machine numerical precision) obtained for the following settings:

- $\theta(p) = (1.7372662352145616, 1.148396070619242)$,
  $\theta(q) = (1.241571556333764, 1.3768479188317202)$,
  $\theta(r) = (1.614143828700357, 1.8451358255393877)$,
  $\alpha_p(q, r) = 90.00000000000001$,

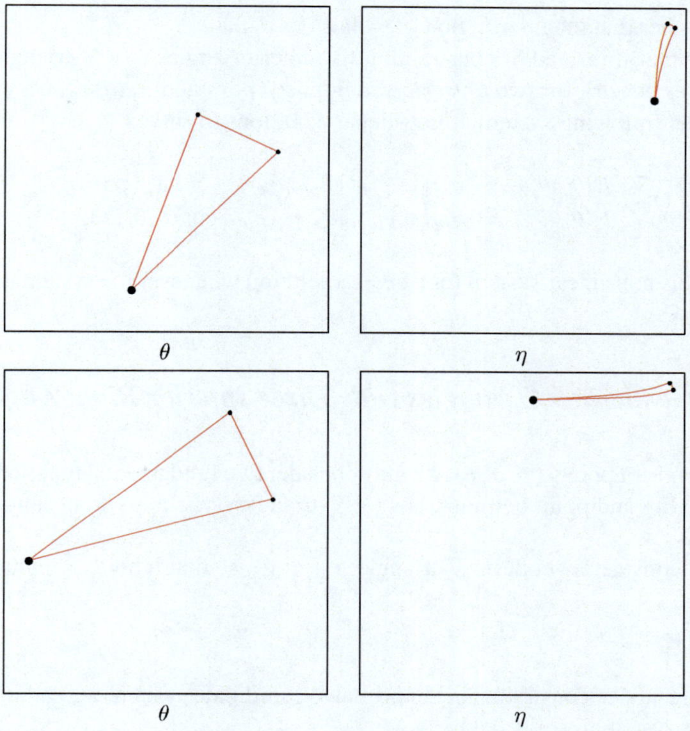

Fig. 7.13 Two examples of ∇-triangles with two interior right angles (the opposite angles to the thicker vertex r)

$\alpha_q(p, r) = 90.0,$
$\alpha_r(p, q) = 12.82764159141668.$

- $\theta(p) = (1.7128340504770114, 1.2510418358297621),$
  $\theta(q) = (1.446857135939727, 1.7930125176801988),$
  $\theta(r) = (1.1177842396781703, 1.5922051785236535),$
  $\alpha_p(q, r) = 90.00000000000001,$
  $\alpha_q(p, r) = 90.00000000000001,$
  $\alpha_r(p, q) = 6.595093466701274.$

Define the following two θ-flat submanifolds:

$$H_q^\theta : \theta_r^\top \nabla^2 F(\theta_q)\theta_{pq} - \theta_q^\top \nabla^2 F(\theta_q)\theta_{pq} = 0, \tag{7.118}$$
$$H_p^\theta : \theta_r^\top \nabla^2 F(\theta_p)\theta_{pq} - \theta_p^\top \nabla^2 F(\theta_p)\theta_{pq} = 0. \tag{7.119}$$

We end up with the following proposition:

Proposition 2 *In a non-Mahalanobis Bregman manifold M, the locii of points r that form a double right angle with the geodesic arc γ_{pq} is $H_q^\theta \cap H_q^\theta$.*

Remember that the intersection of θ-flats is a θ-flat.

Notice that if instead of constraining two interior angles of a ∇-triangle to have right angles, we ask for two *prescribed angles* $\alpha_q(p,r)$ and $\alpha_p(q,r)$, we would end up with the following system of non-linear equations to solve:

$$\begin{cases} \theta_{rq}^\top \nabla^2 F(\theta_q)\theta_{pq} &= \|\theta_{rq}\|_{\nabla^2 F(\theta_q)} \|\theta_{pq}\|_{\nabla^2 F(\theta_q)} \cos\alpha_q(p,r), \\ \theta_{rq}^\top \nabla^2 F(\theta_p)\theta_{pq} &= \|\theta_{rq}|_{\nabla^2 F(\theta_p)} \|\theta_{pq}|_{\nabla^2 F(\theta_p)} \cos\alpha_p(q,r). \end{cases} \tag{7.120}$$

Solving this non-linear system for r gives a solution whenever the system is feasible.

7.3.3 Geodesic ∇-Triangles with Three Interior Right Angles

Given a triple of points (p,q,r), we may consider two fundamental types of triangles (up to duality and point permutations): a ∇-triangle (type ppp) or a triangle of type pdp.

Let us consider a geodesic ∇-triangle $\gamma_{pq}\gamma_{qr}\gamma_{rp}$ so that it holds simultaneously:

$$\gamma_{pq} \perp_p \gamma_{pr}, \quad \gamma_{qr} \perp_q \gamma_{qp}, \quad \gamma_{rp} \perp_r \gamma_{rq}. \tag{7.121}$$

Writing the above constraints in the primal θ-coordinate system, we end up with the following system to solve:

$$\begin{cases} \theta_{qp}^\top \nabla^2 F(\theta_p)\theta_{rp} &= 0 \\ \theta_{rq}^\top \nabla^2 F(\theta_q)\theta_{pq} &= 0 \\ \theta_{pr}^\top \nabla^2 F(\theta_r)\theta_{qr} &= 0. \end{cases} \tag{7.122}$$

Because of the Hessian matrices, this yields in general a non-linear system of equations to solve. The set of feasible solutions define the ∇-triangles with three right angles. In dimension D, we have 3D unknown (the D θ-coordinates of the points p, q, and r) for 3 constraints. That is, the system is underconstrained.

For the 2D Itakura–Saito manifold, the Hessian matrix at p is $\mathrm{diag}\left(\frac{1}{\mathrm{sqr}(\theta_x)}, \frac{1}{\mathrm{sqr}(\theta_y)}\right)$. Thus we get the following system to solve:

$$(\theta_q^x - \theta_p^x)\frac{1}{\mathrm{sqr}(\theta_p^x)}(\theta_r^x - \theta_p^x) + (\theta_q^y - \theta_p^y)\frac{1}{\mathrm{sqr}(\theta_p^y)}(\theta_r^y - \theta_p^y) = 0, \tag{7.123}$$

$$(\theta_r^x - \theta_q^x)\frac{1}{\mathrm{sqr}(\theta_q^x)}(\theta_p^x - \theta_q^x) + (\theta_r^y - \theta_q^y)\frac{1}{\mathrm{sqr}(\theta_q^y)}(\theta_p^y - \theta_q^y) = 0, \tag{7.124}$$

$$(\theta_p^x - \theta_r^x)\frac{1}{\mathrm{sqr}(\theta_r^x)}(\theta_q^x - \theta_r^x) + (\theta_p^y - \theta_r^y)\frac{1}{\mathrm{sqr}(\theta_r^y)}(\theta_q^y - \theta_r^y) = 0. \tag{7.125}$$

Multiplying Eq. 7.123 by $\mathrm{sqr}(\theta_p^x)\mathrm{sqr}(\theta_p^y)$, Eq. 7.124 by $\mathrm{sqr}(\theta_q^x)\mathrm{sqr}(\theta_q^y)$, and Eq. 7.125 by $\mathrm{sqr}(\theta_r^x)\mathrm{sqr}(\theta_r^y)$, we get a system of polynomial equations to solve [30]:

$$(\theta_q^x - \theta_p^x)\mathrm{sqr}(\theta_p^y)(\theta_r^x - \theta_p^x) + (\theta_q^y - \theta_p^y)\mathrm{sqr}(\theta_p^x)(\theta_r^y - \theta_p^y) = 0, \quad (7.126)$$
$$(\theta_r^x - \theta_q^x)\mathrm{sqr}(\theta_q^y)(\theta_p^x - \theta_q^x) + (\theta_r^y - \theta_q^y)\mathrm{sqr}(\theta_q^x)(\theta_p^y - \theta_q^y) = 0, \quad (7.127)$$
$$(\theta_p^x - \theta_r^x)\mathrm{sqr}(\theta_r^y)(\theta_q^x - \theta_r^x) + (\theta_p^y - \theta_r^y)\mathrm{sqr}(\theta_r^x)(\theta_q^y - \theta_r^y) = 0, \quad (7.128)$$

where $\mathrm{sqr}(x) = x^2$. The system of polynomial equations has 3 equations with 6 positive variables defining 3 pairs of distinct points (θ_p^x, θ_p^y), (θ_q^x, θ_q^y), and (θ_r^x, θ_r^y).

When considering the extended Kullback–Leibler manifold, since the Hessian matrix at p is $\mathrm{diag}\left(\frac{1}{\theta_p^x}, \frac{1}{\theta_p^y}\right)$, we get the following system of polynomial equations (after multiplying the first equation by $\theta_p^x\theta_p^y$, the second equation by $\theta_q^x\theta_q^y$ and the third equation by $\theta_r^x\theta_r^y$):

$$\begin{cases} (\theta_q^x - \theta_p^x)\theta_p^y(\theta_r^x - \theta_p^x) + (\theta_q^y - \theta_p^y)\theta_p^x(\theta_r^y - \theta_p^y) = 0, \\ (\theta_r^x - \theta_q^x)\theta_q^y(\theta_p^x - \theta_q^x) + (\theta_r^y - \theta_q^y)\theta_q^x(\theta_p^y - \theta_q^y) = 0, \\ (\theta_p^x - \theta_r^x)\theta_r^y(\theta_q^x - \theta_r^x) + (\theta_p^y - \theta_r^y)\theta_r^x(\theta_q^y - \theta_r^y) = 0. \end{cases} \quad (7.129)$$

Fixing three variables, we get a cubic system of three equations in three unknowns.

We used Wolfram alpha™ to check for potential real solution(s) of the system:

```
a>0, b>0, c>a, x>0, y>c,z>0,
(c-a)*b*(y-a)+(x-b)*a*(z-b)=0,
(y-c)*x*(a-c)+(z-x)*c*(b-x)=0,
(a-y)*z*(c-y)+(b-z)*y*(x-z)=0
```

The system does not admit solution in the positive orthant. We also checked the feasibility of the system for the Itakura–Saito manifold:

```
a>0, b>0, c>a, x>0, y>c,z>0,
(c-a)*b*b*(y-a)+(x-b)*a*a*(z-b)=0,
(y-c)*x*x*(a-c)+(z-x)*c*c*(b-x)=0,
(a-y)*z*z*(c-y)+(b-z)*y*y*(x-z)=0.
```

The system does not admit solution in the positive orthant.

Thus it is an ongoing task to report an example of such a ∇-triangle with three right angles for an asymmetric Bregman divergence, or to prove that such a triangle can never exist.

7.4 Simultaneous Satisfying the Dual Pythagorean Theorems

Fix two points p and q of the Bregman manifold M. We seek for the locii of the third point $r \in M$ such that we have both $\gamma_{pq} \perp_q \gamma^*_{qr}$ and $\gamma^*_{pq} \perp_q \gamma_{qr}$. That is, we need to solve the following system of equations:

$$\begin{cases} (\eta(p) - \eta(q))^\top(\theta(r) - \theta(q)) = 0, \\ (\eta(p) - \eta(q))^\top(\theta(r) - \theta(q)) = 0. \end{cases} \tag{7.130}$$

Notice that finding such triples of points allow one to construct dual geodesic right-angle triangles. When $F(\theta) = \frac{1}{2}\theta^\top Q\theta$ for $Q \succ 0$ a positive-definite matrix, we have B_F that is a squared Mahalanobis distances, and the primal and dual geodesics coincide. Therefore any right-angle triangle is a dually right-angle solution to the problem in Mahalanobis manifolds. Thus we shall consider *asymmetric* Bregman divergences in the remainder since the only symmetric Bregman divergences are squared Mahalanobis distances [5].

7.4.1 Simultaneous Dual Orthogonality: A Geometric Interpretation

The constraint $\gamma_{pq} \perp_q \gamma^*_{qr}$ is equivalent to $(\theta(p) - \theta(q))^\top(\eta(r) - \eta(q)) = 0$. Let $\theta_{pq} = \theta(p) - \theta(q)$, $\eta_r = \eta(r)$ and $\eta_q = \eta(q)$. Then we have the following affine equation in η_r:

$$H^\eta(p, q) : \theta_{pq}^\top \eta_r - \theta_{pq}^\top \eta_q = 0. \tag{7.131}$$

The locii r of points satisfying the above equation is a $(D-1)$-dimensional ∇^*-autoparallel submanifold $H^\eta(p, q)$ (i.e., a $(D-1)$-dimensional η-*flat* [3] or loosely speaking a "∇^*-hyperplane").

Similarly, the constraint $\gamma^*_{pq} \perp_q \gamma_{qr}$ is equivalent to $(\eta(p) - \eta(q))^\top(\theta(r) - \theta(q)) = 0$. Let $\eta_{pq} = \eta(p) - \eta(q)$, $\theta_r = \theta(r)$ and $\theta_q = \theta(q)$. Then we have the following affine equation in θ_r:

$$H^\theta(p, q) : \eta_{pq}^\top \theta_r - \eta_{pq}^\top \theta_q = 0. \tag{7.132}$$

The locii r of points satisfying the above equation is a $(D-1)$-dimensional ∇-autoparallel submanifold $H^\theta(p, q)$ (i.e., a $(D-1)$-dimensional θ-*flat* [3] or ∇-hyperplane).

Notice that point q ought to belong to both $H^\eta(p, q)$ and $H^\theta(p, q)$. Thus to simultaneously satisfy the two dual geodesic orthogonality constraints, the point r should belong to the intersection of a η-flat with a θ-flat:

$$S(p,q) = H^\eta(p,q) \cap H^\theta(p,q). \tag{7.133}$$

We should make sure $S(p,q)$ belongs to the manifold M when solving the equations using either the θ- or η-coordinate system.

Proposition 3 *The locii of points r such that $\gamma_{pq} \perp_q \gamma^*_{qr}$ and $\gamma^*_{pq} \perp_q \gamma_{qr}$ is the intersection of a θ-flat with a η-flat restricted to the manifold: $S(p,q) = H^\eta(p,q) \cap H^\theta(p,q)$.*

In general the intersection of a θ-flat with a η-flat is neither a θ-flat nor a η-flat. In 2D, the submanifolds $H^\eta(p,q)$ and $H^\theta(p,q)$ can be interpreted as a dual geodesic and a primal geodesic, respectively. Thus in 2D, S_{pq} may not be simply connected, and the maximum number of points of the intersection of two geodesics upper bounds the number of solutions for r. The next section illustrates how to build such triples of points for the Itakura–Saito manifold.

7.4.2 *Explicit Construction in the Itakura–Saito Manifold*

Consider the Itakura–Saito manifold described in Sect. 7.2.4.3. Let us exhibit some triple of points (p,q,r) such that $\gamma_{pq} \perp_q \gamma^*_{qr}$ and $\gamma^*_{pq} \perp_q \gamma_{qr}$. To avoid confusion, let us write the 2D θ- and η-coordinates of p by θ^x_p and θ^y_p, and η^x_p and η^y_p.

The second orthogonality constraint equation yields the equation $\eta_{pq}^\top(\theta_r - \theta_q) = \eta^x_{pq}\theta^x_r + \eta^y_{pq}\theta^y_r - \eta_{pq}^\top\theta_q = 0$ which can be rewritten as:

$$\theta^y_r = -\frac{\eta^x_{pq}}{\eta^y_{pq}}\theta^x_r + \frac{\eta_{pq}^\top\theta_q}{\eta^y_{pq}} = a\theta^x_r + b, \tag{7.134}$$

with $a = -\frac{\eta^x_{pq}}{\eta^y_{pq}}$ and $b = \frac{\eta_{pq}^\top\theta_q}{\eta^y_{pq}}$.

The first orthogonality constraint equation yields $(\theta_{pq})^\top(\eta_r - \eta_q)$ with $\eta^x_r = -\frac{1}{\theta^x_r}$ and $\eta^y_r = -\frac{1}{\theta^y_r}$. After multiplying both sides of the equation by $\theta^x_r\theta^y_r$, we find

$$-\theta^x_{pq}\theta^y_r - \theta^y_{pq}\theta^x_r - \theta_{pq}^\top\eta_q\theta^x_r\theta^y_r = 0 \tag{7.135}$$

Letting $\theta^y_r = a\theta^x_r + b$, we get the following quadratic equation to solve:

$$-\theta^x_{pq}(a\theta^x_r + b) - \theta^y_{pq}\theta^x_r - \theta_{pq}^\top\eta_q\theta^x_r(a\theta^x_r + b) = 0. \tag{7.136}$$

We write the former equation into the following canonical form of quadratic equations:

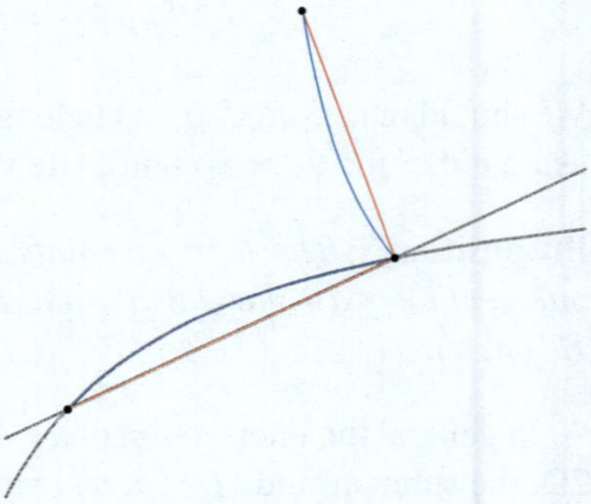

Fig. 7.14 The intersection of $H^\eta(p,q)$ and $H^\theta(p,q)$ (displayed in black) yields two points: point q and the solution point r

$$\underbrace{(-\theta_{pq}^\top \eta_q a)}_{A}(\theta_r^x)^2 + \underbrace{(-\theta_{pq}^x a - \theta_{pq}^y - b\theta_{pq}^\top \eta_q)}_{B}\theta_r^x + \underbrace{(-\theta_{pq}^x b)}_{C} = 0. \quad (7.137)$$

Then we solve the quadratic equation, and get the two solutions: Let $\Delta = B^2 - 4AC > 0$ be the discriminant. We have the two quadratic roots: $\theta_r^{x\prime} = \frac{-B-\sqrt{\Delta}}{2A}$ and $\theta_r^{x\prime\prime} = \frac{-B+\sqrt{\Delta}}{2A}$, and we recover $\theta_r^{y\prime} = a\theta_r^{x\prime} + b$ and $\theta_r^{y\prime\prime} = a\theta_r^{x\prime\prime} + b$. One of the two solutions r' or r'' coincide with point q (with coordinate $\theta(q)$), so the solution point r is the remaining distinct point.

Figure 7.14 displays an example of a triple (p, q, r) with doubly right-angle at q with the pair of constraints $H^\eta(p,q)$ and $H^\eta(p,q)$ passing through point q.

Let us give a numerical example. Set
$\theta(p) = (\theta_p^x, \theta_p^y) = (0.7273955397832663, 0.3279475469672596)$,
$\theta(q) = (\theta_q^x, \theta_q^y) = (0.46251884248040354, 0.3902872167636309)$. Then we solve the quadratic equation and find the two solutions
$\theta(r) = (0.3065847355580658, 0.13822426240588664)$
and
$\theta(r'') = (0.4625188424804033, 0.39028721676363043)$.

Observe that $r'' = q$ so that the solution is $r = r'$. The triple (p, q, r') holds simultaneously the dual Pythagorean theorems at point q.

Figure 7.15 displays three examples of triples of points for which the dual Pythagorean theorems hold simultaneously. The triples of points displayed in Fig. 7.15 are from left to right:

1. $\theta(p) = (0.9704854205553236, 1.4760141668100146)$,
   $\theta(q) = (1.141690604206171, 0.43035569351200803)$ and
   $\theta(r) = (0.22647618241885010.34444830042268043)$.
2. $\theta(p) = (1.3163859900481611, 1.965380252548788)$,
   $\theta(q) = (1.5136826962585432, 1.2440688670072433)$ and
   $\theta(r) = (0.6359397574807304, 0.9494657726625966)$.
3. $\theta(p) = (0.9511702030611633, 1.291145089053253)$,
   $\theta(q) = (0.3277859642409383, 1.906447912395776)$ and
   $\theta(r) = (0.1077217190919158, 0.14622448026891943)$.

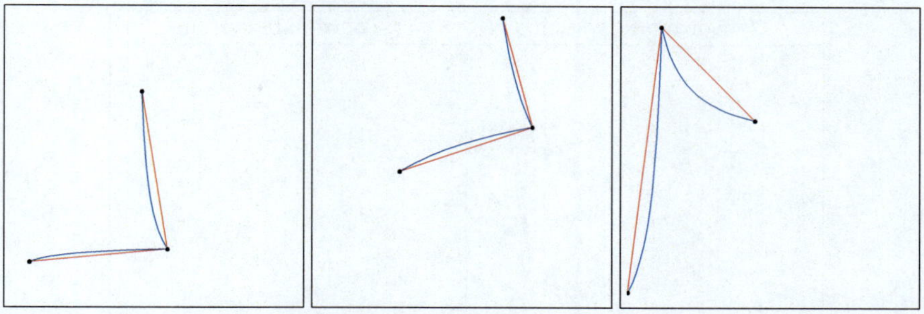

Fig. 7.15 Three examples of triples of points (p, q, r) visualized in the θ-coordinate system for which the dual Pythagorean theorems hold simultaneously at q for the Burg negentropy generator. That is, we have both $\gamma_{pq} \perp_q \gamma^*_{qr}$ and $\gamma^*_{pq} \perp_q \gamma_{qr}$: The two pairs of (red, blue) geodesics form a right-angle at q: A "doubly right-angle"

From these two pairs of dual right angle geodesic arcs at q, we can obtain four geodesic triangles by choosing either the primal or dual geodesic edge for the triangle edge pq: Namely, $\gamma_{pq}\gamma_{qr}\gamma^*_{rp}$, $\gamma^*_{pq}\gamma_{qr}\gamma^*_{rp}$ and $\gamma_{pq}\gamma^*_{qr}\gamma_{rp}$, $\gamma^*_{pq}\gamma^*_{qr}\gamma_{rp}$. These four triangles can be grouped into two dual pairs of dual geodesic triangles which exhibit a *dually right angle* at vertex q: $(\gamma_{pq}\gamma_{qr}\gamma^*_{rp}, \gamma^*_{pq}\gamma^*_{qr}\gamma_{rp})$ and $(\gamma^*_{pq}\gamma_{qr}\gamma^*_{rp}, \gamma_{pq}\gamma^*_{qr}\gamma_{rp})$.

Similarly, solving the dual orthogonality constraint at q for the cubic generator $F(\theta) = \frac{1}{3}\sum_i \theta_i^3$ yields a quadratic equation to solve. However, when considering the extended Shannon negentropy generator $F(\theta) = \sum_{i=1}^{D} \theta^i \log \theta^i - \theta^i$, we get a nonlinear equation (with sum of logarithmic terms) to solve: In 1D, we can easily run numerical optimization to approximate a solution numerically.

7.5 Conclusion

In a dually flat space [3] that we called a Bregman manifold in Sect. 7.2, geodesic triangles can either have angle excesses or angle defects like in arbitrary Riemannian geometry when the manifold is non self-dual (i.e., not of type Mahalanobis). First, we explained in Sect. 7.3 how to build geodesic ∇-triangles with one, two or three right angles provided that the corresponding system of equations is feasible. The system of equations is linear up to two right angles but non-linear when dealing with three right angles. Second, we showed how to build triple of points (p, q, r) such that the dual Pythagorean theorems hold simultaneously at point q yielding a dually right angle at q: two dual pairs of right-angle dual geodesics. It turned out that the locii of such points r for given points p and q is the intersection of a η-flat with a θ-flat. We reported the explicit construction of such triples for the Itakura–Saito manifold in Sect. 7.4.2.

In future work, we shall consider dually flat spaces for symmetric positive-definite matrices [2, 17, 24] where the inner product is the trace of a matrix product. We

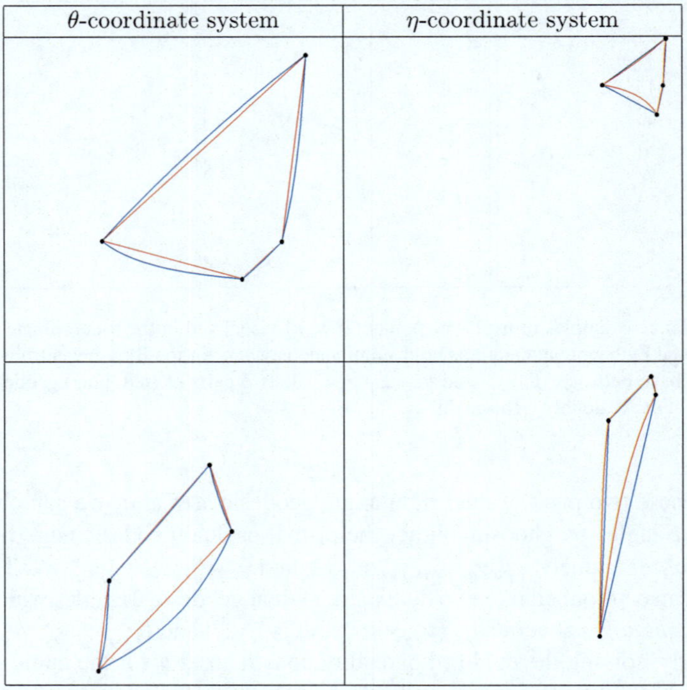

Fig. 7.16 Examples of convex quadrangles in the Itakura–Saito manifold

would also like to prove the following experimental observation: For the 2D Burg negentropy generator, the total sum $\alpha(T)$ of the interior angles of a geodesic ∇-triangle (geodesic triangle with all primal edges) plus the total sum $\beta = \alpha(T^*)$ of the interior angles of a dual geodesic ∇^*-triangle (geodesic triangle with all dual geodesic edges) sum up to 2π. For example, for $\theta(p) = (0.5, 0.5)$, $\theta(q) = (0.75, 0.75)$ and $\theta(r) = (0.95, 0.25)$, we find that $\alpha(T) = 160.19318300825412^o$ (angle defect), $\beta = \alpha(T^*) = 199.80681699174588^o$ (angle excess), and $\alpha + \beta = 360.0^o$. This property seems only to hold for the 2D Itakura–Saito divergence and not in higher dimensions.

We shall also consider an extension of this work to study properties of geodesic convex k-gons instead of geodesic triangles (i.e., 3-gons) in dually flat spaces (2^k such geodesic k-gons). For example, the Lambert quadrilaterals [27] (i.e., 4-gons) have three right angles and the remaining angle which is acute in hyperbolic geometry, obtuse in spherical geometry, and a right angle in Euclidean geometry. In a dually flat space, we have $2^4 = 16$ types of quadrilaterals defining overall $4 \times 4 = 16$ interior angles.[6] Fig. 7.16 displays all pairs of dual geodesics of some convex quadrilaterals in the θ- and η-coordinate systems.

Acknowledgements Figures were programmed using processing.org

[6]A each quadrilateral vertex, we have 4 geodesics defining 6 interior angles between them.

Disclaimer: Views and opinions expressed are those of the authors and do not necessarily represent official positions of their respective companies.

A. Notations

F Strictly convex and C^3 real-valued function
F^* Dual Legendre-Fenchel convex conjugate
$\theta(p) = (\theta^1(p), \ldots, \theta^D(p))$ primal coordinates of point p
$\eta(p) = (\eta_1(p), \ldots, \eta_D(p))$ dual coordinates of point p
$\theta_{ab} = \theta_a - \theta_b$ notational shortcut
$\eta_{ab} = \eta_a - \eta_b$ notational shortcut
$D_F(p : q)$ Divergence between points
$B_F(\theta(p) : \theta(q))$ Bregman divergence
$A_F(\theta(p) : \eta(q))$ Fenchel-Young divergence
(M, g, ∇, ∇^*) Dually flat space (Bregman manifold)
T_p Tangent plane at p
$g_p(u, v)$ inner product between two vectors u and v of T_p
$[g_{ij}$ $=[g(e_i, e_j)]_{ij} = \nabla^{2F}(\theta)]$ Riemannian metric
$[g^{*ij}$ $=[g^*(e^{*i}, e^{*j})]_{ij} = \nabla^2 F^*(\eta)]$ dual Riemannian metric
$\prod_{p,q}(v)$ primal parallel transport of $v \in T_p$ to T_q
$\prod^*_{p,q}(v)$ dual parallel transport of $v \in T_p$ to T_q
$\gamma_{ab}(t)$ Primal geodesic: $\theta(\gamma_{ab}(t)) = (1-t)\theta(a) + t\theta(b)$
$\gamma_{ab}(t)^*$ Dual geodesic: $\eta(\gamma^*_{ab}(t)) = (1-t)\eta(a) + t\eta(b)$
$(v)_B$ vector components in basis B, arranged in a D-tuple
$[v$ $_B]$ vector components in basis B, arranged in a D-dimensional column vector
$B_p = \{e_i = \partial_i = \frac{\partial}{\partial\theta^i}\}$ natural basis at T_p
$B^*_p = \{e^{*i} = \partial^i = \frac{\partial}{\partial\eta_i}\}_i$ reciprocal basis at T_p so that $g(e_i, e^{*j}) = \delta_i^j$
$v_{ab} = \frac{d}{dt}\gamma_{ab}(0) = \dot{\gamma}_{ab}(0)$ tangent vector of $\gamma_{ab}(t)$ at a with contravariant components $\theta(b) - \theta(a)$
$v^*_{ab} = \frac{d}{dt}\gamma^*_{ab}(0) = \dot{\gamma}^*_{ab}(0)$ tangent vector of $\gamma^*_{ab}(t)$ at a with covariant components $\eta(b) - \eta(a)$
$[v^i$ $_B]$ contravariant components of vector v, $v^i = g(v, e^{*i})$
$[v_i$ $_E]$ covariant components of vector v (meaning $[v]_{B^*}$), $v_i = g(v, e_i)$
$g_p(u, v)$ inner product at T_p of two vectors: $g_p(u, v) = u_i v^i = u^i v_i$.

References

1. Akaho, S., Hino, H., Murata, N.: On a convergence property of a geometrical algorithm for statistical manifolds (2019). arXiv:1909.12644

2. Amari, S.: Information geometry of positive measures and positive-definite matrices: decomposable dually flat structure. Entropy **16**(4), 2131–2145 (2014)
3. Amari, S.-I.: Information Geometry and its Applications, vol. 194. Springer, Berlin (2016)
4. Banerjee, A., Merugu, S., Dhillon, I.S., Ghosh, J.: Clustering with Bregman divergences. J. Mach. Learn. Res. **6**, 1705–1749 (2005)
5. Boissonnat, J.-D., Nielsen, F., Nock, R.: Bregman Voronoi diagrams. Discret. Comput. Geom. **44**(2), 281–307 (2010)
6. Crouzeix, J.-P.: A relationship between the second derivatives of a convex function and of its conjugate. Math. Program. **13**(1), 364–365 (1977)
7. Edelsbrunner, H., Wagner, H.: Topological data analysis with Bregman divergences. In: 33rd International Symposium on Computational Geometry (SoCG 2017). Schloss Dagstuhl-Leibniz-Zentrum fuer Informatik (2017)
8. Eguchi, S., et al.: Geometry of minimum contrast. Hiroshima Math. J. **22**(3), 631–647 (1992)
9. Eisenhart, L.P.: Riemannian Geometry. Princeton University Press, Princeton (1997)
10. Kurose, T.: Dual connections and affine geometry. Math. Z. **203**(1), 115–121 (1990)
11. Lauritzen, S.L.: Statistical manifolds. Differ. Geom. Stat. Inference **10**, 163–216 (1987)
12. Mahalanobis, P.C.: On the generalized distance in statistics. Proc. Natl. Inst. Sci. (Calcutta) **2**, 49–55 (1936)
13. Nielsen, F.: Legendre transformation and information geometry (2010)
14. Nielsen, F.: Cramér-Rao lower bound and information geometry. In: Connected at Infinity II, pp. 18–37. Springer, Berlin (2013)
15. Nielsen, F.: An elementary introduction to information geometry (2018). arXiv:1808.08271
16. Nielsen, F., Boltz, S.: The Burbea-Rao and Bhattacharyya centroids. IEEE Trans. Inf. Theory **57**(8), 5455–5466 (2011)
17. Nielsen, F., Bhatia, R.: Matrix Information Geometry. Springer, Berlin (2013)
18. Nielsen, F., Garcia, V. Statistical exponential families: a digest with flash cards (2009). arXiv:0911.4863
19. Nielsen, F., Hadjeres, G.: Monte Carlo information geometry: the dually flat case (2018). arXiv:1803.07225
20. Nielsen, F., Nock, R.: Sided and symmetrized Bregman centroids. IEEE Trans. Inf. Theory **55**(6), 2882–2904 (2009)
21. Nielsen, F., Nock, R. Skew Jensen-Bregman Voronoi diagrams. In: Transactions on Computational Science, vol. XIV, pp. 102–128. Springer, Berlin (2011)
22. Nielsen, F., Nock, R.: On the geometry of mixtures of prescribed distributions. In: IEEE International Conference on Acoustics, Speech and Signal Processing (ICASSP), pp. 2861–2865. IEEE (2018)
23. Nielsen, F., Muzellec, B., Nock, R.: Classification with mixtures of curved Mahalanobis metrics. In: 2016 IEEE International Conference on Image Processing (ICIP), pp. 241–245. IEEE (2016)
24. Nock, R., Magdalou, B., Briys, E., Nielsen, F.: Mining matrix data with Bregman matrix divergences for portfolio selection. In: Matrix Information Geometry, pp. 373–402. Springer, Berlin (2013)
25. Norden, A.P.: On pairs of conjugate parallel displacements in multidimensional spaces. Dokl. Akad. Nauk SSSR **49**(9), 1345–1347 (1945)
26. Pietra, S.F., Pietra, V.D., Lafferty, J.: Inducing features of random fields. IEEE Trans. Pattern Anal. Mach. Intell. **19**(4), 380–393 (1997)
27. Rosenfeld, B.A.: A History of Non-Euclidean Geometry: Evolution of the Concept of a Geometric Space, vol. 12. Springer Science & Business Media (2012)
28. Sen, R.N.: Parallel displacement and scalar product of vectors. In: Proceedings of the National Institute of Sciences of India, vol. 14, p. 45. National Institute of Sciences of India (1948)
29. Shima, H.: The Geometry of Hessian Structures. World Scientific (2007)
30. Sturmfels, B.: Solving Systems of Polynomial Equations, vol. 97. American Mathematical Society (2002)
31. Wolf, J.A.: Spaces of Constant Curvature, vol. 372. American Mathematical Society (1972)
32. Wong, T.-K.L.: Logarithmic divergences from optimal transport and Rényi geometry. Inf. Geom. **1**(1), 39–78 (2018)

Chapter 8
Chain Rule Optimal Transport

Frank Nielsen and Ke Sun

Abstract We define a novel class of distances between statistical multivariate distributions by modeling an optimal transport problem on their marginals with respect to a ground distance defined on their conditionals. These new distances are metrics whenever the ground distance between the marginals is a metric, generalize both the Wasserstein distances between discrete measures and a recently introduced metric distance between statistical mixtures, and provide an upper bound for jointly convex distances between statistical mixtures. By entropic regularization of the optimal transport, we obtain a fast differentiable Sinkhorn-type distance. We experimentally evaluate our new family of distances by quantifying the upper bounds of several jointly convex distances between statistical mixtures, and by proposing a novel efficient method to learn Gaussian mixture models (GMMs) by simplifying kernel density estimators with respect to our distance. Our GMM learning technique experimentally improves significantly over the EM implementation of `sklearn` on the `MNIST` and `Fashion MNIST` datasets.

Keywords Optimal transport · Wasserstein distances · Information geometry · f-divergences · Total Variation and Jensen–Shannon divergence · Bregman divergence · Rényi divergence · Statistical mixtures · Joint convexity

8.1 Introduction and Motivation

Calculating dissimilarities between statistical mixtures is a fundamental operation met in statistics, machine learning, signal processing, and information fusion [6] among others. Minimizing the information-theoretic Kullback–Leibler divergence

F. Nielsen (✉)
Sony Computer Science Laboratories, Tokyo, Japan
e-mail: Frank.Nielsen@acm.org

K. Sun
CSIRO's Data61, Sydney, Australia
e-mail: sunk@ieee.org

F. Nielsen (ed.), *Progress in Information Geometry*,
Signals and Communication Technology,
https://doi.org/10.1007/978-3-030-65459-7_8

(KLD also called relative entropy) between parametric models yields practical learning machines. However, the KLD or in general the Csiszár's f-divergences between statistical mixtures [44] do not admit closed-form formula, and needs in practice to be approximated by costly Monte Carlo stochastic integration. To tackle this computational tractability problem, two research directions have been considered in the literature: ① propose some new distances between mixtures that yield closed-form formula [36, 38] (e.g., the Cauchy-Schwarz divergence, the Jensen quadratic Rényi divergence, the statistical Minkowski distances). ② lower and/or upper bound the f-divergences between mixtures [15, 44]. However, this direction is tricky when considering bounded divergences like the Total Variation (TV) distance or the Jensen–Shannon (JS) divergence that are upper bounded by 1 and log 2, respectively, or when considering high-dimensional mixtures.

When dealing with probability densities, two main classes of statistical distances have been widely studied in the literature: ① The invariant f-divergences of Information Geometry ([2]; IG) characterized as the class of separable distances which are information monotone (i.e., satisfies the partition inequality [63]), and ② The Optimal Transport (OT)/Wasserstein/EMD distance [34, 56] which can be computationally accelerated using entropy regularization [8, 17] (i.e., the Sinkhorn divergence).

In general, computing closed-form formula for the OT between parametric distributions is difficult except in 1D [51]. A closed-form formula is known for elliptical distributions [13] for the 2-Wasserstein metric (including the multivariate Gaussian distributions), and the OT of multivariate continuous distributions can be calculated from the OT of their copulas [22].

The geometry related to these OT/IG distances are different. For example, consider univariate location-scale families (or multivariate elliptical distributions): For OT, the 2-Wasserstein distance between any two members admit the *same* closed-form formula [13, 21] (depending only on the mean and variance parameters, and not on the type of location-scale family). The OT geometry of Gaussian distributions has *positive curvature* [20, 60]. For any smooth f-divergence, the information-geometric manifold has *negative curvature* ([30]; hyperbolic geometry).

In this chapter, we first generalize the work of [33] that proposed a novel family of statistical distances between statistical mixtures (that we term MCOTs, standing for *Mixture Component Optimal Transports*) by solving linear programs between mixture component weights where the elementary distance between any two mixture components is prescribed. Then we propose to learn Gaussian mixture models (GMMs) by simplifying kernel density estimators (KDEs) using our distance.

We describe our main contributions as follows:

- We define the generic *Chain Rule Optimal Transport* (CROT) distance in Definition 8.1, and prove that the CROT distance is a metric whenever the distance between conditional distributions is a metric in Property 8.2. The CROT distance unifies and extends the Wasserstein distances and the MCOT distance [33] between statistical mixtures.
- We report a novel generic upper bound for statistical distances between marginal distributions [45] in Sect. 8.3 (Theorem 8.8) whenever the ground distance is jointly

convex, and introduce its relaxed Sinkhorn distance (SCROT) for fast estimation. Numerical experiments in Sect. 8.4 highlight quantitatively the upper bound performance of the (S)CROT distances for bounding the total variation distance, the Wasserstein W_p metric, and the Rényi α-divergences.

- We design a novel learning algorithm for GMMs by simplifying KDEs with respect to SCROT that yields in that case a closed-form formula (Eq. 8.15) in Sect. 8.5, and demonstrate experimentally better results than the Expectation-Maximization (EM) implementation [11] in `sklearn` [49] on MNIST [32] and Fashion MNIST [64] datasets.

8.2 Chain Rule Optimal Transport

Recall the basic *chain rule* factorization of a joint probability distribution:

$$p(x, y) = p(y)\, p(x|y),$$

where probability $p(y)$ is the *marginal probability*, and probability $p(x|y)$ is the *conditional probability*. Given $p(y)$ and $p(x|y)$ in certain families of simple probability distributions, one can get a density model through marginalization:

$$p(x) = \int p(x, y)\mathrm{d}y.$$

For example, for latent models like statistical mixtures or hidden Markov models [58, 65], x plays the role of the *observed variable* while y denotes the *hidden variable* [16] (unobserved so that inference has to tackle incomplete data, say, using the EM algorithm [11]. Let $\mathcal{X} = \{p(x)\}$ and $\mathcal{Y} = \{p(y)\}$ denote the manifolds of marginal probability densities; let $\mathcal{C} = \{p(x|y)\}$ denote the manifold of conditional probability density. We state the generic definition of the *Chain Rule Optimal Transport* (CROT) distance between the distributions $p(x)$ and $q(x)$ (with $q(x) = \int q(y)q(x|y)\mathrm{d}y$) as follows:

Definition 8.1 (CROT distance) Given two multivariate distributions $p(x, y)$ and $q(x, y)$, we define the *Chain Rule Optimal Transport* as follows:

$$H_D(p, q) := \inf_{r \in \Gamma(p(y), q(z))} E_{r(y,z)} \left[D\Big(p(x|y), q(x|z) \Big) \right], \tag{8.1}$$

where $D(\cdot, \cdot)$ is a ground distance defined on conditional density manifold $\mathcal{C} = \{p(x|y)\}$ (e.g., the Total Variation), $\Gamma(p(y), q(z))$ is the set of all probability measures on $\mathcal{Y}^2$ satisfying the constraints $\int r(y, z)\mathrm{d}z = p(y)$ and $\int r(y, z)\mathrm{d}y = q(z)$, and $E_{r(y,z)}$ denotes the expectation with respect to $r(y, z)$.

When the ground distance D is clear from the context, we write $H(p, q)$ for a shortcut of $H_D(p, q)$. A similar definition was introduced by [55] termed "Markov construction." In our work, the CROT is defined with respect to a distance metric on the manifold $\mathcal{C}$ of conditional densities (information-geometric distance) rather than a section of the distance metric on the space of (x, y).

A key property of CROT is stated as follows:

Property 8.2 (Metric properties) *If $D(\cdot, \cdot)$ is a metric on $\mathcal{C}$, then $H_D(p, q)$ is a metric on $\mathcal{X}$ and a pseudometric on $\mathcal{Y} \times \mathcal{C}$.*

The proof is given in Property 8.2. Notice that H_D is a metric on $\mathcal{X}$ but only a pseudometric (satisfying non-negativity, symmetry, triangle inequality, and $H_D(p, p) = 0$, $\forall p \in \mathcal{Y} \times \mathcal{C}$ instead of the law of indiscernibles of metrics) on the product manifold $\mathcal{Y} \times \mathcal{C}$.

Since $\int r(y, z)\mathrm{d}y\mathrm{d}z = 1$ and since $r(y, z) = p(y)q(z)$ is a feasible transport solution, we get the following upper bounds:

Property 8.3 (Upper bounds)

$$H_D(p, q) \leq \int_y \int_z p(y)q(z)D\Big(p(x|y), q(x|z)\Big)\mathrm{d}y\mathrm{d}z$$
$$\leq \max_{y,z} D\Big(p(x|y), q(x|z)\Big). \tag{8.2}$$

The CROT distances unify and generalize two distances met in the literature:

Remark 8.4 (*CROT generalizes Wasserstein/EMD*) In the case that $p(x|y) = \delta(x - y)$ (Dirac distributions), we recover the Wasserstein distance [60] between point sets (or Earth Mover Distance, EMD; [54]), where $D(\cdot, \cdot)$ is the ground metric distance. Note that point sets can be interpreted as discrete probability measures.

The Wasserstein distance W_p (for $p \geq 1$, with W_1 introduced by [62]) follows from the Kantorovich's [26, 27] relaxation framework of Monge's [34] original optimal mass transport formulation.

Remark 8.5 (*CROT generalizes MCOT*) When both $p(y)$ and $q(z)$ are both (finite) categorical distributions, we recover the distance formerly defined by [33] that we termed the MCOT distance.

CROT is a nontrivial generalization of both the Wasserstein distance and the MCOT, because CROT gives a *flexible* definition on the OT. Given a joint distribution $p(x_1, \ldots, x_n)$, one can consider a *family* of distances, depending on how the random variables $x_1, \ldots, x_n$ split, and how the ground distances D are selected. For example, one can define D to be CROT and we have a nested CROT distance. In the simplest case, let

$$D\left(p(x|y), q(x|z)\right) = \inf_{r'\in\Gamma(p(x|y),q(x'|z))} E_{r'(x,x')}\|(x,y)-(x',z)\|_p, \tag{8.3}$$

then $H_D(p,q)$ becomes a "two-stage optimal transport"

$$H_D(p,q) = \inf_{r\in\Gamma(p(y),q(z))} E_{r(y,z)} \inf_{r'\in\Gamma(p(x|y),q(x'|z))} E_{r'(x,x')}\|(x,y)-(x',z)\|_p, \tag{8.4}$$

We have the following fundamental monotonicity:

Theorem 8.6 *If $D(\cdot,\cdot)$ is given by Eq. 8.3, then we have:*

$$H_D(p,q) \geq \inf_{r\in\Gamma(p(x,y),q(x',z))} E_r\|(x,y)-(x',z)\|_p.$$

The above theorem is true if $\|(x,y)-(x',z)\|_p$ in Eq. 8.3 and the RHS is replaced by any other metric distance. Therefore, through the chain rule factorization of a joint distribution, CROT can give a potentially simpler expression of optimal transport, and its hierarchical structure allows one to use 1D OT problems [4, 9] which enjoys a closed-form solution [51] based on the inverse of the CDFs of the univariate densities:

$$H_D(X,Y) = \left(\int_0^1 c_D(F_X^{-1}(u) - F_Y^{-1}(u))\mathrm{d}u\right),$$

where F_X and F_Y are the cumulative distribution functions (CDFs) of X and Y, respectively, and $D(x,y) := c_D(x-y)$ for a convex and continuous function C_D. Observe that the CROT distance is larger than the optimal transport distance.

Interestingly, the CROT distance provides an upper bound on the marginal distance $D(p(x), q(x))$ provided the base distance D is *jointly convex* [3, 52].

Definition 8.7 (*Jointly convex distance*) A distance $D(\cdot:\cdot)$ on a statistical manifold $\mathcal{M}$ is jointly convex if and only if

$$D((1-\alpha)p_1+\alpha p_2 : (1-\alpha)q_1+\alpha q_2) \leq (1-\alpha)D(p_1:p_2)+\alpha D(p_2:q_2),$$
$$\forall\alpha\in[0,1],\ p_1,p_2\in\mathcal{M}.$$

We write the above inequality more compactly as

$$D((p_1p_2)_\alpha : (q_1q_2)_\alpha) \leq (D(p_1:p_2)D(p_2:q_2))_\alpha, \quad \forall\alpha\in[0,1],$$

where $(ab)_\alpha := (1-\alpha)a+\alpha b$.

Theorem 8.8 (Upper Bound on Jointly Convex Distance, UBJCD) *Given a pair of joint distributions $p(x,y)$ and $q(x,y)$, if $D(\cdot,\cdot)$ is jointly convex, then $D(p(x), q(x)) \leq H_D(p,q)$.*

Notice that $H_D(p, q) \neq H_D(q, p)$ for an asymmetric base distance D.

Let us give some examples of jointly convex distances: ① The f-divergences [47] $I_f(p : q) = \int p(x) f(q(x)/p(x)) \mathrm{d}x$ (for a convex generator $f(u)$ satisfying $f(1) = 0$ and strictly convex at 1); ② The p-powered Wasserstein distances [48] W_p^p; ③ The Rényi divergences [61] for $\alpha \in [0, 1]$; ④ Bregman divergences [5, Exercises 2.3.29 and 2.3.30] provided that the generator F satisfies $\nabla^2 F(y) + \nabla^3 F(y)(y - x) \succeq (\nabla^2 F(x) \nabla^2 F)^{-1}(y)$ where $\succeq$ denotes the Löwner ordering of positive-definite matrices. ⑤ A generalized divergence related to Tsallis divergence [63].

A jointly convex function is separately convex but the converse is false. However, a separately convex bivariate function that is positively homogeneous of degree one is jointly convex (but this result does not hold in higher dimensions; [10]). Conversely, CROT yields a lower bound for jointly concave distances (e.g., fidelity in quantum computing; [46]).

8.3 SCROT: Fast Sinkhorn CROT

Consider two finite statistical mixtures $m_1(x) = \sum_{i=1}^{k_1} \alpha_i p_i(x)$ and $m_2(x) = \sum_{i=1}^{k_2} \beta_i q_i(x)$, not necessarily homogeneous nor of the same type. Let $[k] := \{1, \ldots, k\}$. The MCOT distance proposed by [33] amounts to solve a *Linear Program* (LP) problem.

By defining $U(\alpha, \beta)$ to be set of non-negative matrices $W = [w_{ij}]$ with $\sum_{l=1}^{k_2} w_{il} = \alpha_i$ and $\sum_{l=1}^{k_1} w_{lj} = \beta_j$ (transport polytope; [8]), we get the equivalent compact definition of MCOT (that is a special case of CROT):

$$H_D(m_1, m_2) = \min_{W \in U(\alpha,\beta)} \sum_{i=1}^{k_1} \sum_{j=1}^{k_2} w_{ij} D(p_i, q_j). \tag{8.5}$$

In general, the LP problem (with $k_1 \times k_2$ variables and inequalities, $k_1 + k_2$ equalities whom $k_1 + k_2 - 1$ are independent) delivers an optimal soft assignment of mixture components with exactly $k_1 + k_2 - 1$ nonzero coefficients[1] in matrix $W = [w_{ij}]$. The complexity of linear programming [31] in n variables with b bits using Karmarkar's interior point methods is polynomial, in $O(n^{\frac{7}{2}} b^2)$.

Observe that we necessarily have: $\max_{j \in [k_2]} w_{ij} \geq \frac{\alpha_i}{k_2}$, and similarly that: $\max_{i \in [k_1]} w_{ij} \geq \frac{\beta_j}{k_1}$. Note that $H(m, m) = 0$ since $w_{ij} = D_{ij}$ where D_{ij} denotes the Krönecker symbol: $D_{ij} = 1$ iff $i = j$, and 0 otherwise. We can interpret MCOT as a *Discrete Optimal Transport* (DOT) between (non-embedded) histograms. When $k_1 = k_2 = d$, the transport polytope is the polyhedral set of non-negative $d \times d$ matrices:

$$U(\alpha, \beta) = \{P \in \mathbb{R}_+^{d \times d} : P 1_d = \alpha, P^\top 1_d = \beta\},$$

[1] A LP in d-dimensions has its solution located at a vertex of a polytope, described by the intersection of $d + 1$ hyperplanes (linear constraints).

and

$$H_D(m_1 : m_2) = \min_{P \in U(\alpha,\beta)} \langle P, W \rangle,$$

where $\langle A, B \rangle = \mathrm{tr}(A^\top B)$ is the Fröbenius inner product of matrices, and $\mathrm{tr}(A)$ the matrix trace. This OT can be calculated using the network simplex in $O(d^3 \log d)$ time. Cuturi [8] showed how to relax the objective function in order to get fast calculation using the Sinkhorn divergence:

$$S_D(m_1 : m_2) = \min_{P \in U_\lambda(\alpha,\beta)} \langle P, W \rangle, \tag{8.6}$$

where

$$U_\lambda(\alpha, \beta) := \{P \in U(\alpha, \beta) : \mathrm{KL}(P : \alpha\beta^\top) \leq \lambda\}.$$

The KLD between two $k \times k$ matrices $M = [m_{i,j}]$ and $M' = [m'_{i,j}]$ is defined by

$$\mathrm{KL}(M : M') := \sum_{i,j} m_{i,j} \log \frac{m_{i,j}}{m'_{i,j}},$$

with the convention that $0 \log \frac{0}{0} = 0$. The Sinkhorn divergence is calculated using the equivalent dual Sinkhorn divergence by using matrix scaling algorithms (e.g., the Sinkhorn–Knopp algorithm). Because the minimization is performed on $U_\lambda(\alpha, \beta) \subset U(\alpha, \beta)$, we have

$$H_D(m_1, m_2) \leq S_D(m_1, m_2).$$

Notice that the smooth (dual) Sinkhorn divergence has also been shown experimentally to improve over the EMD in applications (MNIST classification; [8]).

8.3.1 CROT Upper Bounds on Distance Between Statistical Mixtures

First, let us report the basic upper bounds for MCOT mentioned earlier in Property 8.3. The objective function is upper bounded by:

$$H(m_1, m_2) \leq \sum_{i=1}^{k_1} \sum_{j=1}^{k_2} \alpha_i \beta_j D(p_i, q_j) \leq \max_{i \in [k_1], j \in [k_2]} D(p_i, q_j). \tag{8.7}$$

Now, when the conditional density distance D is *separate convex* (i.e., meaning convex in both arguments), we get the following *Separate Convexity Upper Bound*:

$$\text{(SCUB)} \quad D(m_1 : m_2) \leq \sum_{i=1}^{k_1} \sum_{j=1}^{k_2} \alpha_i \beta_j D(p_i : q_j). \tag{8.8}$$

For example, norm-induced distances or f-divergences [42] are separate convex distances. For the particular case of the KLD, we have: $\mathrm{KL}(p : q) := \int p(x) \log \frac{p(x)}{q(x)} \mathrm{d}x$, and when $k_1 = k_2$, we get the following upper bound using the log-sum inequality [12, 43]:

$$\mathrm{KL}(m_1 : m_2) \leq \mathrm{KL}(\alpha : \beta) + \sum_{i=1}^{k} \alpha_i \mathrm{KL}(p_i : q_i), \tag{8.9}$$

Since this holds for *any* permutation of σ of mixture components, we can tight this upper bound by minimizing over all permutations σ:

$$\mathrm{KL}(m_1 : m_2) \leq \min_{\sigma} \mathrm{KL}(\alpha : \sigma(\beta)) + \sum_{i=1}^{k} \alpha_i \mathrm{KL}(p_i : \sigma(q_i)). \tag{8.10}$$

The best permutation σ can be computed using the Hungarian algorithm [23, 24, 53, 59] in cubic time (with cost matrix $C = [c_{ij}]$, and $c_{ij} = \mathrm{kl}(\alpha_i : \beta_j) + \alpha_i \mathrm{KL}(p_i : q_j)$ with $\mathrm{kl}(a : b) = a \log \frac{a}{b}$).

Now, let us further rewrite

$$m_1(x) = \sum_{i=1}^{k_1} \sum_{j=1}^{k_2} w_{i,j} p_i(x)$$

with $\sum_{j=1}^{k_2} w_{i,j} = \alpha_i$, and

$$m_2(x) = \sum_{i=1}^{k_1} \sum_{j=1}^{k_2} w'_{i,j} q_j(x)$$

with $\sum_{i=1}^{k_1} w'_{i,j} = \beta_j$. That is, we can interpret

$$m_1(x) = \sum_{i=1}^{k_1} \sum_{j=1}^{k_2} w_{i,j} p_{i,j}(x)$$

and

$$m_2(x) = \sum_{i=1}^{k_1} \sum_{j=1}^{k_2} w'_{i,j} q_{i,j}(x)$$

as mixtures of $k = k_1 \times k_2$ (redundant) components $\{p_{i,j}(x) = p_i(x)\}$ and $\{q_{i,j}(x) = q_j(x)\}$, and apply the upper bound of Eq. 8.9 for the "best split" of matching mixture components $\sum_{j=1}^{k_2} w_{i,j} p_i(x) \leftrightarrow \sum_{j=1}^{k_1} w'_{j,i} q_i(x)$:

$$\mathrm{KL}(m_1 : m_2) \leq O(m_1 : m_2) \leq \sum_{i=1}^{k_1}\sum_{j=1}^{k_2} w_{i,j} \log\left(\frac{w_{i,j}}{w'_{j,i}}\right) + H_{\mathrm{KL}}(m_1, m_2),$$

where

$$O(m_1 : m_2) = \min_{w \in U(\alpha,\beta)} \sum_{i=1}^{k_1}\sum_{j=1}^{k_2} w_{i,j} \log\left(\frac{w_{i,j}}{w'_{j,i}}\right) + \sum_{i=1}^{k_1}\sum_{j=1}^{k_2} w_{ij} \mathrm{KL}(p_i : q_j). \tag{8.11}$$

Thus CROT allows to upper bound the KLD between mixtures. The technique of rewriting mixtures as mixtures of $k = k_1 \times k_2$ redundant components bears some resemblance with the variational upper bound on the KL divergence between mixtures proposed by [25] that requires to iterate until convergence an update of the variational upper bound. See also [7] for another recent work further pushing that research direction and discussing displacement interpolation and barycenter calculations for Gaussian Mixture Models (GMMs). We note that this framework also applies to or semi-parametric mixtures obtained from Kernel Density Estimators (KDEs; [57]).

8.4 Experiments

We study experimentally the tightness of the CROT upper bound H_D and SCROT upper bound S_D on D between GMMs for the total variation (Sect. 8.4.1), Wasserstein W_p (Sect. 8.4.2) and Rényi distances (Sect. 8.4.3). In Sect. 8.5 we shall further demonstrate how to learn GMMs by minimizing the SCROT distance.

8.4.1 Total Variation Distance

Since TV is a metric f-divergence [28] bounded in [0, 1], so is MCOT. The closed-form formula for the total variation between univariate Gaussian distributions is reported by [37] using the erf function, and the other formula for the total variation between Rayleigh distributions and Gamma distributions are given in [45].

Figure 8.1 illustrates the performances of the various lower/upper bounds on the total variation between mixtures of Gaussian, Gamma, and Rayleigh distributions with respect to the true value which is estimated using Monte Carlo samplings (consistent estimations).

The acronyms of the various bounds are as follows: CELB: Combinatorial Envelope Lower Bound ([44]; applies only for 1D mixtures); CEUB: Combinatorial Envelope Upper Bound ([44]; applies only for 1D mixtures); CGQLB: Coarse-Grained Quantization Lower Bound [44] for 1000 bins (applies only for f-divergences that satisfy the information monotonicity property); CROT: Chain Rule Optimal Transport H_D (this paper); Sinkhorn CROT: Entropy-regularized CROT [8] $S_D \leq H_D$, with $\lambda = 1$ and $\epsilon = 10^{-8}$ (for convergence of the Sinkhorn–Knopp iterative matrix scaling algorithm).

Next, we consider the renown MNIST handwritten digit database [32] of 70,000 handwritten digit 28×28 grey images and the Fashion-MNIST images with exactly the same sample size and dimensions but different image contents [64]. We first use PCA to reduce the original dimensionality $d = 28 \times 28 = 784$ to $D \in \{10, 50\}$. Then we extract two subsets of samples, and estimate respectively two GMMs composed of 10 multivariate Gaussian distributions with a diagonal covariance matrix. The GMMs are learned by the Expectation-Maximization (EM) algorithm implementation of scikit-learn [50]. Notice that we did not use the labels in our estimation, and therefore the mixture components do not necessarily correspond to different digits.

We approximate the TV between D-dimensional GMMs using Monte Carlo by performing stochastic integration of the following integrals:

$$\mathrm{TV}(p, q) := \frac{1}{2}\int |p(x) - q(x)|\mathrm{d}x =$$
$$\frac{1}{2m}\sum_{x_i \sim p(x)} \frac{1 - \exp(r(x_i))}{1 + \exp(r(x_i))} + \frac{1}{2m}\sum_{y_i \sim q(x)} \frac{1 - \exp(r(y_i))}{1 + \exp(r(y_i))},$$

where $\{x_i\}_{i=1}^m$ and $\{y_i\}_{i=1}^m$ are i.i.d. samples drawn from $p(x)$ and $q(x)$, respectively, and $r(x) = |\log p(x) - \log q(x)|$. In our experiments, we set $m = 0.5 \times 10^4$.

To compute the CROT, we use the EMD and Sinkhorn implementations provided by the Python Optimal Transport, POT, library [18]. For Sinkhorn, we set the entropy regularization strength as follows: Sinkhorn (1) means median(M) and Sinkhorn (10) means median$(M)/10$, where M is the metric cost matrix. For example, to compute CROT-TV, M is the pairwise TV distance matrix from all components in the first mixture model to all components in the second mixture. The maximum number of Sinkhorn iterations is 1000, with a stop threshold of 10^{-10}.

To get some intuitions, see Fig. 8.2 for the cost matrix and the corresponding optimal transport matrix, where the cost is defined by TV distance, and the dataset is PCA-processed MNIST. It shows the 10×10 TV distance matrix between mm1's (mixture model) components and mm2's components. Red means a large distance; blue means a small distance. We see that the transportation scheme tries to assign higher weights to small cost pairs (blue region in the cost matrix).

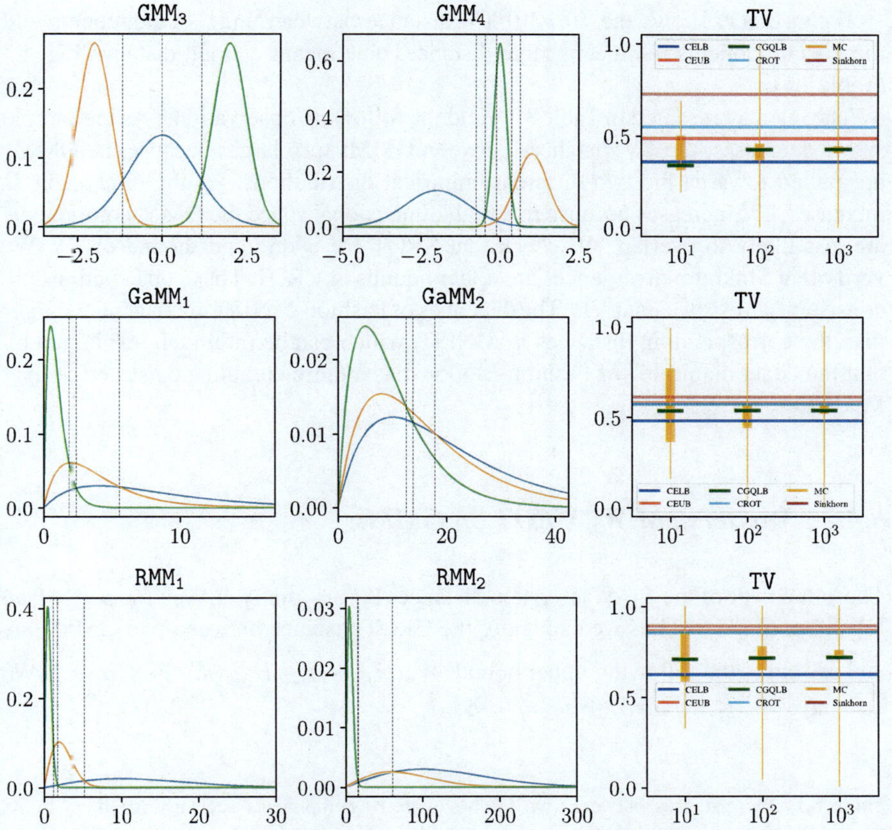

Fig. 8.1 Performance of the CROT distance and the Sinkhorn CROT distance for upper bounding the total variation distance between mixtures of (1) Gaussian, (2) Gamma, and (3) Rayleigh distributions

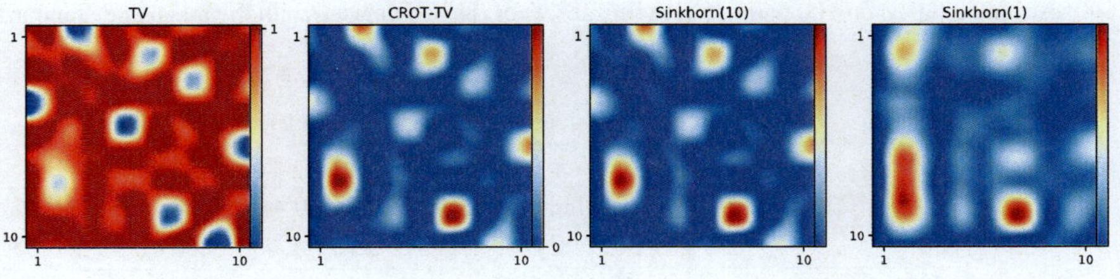

Fig. 8.2 TV distance between two 10-component GMMs estimated on the MNIST dataset: (1) shows the 10×10 matrix TV distance between the first mixture components and the second mixture components (red means large distance and blue means a small distance). (2–4) displays the 10×10 optimal transport matrix W (red means larger weights, blue means smaller weights). The optimal transport matrix is estimated by EMD (2), the Sinkhorn algorithm with weak regularization (3) and the Sinkhorn with strong regularization (4)

Figure 8.3(1) shows the 10×10 TV distance between mm1's components and mm2's components red means large distance, blue means a small distance Fig. 8.3 (Table 8.1).

The results presented in Table 8.1 yield the following observations: As the sample size τ decreases, the TV distances between GMMs turn larger because the GMMs are pulled towards the two different empirical distributions. As the dimension D increases, TV increases because in a high dimensional space the GMM components are less likely to overlap. We check that CROT-TV is an upper bound of TV. We verify that Sinkhorn divergences are upper bounds of CROT. These observations are consistent across two data sets. The distances of Fashion-MNIST are in general larger than the corresponding distances in MNIST, which can be intuitively explained by that the "data manifold" of Fashion-MNIST has a more complicated structure than MNIST.

8.4.2 Wasserstein W_p CROT on GMMs

The pth power of the L_p-Wasserstein distance, W_p^p, is jointly convex for $p \geq 1$ (see Eq. 20, p. 6, [48]). Thus we can apply the CROT distance between two GMMs m_1 and m_2 to get the following upper bound: $W_p(m_1, m_2) \leq H_{W_p^p}^{\frac{1}{p}}(m_1, m_2)$, $\alpha \geq 1$. We also have $W_p \leq W_q$ for $1 \leq p \leq q < \infty$.

Table 8.1 TV distances between two GMMs with 10 components each estimated on PCA-processed images. D is the dimensionality of the PCA. The two GMMs are estimated based on non-overlapping samples, with the parameter $0 < \tau \leq 1$ specifying the relative sample size used to estimated the GMMs. For example, $\tau = 1$ means each GMM is estimated on half of all available images. Sinkhorn (λ) denotes the CROT distance estimated by the Sinkhorn algorithm, where the regularization strength is proportional to $1/\lambda$. For each configuration, the two GMMs are repeatedly estimated based on 100 pairs of random subsets of the full dataset, with the mean and standard deviation reported

Data	D	τ	TV	CROT-TV	Sinkhorn (10)	Sinkhorn (1)
MNIST	10	1	0.16 ± 0.08	0.26 ± 0.14	0.27 ± 0.14	0.78 ± 0.05
	10	0.1	0.29 ± 0.05	0.43 ± 0.08	0.44 ± 0.08	0.84 ± 0.02
	50	1	0.35 ± 0.08	0.43 ± 0.10	0.44 ± 0.10	0.78 ± 0.03
	50	0.1	0.54 ± 0.04	0.64 ± 0.05	0.67 ± 0.06	0.84 ± 0.02
Fashion MNIST	10	1	0.19 ± 0.09	0.23 ± 0.12	0.24 ± 0.12	0.81 ± 0.03
	10	0.1	0.33 ± 0.07	0.40 ± 0.09	0.40 ± 0.09	0.86 ± 0.02
	50	1	0.44 ± 0.11	0.48 ± 0.12	0.50 ± 0.13	0.88 ± 0.03
	50	0.1	0.60 ± 0.07	0.64 ± 0.08	0.67 ± 0.09	0.92 ± 0.02

The OT distance W_2 between Gaussian measures [13, 60] is available in closed-form:

$$W_2(N(\mu_1, \Sigma_1), N(\mu_1, \Sigma_1)) =$$
$$\sqrt{\|\mu_1 - \mu_2\|^2 + \mathrm{tr}(\Sigma_1 + \Sigma_2 - 2(\Sigma_1^{\frac{1}{2}} \Sigma_2 \Sigma_1^{\frac{1}{2}})^{\frac{1}{2}})}.$$

This $H_{W_p^p}^{\frac{1}{p}}$ CROT distance generalizes [7] who considered the W_2 distance between GMMs using discrete OT. They proved that $H_{W_2}(m_1, m_2)$ is a metric, and $W_2(m_1, m_2) \le \sqrt{H_{W_2^2}(m_1, m_2)}$. These results generalize to mixture of elliptical distributions [13]. However, we do not know a closed-form formula for W_p between Gaussian measures when $p \neq 2$.

Given two high-dimensional mixture models m_1 and m_2, we draw respectively n i.i.d. samples from m_1 and m_2, so that $m_1(x) \approx \frac{1}{n} \sum_{i=1}^n D(x_i)$ and $m_2(x) \approx \frac{1}{n} \sum_{j=1}^n D(y_j)$. Then, we have

$$W_p(m_1, m_2) \approx W_p \left(\frac{1}{n} \sum_{i=1}^n D(x_i), \frac{1}{n} \sum_{j=1}^n D(y_j) \right)$$
$$\le H_{W_p^p}^{1/p} \left(\frac{1}{n} \sum_{i=1}^n D(x_i), \frac{1}{n} \sum_{j=1}^n D(y_j) \right). \tag{8.12}$$

Note that $W_p\left(D(x_i), D(x_j)\right) = \|x_i - x_j\|_2$ and therefore the RHS of (8.12) can be evaluated. We use $\mathrm{UB}(W_2)$ to denote this empirical upper bound that will hold if $n \to \infty$. In our experiments $n = 10^3$.

See Table 8.2 for the W_2 distances evaluated on the two investigated data sets. The column $\mathrm{LB}(W_2)$ is a lower bound based on the first and second moments of the mixture models [21]. We can clearly see that $\sqrt{H_{W_2^2}}$ provides a tighter upper bound than $\mathrm{UB}(W_2)$. To compute $\mathrm{UB}(W_2)$ one need to draw a potentially large number of random samples to make the approximation in (8.12), and the computation of the EMD is costly. Therefore one should use $\sqrt{H_{W_2^2}}$ for its better and more efficient approximation.

8.4.3 Rényi CROT Between GMMs

We investigate Rényi α-divergence [40, 41] defined by

$$R_\alpha(p : q) = \frac{1}{1-\alpha} \log \int p(x)^\alpha q(x)^{1-\alpha} \mathrm{d}x,$$

Table 8.2 W_2 distances between two 10-component GMMs estimated on PCA-processed images

Data	D	τ	UB(W_2)	LB(W_2)	$\sqrt{\text{CROT-}W_2^2}$	Sinkhorn (10)	Sinkhorn (1)
MNIST	10	1	1.91 ± 0.02	0.03 ± 0.00	0.84 ± 0.57	0.88 ± 0.58	7.13 ± 0.11
	10	0.1	1.93 ± 0.02	0.09 ± 0.02	1.48 ± 0.38	1.54 ± 0.39	7.29 ± 0.11
	50	1	7.51 ± 0.03	0.07 ± 0.01	2.17 ± 0.93	2.39 ± 0.97	12.02 ± 0.15
	50	0.1	7.53 ± 0.04	0.21 ± 0.02	4.04 ± 0.86	4.33 ± 0.91	12.69 ± 0.22
Fashion MNIST	10	1	1.71 ± 0.05	0.03 ± 0.01	1.19 ± 0.62	1.24 ± 0.63	10.36 ± 0.08
	10	0.1	1.74 ± 0.05	0.10 ± 0.02	1.61 ± 0.63	1.68 ± 0.64	10.43 ± 0.15
	50	1	7.47 ± 0.04	0.07 ± 0.01	3.12 ± 1.01	3.21 ± 1.02	15.31 ± 0.20
	50	0.1	7.50 ± 0.04	0.22 ± 0.02	4.32 ± 1.02	4.45 ± 1.05	15.99 ± 0.29

which encompasses KLD at the limit $\alpha \to 1$. Notice that for multivariate Gaussian densities p and q, $R_\alpha(p : q)$ can be undefined for $\alpha > 1$ as the integral may diverge. In this case the CROT-R_α divergence is undefined. Table 8.3 shows R_α for $\alpha \in \{0.1, 0.5, 0.9\}$ and the corresponding CROT estimated on MNIST and Fashion-MNIST datasets. The observation is consistent with the other distance metrics.

8.5 Learning GMMs with SCROT.KL

This section performs an experimental study to learn mixture models using SCROT. The observed data samples $\{\boldsymbol{x}_i\}_{i=1}^n$ is described by a kernel density estimator (KDE)

$$p(\boldsymbol{x}) = \frac{1}{n}\sum_{i=1}^{n} p_i(\boldsymbol{x}) = \frac{1}{n}\sum_{i=1}^{n} N(\boldsymbol{x}_i, \epsilon I), \tag{8.13}$$

where $\epsilon > 0$ is a hyper parameter. We aim to learn a Gaussian mixture model

$$q(\boldsymbol{x}) = \sum_{i=1}^{m} \alpha_i q_i(\boldsymbol{x}) = \sum_{i=1}^{m} \alpha_i N(\boldsymbol{\mu}_i, \mathrm{diag}(\boldsymbol{\sigma}_i)), \tag{8.14}$$

where $\alpha_i \geq 0$ ($\sum_{i=1}^m \alpha_i = 1$) is the mixture weight of i's component, and diagonal covariance matrices are assumed to reduce the number of free parameters. Minimizing KL$(p : q)$ gives the maximum likelihood estimation [2]. However, the KLD between Gaussian mixture models is known to be not having analytical form [44]. Therefore one has to rely on variational bounds or the re-parametrization trick [29] to bound/approximate KL$(p : q)$. The CROT gives an alternative approach to minimize the KLD by simplifying a KDE [57]. By Theorem 8.8, we have

Table 8.3 Rényi divergences between two 10-component GMMs estimated on PCA-processed images

Data	D	τ	R_α	CROT-R_α	Sinkhorn (10)	Sinkhorn (1)
MNIST $R_{0.1}$	10	1	0.01 ± 0.01	0.07 ± 0.05	0.08 ± 0.05	0.80 ± 0.02
	10	0.1	0.03 ± 0.02	0.15 ± 0.04	0.16 ± 0.04	0.84 ± 0.04
	50	1	0.09 ± 0.06	0.25 ± 0.09	0.29 ± 0.10	1.40 ± 0.07
	50	0.1	0.18 ± 0.09	0.42 ± 0.09	0.46 ± 0.10	1.43 ± 0.09
Fashion MNIST $R_{0.1}$	10	1	0.04 ± 0.03	0.11 ± 0.06	0.12 ± 0.06	1.59 ± 0.05
	10	0.1	0.06 ± 0.03	0.18 ± 0.07	0.19 ± 0.07	1.65 ± 0.07
	50	1	0.12 ± 0.08	0.30 ± 0.11	0.32 ± 0.11	2.37 ± 0.08
	50	0.1	0.20 ± 0.11	0.45 ± 0.10	0.47 ± 0.10	2.41 ± 0.10
MNIST $R_{0.5}$	10	1	0.06 ± 0.05	0.34 ± 0.23	0.37 ± 0.22	4.09 ± 0.12
	10	0.1	0.17 ± 0.05	0.67 ± 0.18	0.72 ± 0.18	4.22 ± 0.10
	50	1	0.31 ± 0.13	1.07 ± 0.41	1.28 ± 0.43	6.73 ± 0.31
	50	0.1	0.69 ± 0.14	1.92 ± 0.40	2.16 ± 0.42	7.01 ± 0.33
Fashion MNIST $R_{0.5}$	10	1	0.17 ± 0.12	0.52 ± 0.29	0.55 ± 0.29	7.54 ± 0.14
	10	0.1	0.28 ± 0.13	0.87 ± 0.28	0.92 ± 0.29	7.79 ± 0.23
	50	1	0.54 ± 0.24	1.45 ± 0.48	1.55 ± 0.48	10.53 ± 0.26
	50	0.1	0.89 ± 0.21	2.16 ± 0.39	2.27 ± 0.40	10.79 ± 0.38
MNIST $R_{0.9}$	10	1	0.14 ± 0.09	0.76 ± 0.42	0.80 ± 0.42	7.18 ± 0.19
	10	0.1	0.31 ± 0.09	1.35 ± 0.37	1.42 ± 0.37	7.53 ± 0.35
	50	1	0.61 ± 0.32	1.90 ± 0.82	2.25 ± 0.85	12.46 ± 0.66
	50	0.1	1.33 ± 0.30	3.51 ± 0.80	3.90 ± 0.82	12.96 ± 0.86
Fashion MNIST $R_{0.9}$	10	1	0.32 ± 0.23	1.07 ± 0.60	1.12 ± 0.61	14.25 ± 0.38
	10	0.1	0.50 ± 0.26	1.69 ± 0.66	1.77 ± 0.67	14.74 ± 0.54
	50	1	1.07 ± 0.43	2.76 ± 0.96	2.93 ± 0.97	21.41 ± 0.78
	50	0.1	1.76 ± 0.45	4.18 ± 1.06	4.40 ± 1.09	22.16 ± 1.02

$H_{\mathrm{KL}}(p : q) \geq \mathrm{KL}(p : q)$. Therefore we minimize the upper bound $H_{\mathrm{KL}}(p : q)$ instead, which can be computed conveniently as the KLD between Gaussian distributions is in closed form. Moreover, because the mixture weights are free parameters, the entropy-regularized optimal transport problem is simplified into

$$\begin{aligned} \min_W \sum_{i=1}^{n}\sum_{j=1}^{m} & \left[w_{ij}\mathrm{KL}\left(p_i, q_j\right) + \frac{1}{\lambda} w_{ij} \log w_{ij} \right], \\ \text{s.t.} \quad & w_{ij} \geq 0, \qquad \forall i, \forall j \\ & \sum_{j=1}^{m} w_{ij} = \frac{1}{n}, \end{aligned}$$

where $\lambda > 0$ is a regularization strength parameter (same as the Sinkhorn algorithm). By a similar analysis [8], the optimal weights $w^\star_{ij}$ must satisfy

$$w^\star_{ij} = \frac{1}{n} \frac{\exp(-\lambda \mathrm{KL}(p_i, q_j))}{\sum_{j=1}^{m} \exp(-\lambda \mathrm{KL}(p_i, q_j))}, \tag{8.15}$$

We therefore minimize $\sum_{i=1}^{n'} \sum_{j=1}^{m} w^\star_{ij} \mathrm{KL}(p_i, q_j)$ based on gradient descent on mini-batches of n' samples. We set empirically the hyper-parameter $m = 10$ (number of components), $\lambda = 0.005$ (Sinkhorn regularization parameter) and $\epsilon = 10^{-6}$ (KDE bandwidth). Fine tuning them can potentially yields better results. We use the training dataset to learn the q distribution (GMM) and estimate the testing error based on its distance with $\hat{p}$, a KDE w.r.t. the testing datasets.

Figure 8.3 shows the learning curves when estimating a 10-component-GMM on MNIST (left) and Fashion MNIST (right). One can observe that SCROT.KL is indeed an upper bound of KL. Minimizing SCROT.KL can effectively learn a mixture model on these two datasets. The resulting model achieves better testing error as compared to `sklearn`'s EM algorithm [49]. This is because we use KDE as the data distribution, which better describes the data as compared to the empirical distribution. Comparatively, the KLD is larger on the Fashion MNIST dataset, where the data distribution is more complicated and cannot be well described by the GMM. EM takes 2 min. SCROT is implemented in Tensorflow [1] using gradient descent (Adam), and takes around 20 min for 100 epochs on an Intel `i5-7300U` CPU.

In order to efficiently estimate the KLD (corresponding to "KL" and "KL(EM)" in the figure), we use the information-theoretical bound $H(X, Y) \leq H(X) + H(Y)$, where H denotes Shannon's entropy. Therefore $\mathrm{KL}(p : q) = -H(p) - \int p(\boldsymbol{x})$

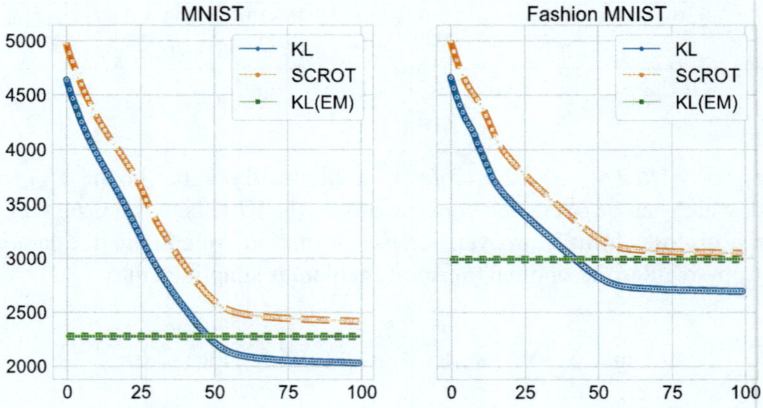

Fig. 8.3 Testing error against the number of epochs on MNIST (left) and Fashion-MNIST (right). The curve "KL" shows the estimated KLD between the data distribution (KDE based on the testing dataset) and the learned GMM. The curve "SCROT" shows the SCROT distance (the learning cost function). The curve "KL(EM)" shows the KLD between the data distribution and a GMM learned using `sklearn`'s EM algorithm

$\log q(x)\mathrm{d}x \geq -H(U) - H(p_i) - \int p(\boldsymbol{x}) \log q(\boldsymbol{x})\mathrm{d}x$, where $U = (1/n, \ldots, 1/n)$ is the uniform distribution, and the integral $\int p(\boldsymbol{x}) \log q(\boldsymbol{x})\mathrm{d}x$ is estimated by Monte-Carlo sampling.

8.6 Conclusion

We defined the generic *Chain Rule Optimal Transport* (CROT) distance (Definition 8.1) H_D for any ground distance D. CROT unifies and generalizes the Wasserstein/EMD distance between discrete measures [54] and the *Mixture Component Optimal Transport* [33] distance. We proved that H_D is a metric whenever D is a metric (Property 8.2). We then dealt with statistical mixtures, and showed that $H_D(m_1, m_2) \geq D(m_1, m_2)$ (Theorem 8.8) whenever D is jointly convex, and considered the smooth Sinkhorn CROT distance $S_D(m_1, m_2)$ (SCROT) for fast calculations of $H_D(m_1, m_2)$ via matrix scaling algorithms (Sinkhorn–Knopp algorithm) so that $D(m_1, m_2) \leq H_D(m_1, m_2) \leq S_D(m_1, m_2)$. These bounds hold in particular for statistical f-divergences $I_f(p : q) = \int p(x) f(q(x)/p(x))\mathrm{d}x$ which includes the Kullback–Leibler divergence). Finally, we proposed a novel efficient method to learn Gaussian mixture models from a semi-SCROT distance that bypasses Sinkhorn iterations and uses a simple normalization (Eq. 8.15). Our learning method by KDE simplification is shown to outperform the EM algorithm of `sklearn` for the MNIST and Fashion MNIST datasets.

Acknowledgements Frank Nielsen thanks Professor Steve Huntsman for pointing out reference [33] to his attention. The authors are grateful to Professor Patrick Forré (University of Amsterdam) for letting us know of an earlier error in the definition of CROT, and to Professor Rüschendorf for sending us his work [55].

Disclaimer: Views and opinions expressed are those of the authors and do not necessarily represent official positions of their respective companies.

A. Proof of CROT Metric (Property 8.2)

Proof We prove that $H(p, q)$ satisfies the following axioms of metric distances:

Non-negativity. As $D\Big(p(x|y), q(x|z)\Big) \geq 0$, we have by definition that $H_D(p, q) \geq 0$.

Law of indiscernibles. If $H_D(p, q) = 0$, then $\forall \epsilon > 0, \exists r^\star \in \Gamma(p(y), q(z))$, such that

$$E_{r^\star(y,z)} D\left(p(x|y), q(x|z)\right) < \epsilon.$$

As $D(\cdot,\cdot)$ is a metric, the density $r^\star(y,z)$ is concentrated on the region $p(x|y) = q(x|z)$ so that

$$\int r^\star(y,z)p(x|y)\mathrm{d}y\mathrm{d}z = \int r^\star(y,z)q(x|z)\mathrm{d}y\mathrm{d}z.$$

We therefore have

$$\begin{aligned} p(x) &= \int p(y)p(x|y)\mathrm{d}y = \int r^\star(y,z)\mathrm{d}z p(x|y)\mathrm{d}y = \int r^\star(y,z)p(x|y)\mathrm{d}y\mathrm{d}z \\ &= \int r^\star(y,z)q(x|z)\mathrm{d}y\mathrm{d}z = \int r^\star(y,z)\mathrm{d}y q(x|z)\mathrm{d}z = \int q(z)q(x|z)\mathrm{d}z \\ &= q(x). \end{aligned}$$

Symmetry.

$$\begin{aligned} H_D(p,q) &= \inf_{r\in\Gamma(p(y),q(z))} \int r(y,z) D\Big(p(x|y), q(x|z)\Big)\,\mathrm{d}y\mathrm{d}z \\ &= \inf_{r\in\Gamma(p(y),q(z))} \int r(y,z) D\Big(q(x|z), p(x|y)\Big)\,\mathrm{d}y\mathrm{d}z \\ &= \inf_{R\in\Gamma(q(z),p(y))} \int R(z,y) D\Big(q(x|z), p(x|y)\Big)\,\mathrm{d}z\mathrm{d}y \\ &= H_D(q,p), \end{aligned}$$

where $R(z,y) = r(y,z)$ s.t. $\int R(z,y)\mathrm{d}y = q(z)$ and $\int R(z,y)\mathrm{d}z = p(y)$.
Triangle inequality. Denote

$$\begin{aligned} r_{12} &= \arg\min\nolimits_{r\in\Gamma(p_1(y_1),\ p_2(y_2))} E_{r(y_1,y_2)} D(p_1(x|y_1), p_2(x|y_2)), \\ r_{23} &= \arg\min\nolimits_{r\in\Gamma(p_2(y_2),\ p_3(y_3))} E_{r(y_2,y_3)} D(p_2(x|y_2), p_3(x|y_3)). \end{aligned}$$

$$\begin{aligned} & H_D(p_1,p_2) + H_D(p_2,p_3) \\ =& E_{r_{12}(y_1,y_2)} D(p_1(x|y_1), p_2(x|y_2)) + E_{r_{23}(y_2,y_3)} D(p_2(x|y_2), p_3(x|y_3)) \\ \ge& \inf_s E_{s(y_1,y_2,y_3)} \left[D(p_1(x|y_1), p_2(x|y_2)) + D(p_2(x|y_2), p_3(x|y_3))\right] \\ \ge& \inf_s E_{s(y_1,y_2,y_3)} D(p_1(x|y_1), p_3(x|y_3)) \\ =& \inf_r E_{r(y,z)} D(p_1(x|y), p_3(x|z)) \\ =& H_D(p_1,p_3), \end{aligned}$$

where $s(y_1,y_2,y_3)$ denotes the set of all probability measures on $\mathcal{Y}^3$ with marginals p_1, p_2 and p_3. Clearly, $\frac{r_{12}(y_1,y_2)r_{23}(y_2,y_3)}{p_2(y_2)} \in s(y_1,y_2,y_3)$.

B. Proof of Upper Bound of H_D

Without loss of generality we assume p and q are mixture models. The proof for the general case is similar.

Proof

$$
\begin{aligned}
D(m_1 : m_2) &= D\left(\sum_{i=1}^{k_1} \alpha_i p_i, \sum_{j=1}^{k_2} \beta_j q_j\right)\\
&= D\left(\sum_{i=1}^{k_1}\sum_{j=1}^{k_2} w_{i,j} p_{i,j} : \sum_{i=1}^{k_1}\sum_{j=1}^{k_2} w_{i,j} q_{i,j}\right)\\
&\leq \sum_{i=1}^{k_1}\sum_{j=1}^{k_2} w_{i,j} D(p_{i,j} : q_{i,j}),\\
&\leq \sum_{i=1}^{k_1}\sum_{j=1}^{k_2} w_{i,j} D(p_i : q_j) =: H_D(m_1, m_2).
\end{aligned}
$$

C. Upper Bounding f-Divergences

First, let us start by proving the following lemma for the Kullback–Leibler divergence:

Lemma 8.9 *The Kullback–Leibler divergence between two Radon–Nikodym p and q with respect to μ is upper bounded as follows:* $\mathrm{KL}(p : q) \leq \int \frac{p(x)^2}{q(x)} \mathrm{d}\mu(x) - 1$.

Proof Consider a strictly convex and differentiable function $F(x)$ on $(0, \infty)$. Then we have

$$F(b) - F(a) \geq F'(a)(b - a), \tag{8.16}$$

for any $a, b \in (0, \infty)$, with equality iff. $a = b$. Indeed, this inequality is related to the non-negativeness of the scalar Bregman divergence $B_F(b, a) = F(b) - F(a) - (b - a)F'(a) \geq 0$.

Plugging $F(x) = -\log x$ (with $F'(x) = -\frac{1}{x}$ and $F''(x) = \frac{1}{x^2} > 0$), $a = q(x)$ and $b = p(x)$ in Eq. 8.16, we get

$$\log q(x) - \log p(x) \geq \frac{q(x) - p(x)}{q(x)}.$$

Multiplying both sides of the inequality by $-p(x) < 0$ (and reversing the inequality), we end up with

$$p(x)\log\frac{p(x)}{q(x)} \leq \frac{p^2(x)}{q(x)} - p(x).$$

Then taking the integral over the support $\mathcal{X}$ of the distributions yields:

$$\mathrm{KL}(p:q) \leq \int_{\mathcal{X}} \frac{p(x)^2}{q(x)}\mathrm{d}\mu(x) - 1,$$

with equality when $p(x) = q(x)$ almost everywhere. Notice that the right-hand side integral $\int_{\mathcal{X}} \frac{p(x)^2}{q(x)}\mathrm{d}\mu(x)$ may diverge (e.g., when KL is infinite).

Now, let us consider two mixtures $m(x) = \sum_{i=1}^k w_i p_i(x)$ and $m'(x) = \sum_{i=1}^{k'} w_i' p_i'(x)$. Apply Lemma 8.9 to get

$$\mathrm{KL}(m:m') \leq \sum_{i,j}\int w_i w_j \frac{p_i(x)p_j(x)}{m'(x)}\mathrm{d}\mu(x) - 1.$$

Let us upper bound $A_{ij} = \int \frac{p_i(x)p_j(x)}{m'(x)}\mathrm{d}\mu(x)$ to upper bound

$$\mathrm{KL}(m:m') \leq \sum_{i,j} w_i w_j A_{ij} - 1.$$

For bounding the terms A_{ij}, we interpret the mixture density as an arithmetic weighted mean that is greater or equal than a geometric mean (AGM inequality). Therefore we get:

$$\int \frac{p_i(x)p_j(x)}{m'(x)}\mathrm{d}\mu(x) \leq \int \frac{p_i(x)p_j(x)}{\prod_{l=1}^{k'} w_l' p_l'(x)}\mathrm{d}\mu(x).$$

When the mixture components belong to a same exponential family [39], we get a closed-form upper bound since $\theta_i + \theta_j - \sum_{l=1}^{k'} w_l'\theta_l' \in \Theta$: Let $\bar{\theta}' = \sum_{l=1}^{k'} w_l'\theta_l'$ denote the barycenter of the natural parameters of the mixture components of m'. We have:

$$\frac{p(x;\theta_i)p(x;\theta_j)}{\prod_{l=1}^{k'} w_l' p(x;\theta_l')} = \exp\left(\left(\theta_i + \theta_j - \bar{\theta}'\right)^\top t(x) - F(\theta_i) - F(\theta_j) + \sum_{l=1}^{k'} w_l' F(\theta_l') + k(x)\right).$$

Taking the integral over the support we find that

$$A_{ij} \leq \exp\left(F\left(\theta_i + \theta_j - \bar{\theta}'\right) - F(\theta_i) - F(\theta_j) + \sum_{l=1}^{k'} w_l' F(\theta_l')\right).$$

Overall, we get the upper bound:

$$\mathrm{KL}(m : m') \leq \left(\sum_{i,j} w_i w_j \exp\left(F\left(\theta_i + \theta_j - \bar{\theta}'\right) - F(\theta_i) - F(\theta_j) + \sum_{l=1}^{k'} w_l' F(\theta_l')\right)\right) - 1. \tag{8.17}$$

In general, we have the following upper bound for f-divergences [14]:

Property 8.10 (f-divergence upper bound) *The f-divergence between two densities p and q with respect to μ is upper bounded as follows: $I_f(p : q) \leq \int (q(x) - p(x)) f'\left(\frac{q(x)}{p(x)}\right) \mathrm{d}\mu(x)$.*

Proof Let us use the non-negative property of scalar Bregman divergences:

$$B_F(a : b) = F(a) - F(b) - (a - b)F'(b) \geq 0.$$

Let $F(x) = f(x)$ (with $F(1) = f(1) = 0$), and $a = 1$ and $b = \frac{q}{p}$. It follows that

$$B_F\left(1 : \frac{q}{p}\right) = -f\left(\frac{q}{p}\right) - \left(1 - \frac{q}{p}\right) f'\left(\frac{q}{p}\right) \geq 0.$$

That is,

$$pf\left(\frac{q}{p}\right) \leq p\left(\frac{q}{p} - 1\right) f'\left(\frac{q}{p}\right).$$

Taking the integral over the support, we get

$$I_f(p : q) \leq \int (q - p) f'\left(\frac{q}{p}\right) \mathrm{d}\mu.$$

For example, when $f(u) = -\log u$ (with $f'(u) = -\frac{1}{u}$), we recover the former upper bound:

$$\mathrm{KL}(p : q) \leq \int (p - q)\frac{p}{q} \mathrm{d}\mu = \int \frac{p^2}{q} \mathrm{d}\mu - 1.$$

Notice that $\int \frac{p^2}{q} \mathrm{d}\mu - 1$ is a f-divergence for the generator $f(u) = \frac{1}{u} - 1$.

D. Square Root of the Symmetric α-Jensen–Shannon Divergence

TV is bounded in [0, 1] which makes it difficult to appreciate the quality of the CROT upper bounds in general. We shall consider a different parametric distance D_α that is upper bounded by an arbitrary bound: $D_\alpha(p, q) \le C_\alpha$.

It is well known that the square root of the Jensen–Shannon divergence is a metric [19] satisfying the triangle inequality. In [35], a generalization of the Jensen–Shannon divergence was proposed, given by

$$\mathrm{JS}_\alpha(p : q) := \frac{1}{2}\mathrm{KL}(p : (pq)_\alpha) + \frac{1}{2}\mathrm{KL}(q : (pq)_\alpha), \tag{8.18}$$

where $(pq)_\alpha := (1-\alpha)p + \alpha q$. JS_α unifies (twice) the Jensen–Shannon divergence (obtained when $\alpha = \frac{1}{2}$) with the Jeffreys divergence ($\alpha = 1$; [35]). A nice property is that the skew K-divergence is upper bounded as follows:

$$\mathrm{KL}(p : (pq)_\alpha) \le \int p \log \frac{p}{(1-\alpha)p} \le -\log(1-\alpha)$$

for $\alpha \in (0, 1)$, so that $\mathrm{JS}_\alpha[p : q] \le -\frac{1}{2}\log(1-\alpha) - \frac{1}{2}\log\alpha$ for $\alpha \in (0, 1)$.

Thus, we have the square root of the symmetrized α-divergence that is upper bounded by

$$\sqrt{\mathrm{JS}_\alpha(p : q)} \le C_\alpha = \sqrt{-\frac{1}{2}\log(1-\alpha) - \frac{1}{2}\log\alpha}.$$

However, $\sqrt{\mathrm{JS}_\alpha[p : q]}$ is not a metric in general [47]. Indeed, in the extreme case of $\alpha = 1$, it is known that any positive power of the Jeffreys divergence does not yield a metric.

Observe that JS_α is a f-divergence since $K_\alpha(p : q) := \mathrm{KL}(p : (pq)_\alpha)$ is a f-divergence for the generator $f(u) = -\log((1-\alpha) + \alpha u)$, and we have $\mathrm{KL}(q : (pq)_\alpha) = K_{1-\alpha}(q : p)$. Since $I_f(q : p) = I_{f^\diamond}(p : q)$ for $g(u) = uf(1/u)$, it follows that the f-generator f_{JS_α} for the JS_α divergence is:

$$f_{\mathrm{JS}_\alpha}(u) = -\log\left((1-\alpha) + \alpha u\right) - \log\left(\alpha + \frac{1-\alpha}{u}\right). \tag{8.19}$$

Figure 8.4 and Table 8.4 display the experimental results obtained for the α-JS divergences. One can have similar observations with the TV results.

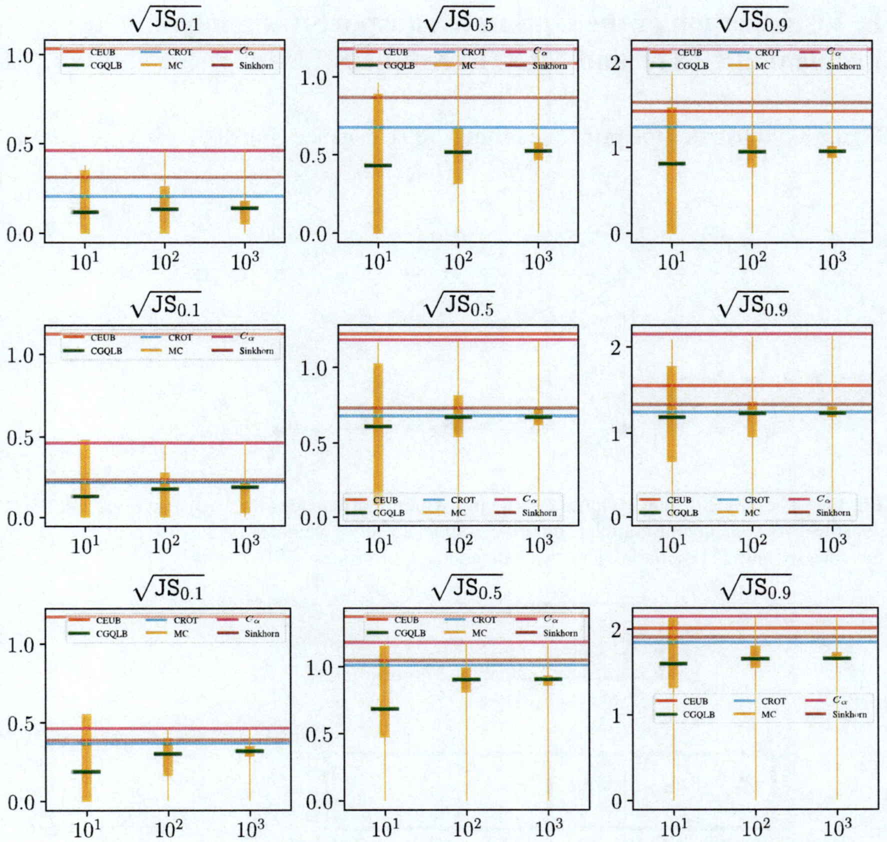

Fig. 8.4 Performance of the CROT distance and the Sinkhorn CROT distance for upper bounding the square root of the α-Jensen–Shannon distance between mixtures of (1) Gaussian, (2) Gamma, and (3) Rayleigh distributions

Table 8.4 Square root of the Jensen–Shannon divergence between two 10-component GMMs estimated on PCA-processed images

Data	D	τ	$\sqrt{\mathrm{JS}_{0.5}}$	CROT-$\sqrt{\mathrm{JS}_{0.5}}$	Sinkhorn (10)	Sinkhorn (1)
MNIST	10	1	0.25 ± 0.11	0.36 ± 0.17	0.37 ± 0.17	0.94 ± 0.05
	10	0.1	0.39 ± 0.05	0.55 ± 0.07	0.56 ± 0.08	1.00 ± 0.02
	50	1	0.51 ± 0.11	0.54 ± 0.12	0.56 ± 0.13	0.93 ± 0.04
	50	0.1	0.69 ± 0.05	0.76 ± 0.07	0.79 ± 0.07	1.00 ± 0.03
Fashion MNIST	10	1	0.33 ± 0.15	0.31 ± 0.13	0.33 ± 0.14	0.96 ± 0.04
	10	0.1	0.46 ± 0.09	0.48 ± 0.09	0.49 ± 0.10	1.01 ± 0.03
	50	1	0.60 ± 0.12	0.57 ± 0.14	0.59 ± 0.15	1.03 ± 0.04
	50	0.1	0.75 ± 0.07	0.76 ± 0.09	0.80 ± 0.10	1.08 ± 0.02

E. Visualization of the Optimal Transport Assignment Problem of CROT and MCOT Distances

Figure 8.5 illustrates the principle of the CROT distance (Fig. 8.6).

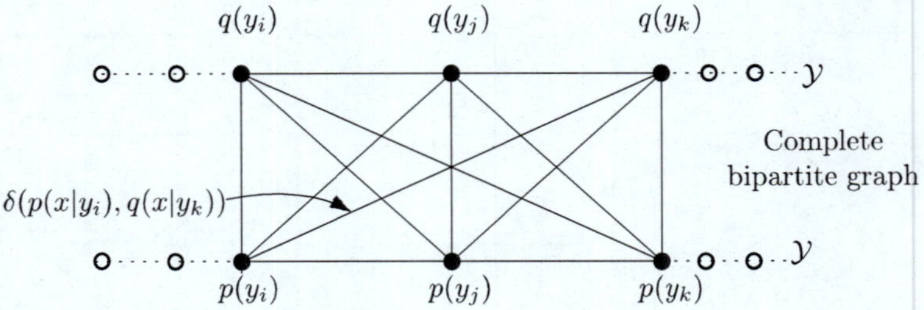

Fig. 8.5 The CROT distance: Optimal matching of marginal densities w.r.t. a distance on conditional densities. We consider the complete bipartite graph with edges weighted by the distances D between the corresponding conditional densities defined at edge vertices

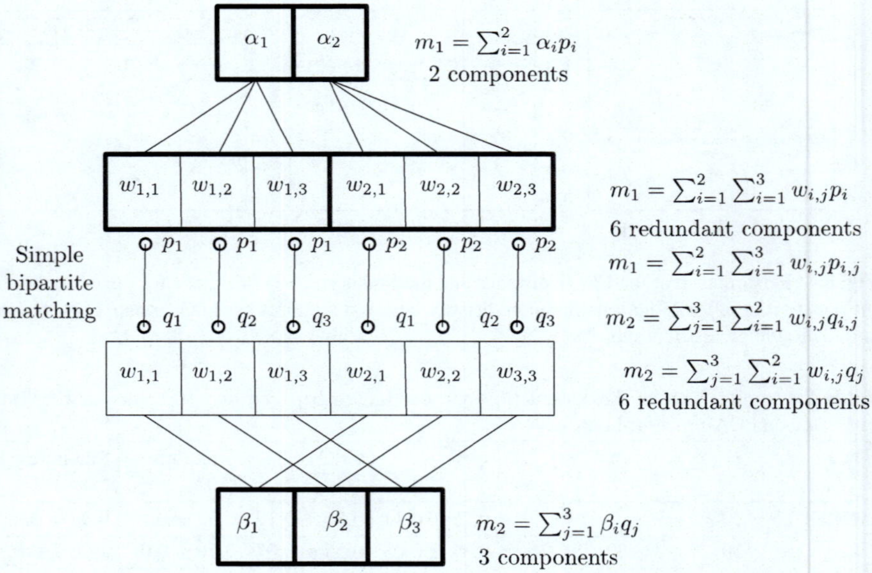

Fig. 8.6 An interpretation of CROT by rewriting the mixtures $m_1 = \sum_{i=1}^{k_1}\sum_{j=1}^{k_2} w_{i,j} p_{i,j}$ and $m_2 = \sum_{i=1}^{k_1}\sum_{j=1}^{k_2} w_{i,j} q_{i,j}$ with $p_{i,j} = p_i$ and $q_{i,j} = q_j$ and using the joint convexity of the base distance D

References

1. Abadi, M., Barham, P., Chen, J., Chen, Z., Davis, A., Dean, J., Devin, M., Ghemawat, S., Irving, G., Isard, M., et al.: Tensorflow: a system for large-scale machine learning. In: 12th USENIX Symposium on Operating Systems Design and Implementation (OSDI 16), pp. 265–283 (2016)
2. Amari, S.-I.: Information Geometry and Its Applications. Applied Mathematical Sciences. Springer, Japan (2016)
3. Bauschke, H.H., Borwein, J.M.: Joint and separate convexity of the Bregman distance. In: Studies in Computational Mathematics, vol. 8, pp. 23–36. Elsevier (2001)
4. Bonneel, N., Rabin, J., Peyré, G., Pfister, H.: Sliced and radon Wasserstein barycenters of measures. J. Math. Imaging Vis. **51**(1), 22–45 (2015)
5. Borwein, J.M., Vanderwerff, J.D.: Convex Functions: Constructions, Characterizations and Counterexamples, vol. 109. Cambridge University Press, Cambridge (2010)
6. Chang, K.-C., Sun, W.: Scalable fusion with mixture distributions in sensor networks. In: 11th International Conference on Control Automation Robotics & Vision (ICARCV), pp. 1251–1256 (2010)
7. Chen, Y., Georgiou, T.T., Tannenbaum, A.: Optimal transport for Gaussian mixture models. IEEE Access **7**, 6269–6278 (2019)
8. Cuturi, M.: Sinkhorn distances: lightspeed computation of optimal transport. In: NIPS, pp. 2292–2300 (2013)
9. Cuturi, M., Teboul, O., Vert, J.: Differentiable sorting using optimal transport: the Sinkhorn CDF and quantile operator (2019). CoRR arXiv:abs/1905.11885
10. Dacorogna, B., Maréchal, P.: The role of perspective functions in convexity, polyconvexity, rank-one convexity and separate convexity. J. Convex Anal. **15**(2), 271 (2008)
11. Dempster, A.P., Laird, N.M., Rubin, D.B.: Maximum likelihood from incomplete data via the EM algorithm. J. R. Stat. Soc. Ser. B (Methodol.), pp. 1–38 (1977)
12. Do, M.N.: Fast approximation of Kullback–Leibler distance for dependence trees and hidden Markov models. IEEE Signal Process. Lett. **10**(4), 115–118 (2003)
13. Dowson, D.C., Landau, B.: The Fréchet distance between multivariate normal distributions. J. Multivar. Anal. **12**(3), 450–455 (1982)
14. Dragomir, S.S.: Inequalities for Csiszár f-divergence in information theory. Victoria University, Melbourne, Australia (2000)
15. Durrieu, J.-L., Thiran, J.-P., Kelly, F.: Lower and upper bounds for approximation of the Kullback–Leibler divergence between Gaussian mixture models. In: 2012 IEEE International Conference on Acoustics, Speech and Signal Processing (ICASSP), pp. 4833–4836. IEEE (2012)
16. Everett, B.: An Introduction to Latent Variable Models. Springer Science & Business Media (2013)
17. Feydy, J., Séjourné, T., Vialard, F.-X., Amari, S.-I., Trouvé, A., Peyré, G.: Interpolating between optimal transport and MMD using Sinkhorn divergences (2018). arXiv:1810.08278
18. Flamary, R., Courty, N.: POT python optimal transport library (2017)
19. Fuglede, B., Topsoe, F.: Jensen-Shannon divergence and Hilbert space embedding. In: International Symposium on Information Theory (ISIT 2004), p. 31. IEEE (2004)
20. Gangbo, W., McCann, R.J.: The geometry of optimal transportation. Acta Math. **177**(2), 113–161 (1996)
21. Gelbrich, M.: On a formula for the L2 Wasserstein metric between measures on Euclidean and Hilbert spaces. Mathematische Nachrichten **147**(1), 185–203 (1990)
22. Ghaffari, N., Walker, S.: On multivariate optimal transportation (2018)
23. Goldberger, J., Aronowitz, H.: A distance measure between GMMs based on the unscented transform and its application to speaker recognition. In: INTERSPEECH European Conference on Speech Communication and Technology, pp. 1985–1988 (2005)
24. Goldberger, J., Gordon, S., Greenspan, H.: An efficient image similarity measure based on approximations of KL-divergence between two Gaussian mixtures. In: IEEE International Conference on Computer Vision (ICCV), p. 487. IEEE (2003)

25. Hershey, J.R., Olsen, P.A.: Approximating the Kullback-Leibler divergence between Gaussian mixture models. In: ICASSP, vol. 4, pp. IV–317. IEEE (2007)
26. Kantorovich, L.: On the transfer of masses. Doklady Akademii Nauk **37**(2), 227–229 (1942). (in Russian)
27. Kantorovitch, L.: On the translocation of masses. Manag. Sci. **5**(1), 1–4 (1958)
28. Khosravifard, M., Fooladivanda, D., Gulliver, T.A.: Confliction of the convexity and metric properties in f-divergences. IEICE Trans. Fundam. Electron. Commun. Comput. Sci. **90**(9), 1848–1853 (2007)
29. Kingma, D.P., Welling, M.: Auto-encoding variational Bayes. In: ICLR (2014)
30. Komaki, F.: Bayesian prediction based on a class of shrinkage priors for location-scale models. Ann. Inst. Stat. Math. **59**(1), 135–146 (2007)
31. Korte, B., Vygen, J.: Linear programming algorithms. In: Combinatorial Optimization, pp. 75–102. Springer (2018)
32. LeCun, Y., Bottou, L., Bengio, Y., Haffner, P.: Gradient-based learning applied to document recognition. Proc. IEEE **86**(11), 2278–2324 (1998)
33. Liu, Z., Huang, Q.: A new distance measure for probability distribution function of mixture type. In: ICASSP, vol. 1, pp. 616–619. IEEE (2000)
34. Monge, G.: Mémoire sur la théorie des déblais et des remblais. Imprimerie Royale (1781)
35. Nielsen, F.: A family of statistical symmetric divergences based on Jensen's inequality (2010). arXiv:1009.4004
36. Nielsen, F.: Closed-form information-theoretic divergences for statistical mixtures. In: 2012 21st International Conference on Pattern Recognition (ICPR), pp. 1723–1726. IEEE (2012)
37. Nielsen, F.: Generalized Bhattacharyya and Chernoff upper bounds on bayes error using quasi-arithmetic means. Pattern Recognit. Lett. **42**, 25–34 (2014)
38. Nielsen, F.: The statistical Minkowski distances: closed-form formula for Gaussian mixture models (2019). arXiv:1901.03732
39. Nielsen, F., Garcia, V.: Statistical exponential families: a digest with flash cards (2009). arXiv:0911.4863
40. Nielsen, F., Nock, R.: A closed-form expression for the Sharma-Mittal entropy of exponential families. J. Phys. A: Math. Theor. **45**(3), 032003 (2011)
41. Nielsen, F., Nock, R.: On Rényi and Tsallis entropies and divergences for exponential families (2011). arXiv:1105.3259
42. Nielsen, F., Nock, R.: On the chi square and higher-order chi distances for approximating f-divergences. IEEE Signal Process. Lett. **21**(1), 10–13 (2014)
43. Nielsen, F., Nock, R.: On w-mixtures: finite convex combinations of prescribed component distributions (2017). CoRR arXiv:abs/1708.00568
44. Nielsen, F., Sun, K.: Guaranteed bounds on information-theoretic measures of univariate mixtures using piecewise log-sum-exp inequalities. Entropy **18**(12), 442 (2016)
45. Nielsen, F., Sun, K.: Guaranteed deterministic bounds on the total variation distance between univariate mixtures. IEEE Mach. Learn. Signal Process. (MLSP) 1–6 (2018)
46. Nielsen, M.A., Chuang, I.: Quantum computation and quantum information (2002)
47. Österreicher, F., Vajda, I.: A new class of metric divergences on probability spaces and its applicability in statistics. Ann. Inst. Stat. Math. **55**(3), 639–653 (2003)
48. Ozawa, R., Yokota, T.: Stability of RCD condition under concentration topology. J. Phys. A: Math. Theor. **45**(3), 032003 (2011)
49. Pedregosa, F., Varoquaux, G., Gramfort, A., Michel, V., Thirion, B., Grisel, O., Blondel, M., Prettenhofer, P., Weiss, R., Dubourg, V., et al.: Scikit-learn: Machine learning in python. J. Mach. Learn. Res. **12**, 2825–2830 (2011)
50. Pedregosa, F., Varoquaux, G., Gramfort, A., Michel, V., Thirion, B., Grisel, O., Blondel, M., Prettenhofer, P., Weiss, R., Dubourg, V., et al.: Scikit-learn: machine learning in Python. J. Mach. Learn. Res. **12**, 2825–2830 (2011)
51. Peyré, G., Cuturi, M., et al.: Computational optimal transport. Found. Trends® in Mach. Learn. **11**(5–6), 355–607 (2019)

52. Pitrik, J., Virosztek, D.: On the joint convexity of the Bregman divergence of matrices. Lett. Math. Phys. **105**(5), 675–692 (2015)
53. Reynolds, D.A., Quatieri, T.F., Dunn, R.B.: Speaker verification using adapted Gaussian mixture models. Digital Signal Process. **10**(1–3), 19–41 (2000)
54. Rubner, Y., Tomasi, C., Guibas, L.J.: The earth mover's distance as a metric for image retrieval. Int. J. Comput. Vis. **40**(2), 99–121 (2000)
55. Rüschendorf, L.: The Wasserstein distance and approximation theorems. Probab. Theory Relat. Fields **70**, 117–129 (1985)
56. Santambrogio, F.: Optimal Transport for Applied Mathematicians, pp. 99–102. Birkäuser, NY (2015)
57. Schwander, O., Nielsen, F.: Learning mixtures by simplifying kernel density estimators. In: Matrix Information Geometry, pp. 403–426. Springer (2013)
58. Silva, J., Narayanan, S.: Upper bound Kullback-Leibler divergence for hidden Markov models with application as discrimination measure for speech recognition. In: IEEE International Symposium on Information Theory (ISIT), pp. 2299–2303. IEEE (2006)
59. Singer, Y., Warmuth, M.K.: Batch and on-line parameter estimation of Gaussian mixtures based on the joint entropy. In: NIPS 578–584 (1999)
60. Takatsu, A., et al.: Wasserstein geometry of Gaussian measures. Osaka J. Math. **48**(4), 1005–1026 (2011)
61. Van Erven, T., Harremos, P.: Rényi divergence and Kullback-Leibler divergence. IEEE Trans. Inf. Theory **60**(7), 3797–3820 (2014)
62. Vaserstein, L.N.: Markov processes over denumerable products of spaces, describing large systems of automata. Probl. Peredachi Informatsii **5**(3), 64–72 (1969)
63. Vigelis, R.F., De Andrade, L.H., Cavalcante, C.C.: Properties of a generalized divergence related to Tsallis generalized divergence. IEEE Trans. Inf. Theory **66**(5), 2891–2897 (2019)
64. Xiao, H., Rasul, K., Vollgraf, R.: Fashion-MNIST: a novel image dataset for benchmarking machine learning algorithms. Technical report, Zalando Research, Berlin, Germany (2017). arXiv:cs.LG/1708.07747
65. Xie, L., Ugrinovskii, V.A., Petersen, I.R.: Probabilistic distances between finite-state finite-alphabet hidden Markov models. IEEE Trans. Autom. Control. **50**(4), 505–511 (2005)

Chapter 9
Towards the "Shape" of Cosmological Observables and the String Theory Landscape with Topological Data Analysis

Alex Cole and Gary Shiu

Abstract Persistent homology is a technique from Topological Data Analysis that computes the multiscale "shape" of a data set. We review the basic formalism of persistent homology as well as several applications to cosmology and string theory. We describe how persistent homology provides efficient and robust morphological summaries of cosmological observables including the Cosmic Microwave Background and Large-Scale Structure, and how statistical pipelines involving persistent homology can potentially constrain cosmological parameters. We also review the string theory landscape as an interesting data set displaying complex features in high-dimensional spaces, and thus an ideal setting for persistent homology. We describe the characterization of distributions of flux vacua using persistent homology.

Keywords Topological Data Analysis · Persistent homology · Inflationary cosmology · Cosmic microwave background · Large scale structure · Non-gaussianity · String landscape

---

This research was supported in part by the DOE grant DE-SC0017647, the Heising-Simons Foundation, the Simons Foundation, National Science Foundation Grant No. NSF PHY-1748958, and the Kellett Award of the University of Wisconsin. GS gratefully acknowledges the hospitality of the Kavli Institute for Theoretical Physics during the final stage of this work. AC is grateful to have been supported by a Straka Fellowship at UW-Madison.

A. Cole
GRAPPA and ITFA, Institute of Physics, University of Amsterdam, 1090 Amsterdam, WX,, Netherlands

A. Cole · G. Shiu (✉)
University of Wisconsin, Madison, WI 53706, USA
e-mail: shiu@physics.wisc.edu

Kavli Institute for Theoretical Physics, University of California, Santa Barbara, CA 93106, USA

F. Nielsen (ed.), *Progress in Information Geometry*, Signals and Communication Technology,
https://doi.org/10.1007/978-3-030-65459-7_9

9.1 Introduction

Physics and information theory have long had a rich exchange. Many concepts in information theory admit an understanding via statistical physics. Moreover, information theory contributes a great deal to both experimental and theoretical physics. The experimental physicist seeks to understand the information content of various experimental observations and the constraining power of future experiments, making extensive use of the Fisher information. On the theoretical side, the "it from qubit" program has revealed that many aspects of gravity seem to emerge from information-theoretic quantities such as entanglement between different statistical systems. A similar relationship exists more broadly between machine learning/data science and physics, with physically-motivated architectures and algorithms such as the Boltzmann machine [3] and importance sampling and a growing presence of deep learning in both theoretical and experimental physics. This symbiotic relationship between fields shows no signs of slowing down.

In this chapter, we review the application of *persistent homology* to several physical data sets. Persistent homology forms the main pillar of Topological Data Analysis (TDA), and formalizes a notion of "shape" for discrete data sets. As such, we can regard it as a powerful pattern recognition tool with potential to identify signatures of subtle physical phenomena. In particular, we are motivated by problems in cosmology and string theory. In cosmology, we would like to understand how fundamental physics affects the complex structure (such as the spatial distribution of galaxies) we observe in our universe. For string theory, we are interested in the problem of navigating the so-called string landscape, the vast set of ground states of the theory.

The organization of this chapter is as follows. In Sect. 9.2 we briefly review the basics of persistent homology as a general pattern recognition tool. We emphasize its stability theorems and describe how one may derive coarser statistics than the persistence diagram. In Sect. 9.3 we note how topological statistics have previously appeared in the cosmology literature, and describe how persistent homology techniques make contact with these statistics. We describe how persistent homology provides novel statistical descriptions of the topology of cosmological observables including the Cosmic Microwave Background and Large-Scale Structure. Section 9.4 motivates the use of data-centric approaches to the string theory landscape. After reviewing relevant aspects of a construction of string vacua supported by fluxes, we discuss how computational topology allows one to characterize these data sets, recovering previously known features and discovering new structure.

9.2 A Short Review of Persistent Homology

In this section, we review persistent homology as a tool allowing one to formalize a notion of "shape" for discrete data sets. We begin by defining simplicial complexes, the discrete mathematical structures in which we embed our data. We describe how

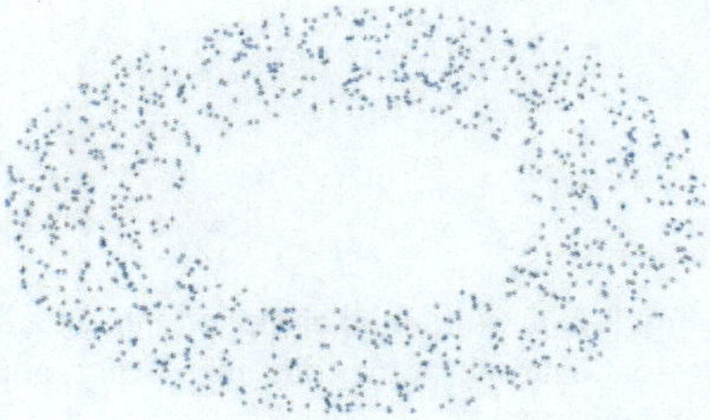

Fig. 9.1 Points randomly sampled from an annulus. Persistent homology quantifies the presence of the "hole" in the middle, as well as topological aspects of the distribution's small scale structure

the topology of simplicial complexes is described by simplicial homology. We then confront problems of representational ambiguity and instability. It turns out that to resolve these problems, it is helpful to introduce a notion of *persistent* homology, in which a data set is represented by a family of simplicial complexes and their homologies. After defining persistent homology, we review its stability results and detail how one can use persistence to develop topological statistics and inference. For more in-depth reviews and more complete references, see [21, 44, 83, 103].

In its classical formulation (see e.g.. [57, 82]), topology is concerned with the deformation-invariant features of a space. The rules of topology identify two spaces if one can be deformed into the other. In other words, we may stretc.h or squeeze a space, but not tear it apart or attach new components. For our purposes, the deformation-invariant features we will consider are "holes" of various dimensions. Allowed deformations cannot create or destroy these holes. Topology is then a rather coarse-grained summary of a space, describing its gross aspects.

Persistent homology allows one to extend this framework to discrete data sets, for which classical methods cannot say much. One reason to do this is that by restricting to a toplogical description, we might hope to receive some benefits in terms of robustness. Consider the set of points (henceforth a "point cloud") in Fig. 9.1. The distribution of points has a clear shape whose characterization does not fall within the purview of classical topology. To describe the point cloud's shape, we embed it in the structure of a *simplicial complex*.

9.2.1 Simplicial Homology

Simplicial complexes are made of simplices. Simplices are vertices (0-simplices), edges (1-simplices), triangles (2-simplices), and so on. We will refer to a simplex of arbitrary dimension as a p-simplex. Note that an edge has two vertices as faces, a triangle has three edges as faces, and so on. We will sometimes write a p-simplex σ whose vertices are $v_0, \ldots, v_p$ as $\sigma = [v_0 \ldots v_p]$. A simplicial complex is a collection S of simplices such that

1. $\sigma \cap \tau \in S$ for all $\sigma, \tau \in S$;
2. $\sigma \subset \tau$ implies $\sigma \in S$ for all $\tau \in S$.

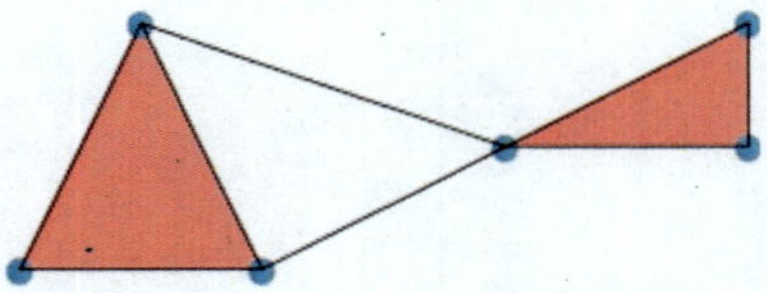

Fig. 9.2 A simplicial complex. Persistent homology uses families of simplicial complexes to represent data sets across different scales, and tracks when "holes" of various dimensions are created and destroyed

In other words, simplicial complexes are closed under taking faces and intersections of simplices. These requirements ensure the mathematical operations we will perform are well-defined. A graphical representation of a simplicial complex is shown in Fig. 9.2. Note that we may specify a simplicial complex as a finite collection S of sets such that $\sigma \in S$ and $\tau \subseteq \sigma$ implies $\tau \in S$, without reference to a geometric embedding. A geometric embedding always exists, as can be easily shown. The following discussion does not require we make a distinction between geometric and abstract simplicial complexes. There are other topics, for example the consideration of simplicial maps (the simplicial analogues of continuous maps between topological spaces) for which it is more natural to consider geometric embeddings of simplicial complexes.

We are interested in identifying "holes" of various dimensions in a simplicial complex. To do this formally, we first make some definitions. Given a simplicial complex S, a p-chain is a collection of p-simplices in S. We can formally write these as sums $\sum_i a_i \sigma_i$, where $a_i \in \mathbb{Z}_2$ and i runs over p-simplices in S.[1] The p-chains form a group under addition, which we denote C_p.

There are two special subgroups of C_p. To define them, we first define the boundary operator $\partial_p : C_p \to C_{p-1}$. Writing a p-simplex in terms of its vertices as $[v_0 \ldots v_p]$, the action of the boundary operator is defined by

$$\partial_p[v_0 \ldots v_p] = \sum_{j=0}^{p} (-1)^p [v_0 \ldots \hat{v}_j \ldots v_p] \tag{9.1}$$

and linear extension. Here the hatted vertex is omitted. The boundary operator deserves its name; it takes a p-chain to its boundary. For example, the boundary of a 1-simplex $[v_0 v_1]$ is given by $\partial_1[v_0 v_1] = [v_0] + [v_1]$. Moreover, the boundary operator defines a homomorphism from C_p to C_{p-1}. Accordingly, its image and kernel can be used to define subgroups.

We call a p-chain σ a p-cycle if its boundary vanishes, $\partial_p \sigma = 0$. As a simple example, a 1-chain that "loops" is a 1-cycle, $\partial_1([v_0 v_1] + [v_1 v_2] + \cdots + [v_n v_0]) =$

[1] One may also consider (persistent) homology defined over other coefficient fields. For simplicity, we restrict to $\mathbb{Z}_2$. In fact, this is the most efficient choice if the underlying space does not have torsion, and the results computed for $\mathbb{Z}_2$ can be translated to homology over other coefficient fields. See Sect. 4.3 of [103] for a discussion.

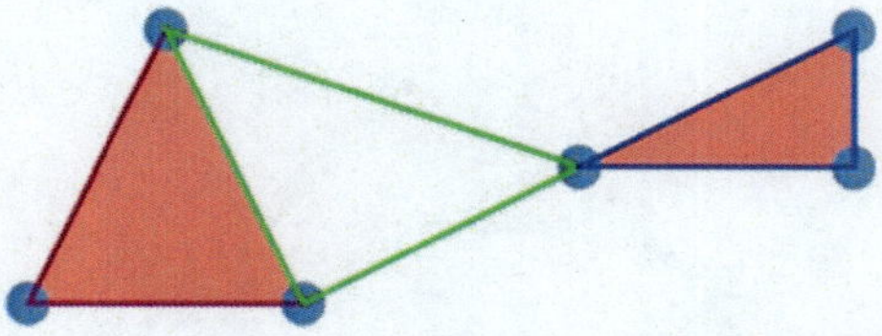

Fig. 9.3 Two 1-cycles are highlighted in green and blue. The green 1-cycle is not the boundary of any 2-chain, and therefore represents a nontrivial 1-homology class. The blue 1-cycle is the boundary of the 2-chain comprised of the interior triangle, and is therefore homologous to the empty cycle. The 1-cycle that uses two of the green 1-cycle's edges and two of the purple edges is homologous to the green 1-cycle

$[v_0] + [v_1] + [v_1] + \cdots + [v_n] + [v_n] + [v_0] = 0$. Since ∂_p is linear, the p-cycles form a subgroup of C_p, which we denote Z_p. In other words, $Z_p = \ker \partial_p$. Analogously, we call a p-cycle σ a p-boundary if it is the boundary of a $(p+1)$-cycle, i.e. $\sigma = \partial_p \tau$ for some $(p+1)$-cycle τ. The p-boundaries also form a group, which we denote B_p. A crucial observation is that the boundary of a boundary is always empty, $\partial_p \partial_{p+1} \sigma = 0$ for all σ. This is called the Fundamental Lemma of Homology. In other words B_p is a subgroup of Z_p. This observation leads us to define the p-th homology group as

$$H_p = Z_p / B_p \tag{9.2}$$

In other words, elements of the p-th homology group are p-cycles modulo the equivalence relation $\sigma \simeq \sigma + \partial_{p+1}\tau$. Nontrivial homology elements are cycles that are not boundaries of higher dimensional cycles. These cycles are homologous to the empty p-cycle. See Fig. 9.3 for examples.

Recall that we are interested in identifying "holes" of various dimensions in a simplicial complex. The reason we have gone through the preceding mathematical definitions is that *a hole of dimension p corresponds to a nontrivial element of the p-th homology group.*[2] Moreover, since the p-th homology group is defined in terms of subspaces of the linear boundary operator ∂_p, its rank (counting the number of independent p-holes) can be calculated via simple linear algebra. These ranks are called the *Betti numbers*, which we denote b_p. They count the number of "holes" of dimension p.

9.2.2 Persistent Simplicial Homology

The machinery of simplicial homology allows us to compute topological invariants of simplicial complexes. These invariants are preserved by simplicial homeomorphisms. As in the continuous case, we might hope that in the context of data analysis we

[2] Strictly speaking, the 0-th homology group counts the independent components of the simplicial complex, i.e. disconnected clusters.

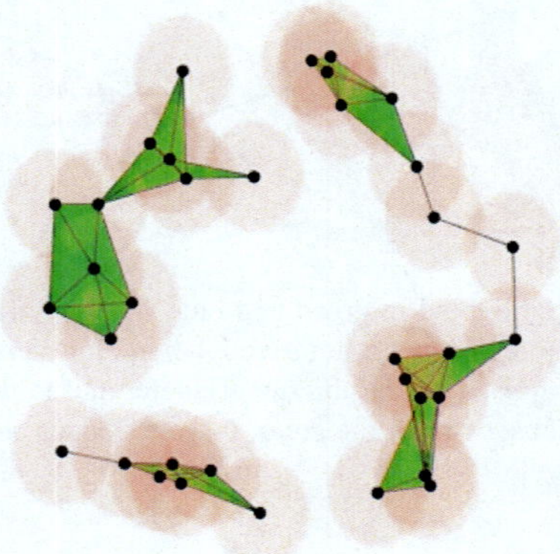

Fig. 9.4 A Vietoris-Rips complex. Vertices are connected by edges if their separation is less than some distance r (visualized via overlapping balls). Triangles are shaded in if all required edges are present. In persistent homology, the coarsening scale r is increased, and the births and deaths of individual topological features are tracked

have earned some robustness. This robustness in hand, we should then represent our data set by a simplicial complex and proceed to compute the simplicial homology. However, taking this approach suffers from immediate complications.

First of all, going from a point cloud to a single simplicial complex is not straightforward. It is natural to associate each point in the point cloud with a vertex, but when it comes to choosing higher-dimensional simplices, choices must be made. Choosing a single simplicial complex to represent our data would be asking for trouble, since different choices can lead to different topological invariants. Moreover, the presence of noise in a data set would cause similar problems. We wouldn't have gained much in the way of stability by resorting to a topological description.

We will work around these issues by instead representing our data set with a *filtration*, an increasing family of simplicial complexes $S_1 \subset S_2 \subset \cdots \subset S_n$. A filtration is often parameterized by some coarsening scale. For example, the Vietoris-Rips filtration at scale r is defined by including edges between two vertices v_0, v_1 if the distance between them is less than r, $d(v_0, v_1) < r$, and including higher dimensional simplices if all of their faces are present. An example of an intermediate-scale Vietoris-Rips complex is shown in Fig. 9.4. The sequence of simplicial complexes induces a sequence of homology groups $H_p(S_1) \to H_p(S_2) \to \cdots \to H_p(S_n)$. As the filtration parameter increases, nontrivial homology cycles are "born" (e.g.. edges are included to make holes) and "die" (e.g.. triangles fill in holes). When two formerly disconnected components (or elements of higher homology groups) merge, we adopt the *elder rule*: the component that was born earlier survies, while the other component dies. In persistent homology, one is able to track the lifetimes of *individual* homology classes. Intuitively, features that *persist* for a long time in the filtration are robust aspects of the point cloud. For example, for the Vietoris-Rips filtration in Fig. 9.4, we expect a long-lived 1-cycle, corresponding to the large "hole" in the middle. Persistence allows us to assign a notion of significance to different topological features. This can be exploited to define distances between persistence diagrams that enjoy nice stability properties.

In can be shown that when each p-simplex in a filtration is added to the simplicial complex, it either creates a p-cycle or destroys a $(p-1)$-cycle. In persistent homology, we track individual topological features by computing the *persistence pair* (σ, τ) that create and destroy a particular homology class. The homology class is then "born" when σ is added to the filtration and "dies" when τ is added. Given a filtration containing N simplices, one can compute the persistence pairing in $O(N^3)$ time, often quite faster, via linear algebra operations on the boundary operators [44].

9.2.3 Persistence Diagrams and Other Representations

One useful way to represent and interpret the result of of a persistent homology computation is the persistence diagram. Persistence diagrams (PDs) are scatter plots of the "birth" and "death" times of individual nontrivial homology cycles. Naturally, features that "persist" for a long time fall far from the $\nu_{\text{birth}} = \nu_{\text{death}}$ line. In general one can plot different PDs for cycles of different dimension, or can use a color-coding scheme to plot cycles of different dimension on the same diagram. An example PD is shown in Fig. 9.5.

PDs can be processed further to give one dimensional functions. The *persistent Betti numbers* are one dimensional functions that count the number of "living" p-cycles at a given filtration parameter. They are simply the Betti numbers of simplicial complexes in the filtration, $b_p(\nu) = \text{rank} H_p(S_\nu)$. In equations, the p-th persistent Betti number at filtration parameter ν can be calculated from a PD as

$$b_p(\nu) = \sum_{b \leq \nu, d > \nu} O_p(b, d) \tag{9.3}$$

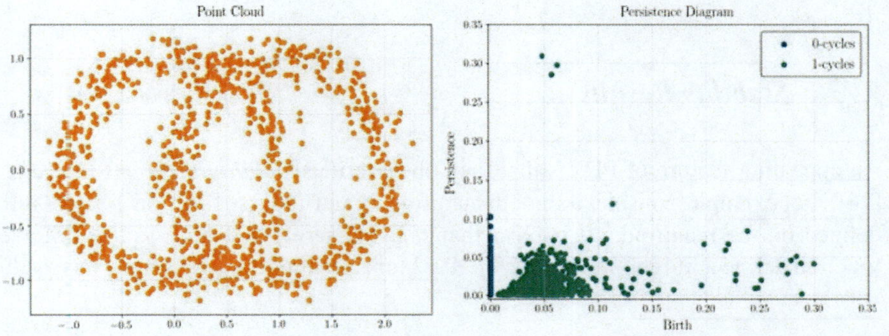

Fig. 9.5 Left: input point cloud corresponding to two overlapping rings. Right: the corresponding PD. There are three 1-cycles of large persistence

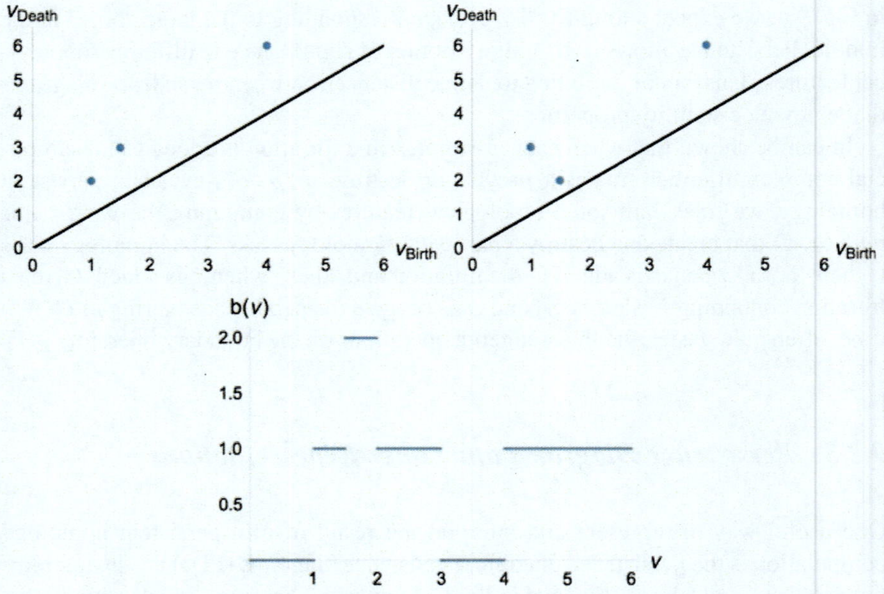

Fig. 9.6 Distinct persistence diagrams (PDs, above) can give rise to the same persistent Betti numbers (below). Since PDs track individual cycles, they contain more information than persistent Betti numbers, which merely count cycles

Here $O_p(b, d)$ counts the number of p-cycles that are born at b and die at d. Clearly, the transformation from a PD to a persistent Betti number is *not invertible*. In other words, PDs contain strictly more information than persistent Betti numbers. This is because persistent Betti numbers merely count topological features, while PDs track individual cycles. An example of two distinct PDs giving rise to the same persistent Betti number is shown in Fig. 9.6. In addition to the persistent Betti numbers, one can slice a persistence diagram in many ways to derive other one-dimensional statistics, for example counting the number of cycles born before a particular filtration time, etc.

9.2.4 Stability Results

An appealing feature of PDs is that they obey various *stability theorems* (see e.g.. [26]). For example, consider a smooth manifold M and a Morse function $f : M \to \mathbb{R}$ defined on the manifold.[3] It is clear that *sublevel sets* $f^{-1}(-\infty, \nu]$ of the Morse function define a filtration. See [44, 77, 103] for relevant details on Morse theory. In

[3]Strictly speaking, to use the simplcial machinery we have considered so far, M and f are suitably discretized.

fact, one can make contact to filtrations defined via point sets $X \subset \mathbb{R}^m$ by considering the *distance function*

$$f_X(y) = \inf_{x \in X} \|x - y\| \tag{9.4}$$

Now letting $PD_n(f)$ represent the persistence diagram corresponding to the n-dimensional persistent homology of the sublevel filtration of f, one can show that

$$d_B(PD_n(f), PD_n(g)) \leq \|f - g\|_\infty \tag{9.5}$$

where d_B is the so-called *bottleneck distance* between persistence diagrams, defined by

$$d_B(PD_n(f), PD_n(g)) = \inf_\eta \sup_{x \in PD_n(f)} \|x - \eta(x)|_\infty \tag{9.6}$$

where η is a bijection between $PD_n(f)$ and $PD_n(g)$, and the finite number of nonzero persistence cycles in a given persistence diagram is supplemented by an infinite number of zero-persistence cycles along the "birth=death" line of the persistence diagram. In other words, if the input data of a filtration is perturbed, the change in the PD is provably small. Intuitively, this is because small noise will not create or destroy significant topological features. Thus we see how the introduction of persistence leads to significant robustness.

This guarantee of robustness might be compared with the problem of adversarial perturbations for machine learning frameworks [97]. Here, imperceptible noise can cause neural networks and other ML architectures to e.g.. misclassify images in dramatic ways. One might hope that a machine learning pipeline utilizing persistent homology can be made more robust to such perturbations. Note also that slicings of PDs like the persistent Betti numbers do not inherit stability from PDs.

9.2.5 Coupling to Machine Learning, Statistical Inference

As scatter plots, PDs do not immediately couple nicely to machine learning pipelines. However, there are several routes one can take to define nicer mathematical objects. One derived statistic that does inherit stability is the persistence image [4], which can be used for statistical learning. Roughly speaking, the persistence image is a persistence diagram where individual cycles have been smoothed via some specified kernel and the resulting scalar function has been binned as a 2-dimensional histogram. To inherit the stability of PDs, one also multiplies the kernel of a given cycle by a factor that depends on the cycle's persistence, for example $\log(1 + \nu_{\text{death}} - \nu_{\text{birth}})$. The resulting object lives in a vector space and as such naturally couples to statistical learning pipelines. Another persistence-based object suitable for statistical inference is the persistence landscape [16]. Coupling a differentiable persistence computation to a neural network architecture and proving stability bounds is an active area of research, see e.g.. [15, 20, 62, 63, 65, 84].

9.3 TDA for Cosmology

Cosmology is marching into an era of Big Data. For example, ongoing and upcoming surveys promise to generate huge numbers of new observations of galaxies, giving us greater access to the history of the universe than ever before. To fully leverage this flood of data, we would benefit greatly from a general pattern recognition tool. However, to sufficiently trust such a tool, we should understand the pipeline in terms of familiar concepts. For example, for a given N-body simulation of Large-Scale Structure, the simulation's trustworthiness breaks down before the simulation's grid size. (See [92] for a comparison of matter power spectrum predictions with various simulation methods.) We would like to be able to formulate a data science pipeline that does not rely on simulation details at this scale. It appears that persistent homology could provide such a general tool.

For example, we would like to learn about inflation. Inflation is a postulated period of exponential expansion in the early universe that solves the flatness, horizon, and monopole problems of Big Bang cosmology [6, 52, 72, 94]. During this inflationary phase, quantum fluctuations are stretc.hed to superhorizon scales, where they freeze and classicalize. When these fluctuations reenter the horizon, they evolve under gravity, with overdense regions collapsing to form the cosmic structure we observe today. Generically, inflation predicts an almost scale-invariant primordial density perturbation spectrum [11, 53, 58, 79–81, 95]. These perturbations are not exactly Gaussian, but their deviation from Gaussianity is typically small for single field slow-roll inflation [73]. A broader class of inflationary models, even within the single field context, allows significant levels of non-Gaussianity with distinctive "shapes" (functional dependences on momenta) [22]. These shapes reflect the underlying physics of the model, e.g.. a large non-Gaussianity in the so-called *squeezed limit* can only by achieved by multi-field inflation. We therefore aim to constrain models of inflation by studying the statistics of cosmological observables and reverse engineering the quantum perturbations and primordial physics that generated them. Further non-Gaussianities develop via the nonlinearity of gravitational evolution, so that we can also access the dynamics of cosmological expansion (and e.g.. the dark energy equation of state) via non-Gaussianities. Observables allowing us to do this include the Cosmic Microwave Background (CMB) and the "cosmic web" of Large-Scale Structure (LSS).

However, directly detecting primordial non-Gaussianity via higher-order moments proves difficult in practice, as it relies on the measurement of rare events. Moreover, since a random variable can deviate from Gaussianity in infinitely many ways, there is no single signature of primordial non-Gaussianity. So far, primordial non-Gaussianity in the CMB has proven elusive, with the primordial non-Gaussian component of data dominated by observational effects [5]. We must therefore employ complementary approaches to data analysis. Current approaches to measuring non-Gaussianity fall into two categories. The geometric approach aims to quantify the structure of a cosmological observable in position-space. Another approach, which generally takes place in momentum-space, fits higher-order correlation functions against templates for different well-motivated models [9, 71].

In this section we review aspects of the geometric approach, focusing on recent developments and going into more detail regarding the present authors' contributions.

9.3.1 Morphology of Cosmological Observables: Recent Progress

Roughly speaking, there are two ways one can gather geometric information from cosmological observables. The primordial perturbations set up by inflation are naturally described in the language of random fields. For "field-like" observables, one can consider geometric aspects of excursion sets of the field. For example, primordial perturbations are imprinted on the Cosmic Microwave Background as temperature and polarization anisotropies. In the language of Sect. 9.2 one regards e.g.. the temperature anisotropy as a Morse function on the sphere, $\frac{\Delta T}{T_{\text{avg}}} : S^2 \to \mathbb{R}$. Geometric aspects of the sublevel sets $\left(\frac{\Delta T}{T_{\text{avg}}}\right)^{-1}(-\infty, \nu]$ with varying ν are then studied.

On the other hand, we do not observe the density perturbation directly as an instance of a random field. Rather, we observe *tracers* of the underlying distribution, such as galaxies. The organization of this distribution into an intricate hierarchical network of interlocking filaments, voids, and clusters is sometimes called the cosmic web [13, 68]. In this context, it is more natural to take a "point-like" viewpoint. As we saw in Sect. 9.2, persistent homology naturally applies to both contexts.[4]

One class of geometric observables in the "field-like" category are the Minkowski Functionals (MFs) [76, 90, 91, 102]. Analytic expressions for the MFs are known in the weakly non-Gaussian case [48, 61, 75]. In particular, for the CMB, there are three MFs: area, length, and "genus" of excursion sets.

The "genus" is in fact, up to normalization, the Euler character of these sets. This can be related to the Betti numbers via the Euler-Poincaré formula

$$\chi = \sum_p (-1)^p b_p \tag{9.7}$$

Here we should regard χ and b_p as functions of the threshold temperature ν. The non-invertibility of the transformation $b_p \to \chi$ tells us that the persistent Betti numbers

[4] As was also described in Sect. 9.2, the distinction between "point-like" and "field-like" observables is somewhat artificial, since collections of points can be used to define "field-like" observables and vice versa. In practice, however, transforming between these viewpoints may require nontrivial assumptions about the underlying physics, for examples regarding the bias of various tracers.

contain more information about a set than the Euler character. This was exploited in [23], which showed on a set of CMB simulations that the Betti numbers were quantitatively more sensitive to a particular form of primordial non-Gaussianity than the Euler character. Moreover, in Sect. 9.2 we saw that the persistence diagram in fact contains more information than the persistent Betti numbers. One might then ask how much of the information ignored by persistent Betti numbers is useful for statistical discrimination. As such, in [29] the present authors studied the ability of persistent homology to identify primordial local non-Gaussianity. More recently, [85] computed the persistent Betti numbers of Planck data and, comparing to the persistent Betti numbers of Planck best-fit simulations, identified some mild anomalies.

An important theme in recent work is that persistence-based statistics would be much more powerful with a strong analytic foothold. While analytic expressions for the MFs are known for mildly non-Gaussian fields [49, 75], we do not have analogous expressions for even the persistent Betti numbers of a Gaussian random field. These quantities were studied numerically in [87]. Additionally [47] proposed a semi-analytic formula for the persistent Betti numbers of a Gaussian random field. A full derivation, however, is still lacking. A higher aim would be predictions for the persistence diagrams of these random fields.

Some analytic progress for other geometric observables has been made in [25], which used DisPerSE [93] to extract the "persistent skeleton" (i.e. network connecting maxima and saddle points of a 3D field) of Gaussian random fields and cosmological simulations, and found agreement regarding the connectivity of this skeleton with predictions made via peak theory.

For progress from the "point-like" perspective, [86] computed persistence diagrams and persistent Betti numbers for simulations of some toy cosmological models, see also [86, 101]. Similar methods were applied to SDSS data in [66]. One intriguing aspect of this approach is one can probe the dynamics of cosmological evolution (including for example the dark energy equation of state) by studying LSS at different redshifts. More recently, [12] initiated a program for detecting primordial non-Gaussianity and constraining cosmological parameters via the persistent homology of LSS, taking the first steps towards making persistent homology a standard tool for computational cosmologists.

9.3.2 *TDA for CMB [29]*

In this subsection, we describe in more detail some relevant aspects of applying persistent homology to cosmology, following [29]. We review the ansatz of primordial local non-Gaussianity and describe how this primordial imprint propagates to cosmological observables. We describe some publicly available simulations [46] and the statistical sensitivity of a persistence-based pipeline to non-Gaussianity in these simulations. We find that a combination of the persistence diagrams is nearly twice as sensitive as the persistent Betti numbers to non-Gaussianity in these simulations. Throughout, we emphasize several pitfalls to avoid when studying the persistent homology of cosmologically-relevant non-Gaussian fields.

9.3.2.1 Local Non-Gaussianity

A phenomenological parameterization of primordial non-Gaussianity called local non-Gaussianity takes the ansatz [69]

$$\Phi(\vec{x}) = \Phi^{\mathrm{G}}(\vec{x}) + f_{\mathrm{NL}}^{\mathrm{loc}}\left(\Phi^{\mathrm{G}}(\vec{x})^2 - \langle\Phi^{\mathrm{G}}(\vec{x})^2\rangle\right) \tag{9.8}$$

for the primordial gravitational potential Φ, where Φ^{G} is a random Gaussian field and $f_{\mathrm{NL}}^{\mathrm{loc}}$ is a free parameter. Propagating the primordial gravitational potential to CMB temperature anisotropies and expanding in spherical harmonics, one has $\frac{\Delta T}{T_{\mathrm{avg}}} = \sum_{\ell m} a_{\ell m} Y_{\ell m}$ with

$$a_{\ell m} = 4\pi(-i)^{\ell} \int \frac{d^3 k}{(2\pi)^3} \Phi(\vec{k}) \Delta_{\ell}(k) Y_{\ell m}^*(\hat{k}) \tag{9.9}$$

where $\Delta_{\ell}(k)$ is the transfer function for temperature in momentum space. Note that since $\Phi \sim 10^{-5}$, the three-point function in these models is correspondingly suppressed relative to the two-point function, so that even for $f_{\mathrm{NL}}^{\mathrm{loc}} \sim 100$ (which is already well-excluded by current experimental bounds), the non-Gaussianity is still weak. In other words, the relevant expansion parameter is $f_{\mathrm{NL}}\Phi$, not just f_{NL}. Thus if one normalizes the field to have unit variance, $f_{\mathrm{NL}}^{\mathrm{loc}}$ must be suitably rescaled to remain in the regime of weak non-Gaussianity. It is also important to note that local ansatz applies to primordial field, not the temperature anisotropy. This makes simulating the CMB with primordial non-Gaussianity a more involved procedure than taking $\Delta T = \Delta T_{\mathrm{G}} + f_{\mathrm{NL}}^{\mathrm{loc}}\left(\Delta T_{\mathrm{G}}^2 - \langle \Delta T_{\mathrm{G}}\rangle^2\right)$.

Publicly available simulations of local non-Gaussianity have been provided by Elsner and Wandelt [46]. The simulations are generated using a seed Gaussian field Φ_{G}. The spherical harmonic coefficients are then calculated by combining the ansatz (9.8) with (9.9). The simulations provide 1000 sets of Gaussian spherical harmonic coefficients $a_{\ell m}^{L}$ and corresponding non-Gaussian $a_{\ell m}^{NL}$. One can tune the amount of non-Gaussianity in a particular map by using $a_{\ell m} = a_{\ell m}^{\mathrm{L}} + f_{\mathrm{NL}}^{\mathrm{loc}} a_{\ell m}^{\mathrm{NL}}$.

In [29] each simulation was preprocessed by cutting out squares, 2 per simulation. The persistent homology of a sublevel filtration for the temperature anisotropy of each square was then computed. From the PDs, two derived statistics were considered, the PDs binned as two-dimensional histograms and the persistent Betti numbers.

9.3.2.2 Statistical Power

The constraining power of a statistic may be estimated by computing $P(\mathbf{p}|\mathbf{d})$. Here P gives the probability that some data being measured implies the model under consideration has parameters $\mathbf{p}$. Via Bayes' theorem, one has

Table 9.1 Relative information content of various topological statistics

Statistic	$\Delta f_{\mathrm{NL}}^{\mathrm{loc}}$
b_0	67.4
b_1	66.1
$b_0 + b_1$	60.6
PD_0	39.1
PD_1	37.4
$PD_0 + PD_1$	35.8

$$P(\mathbf{p}|\mathbf{d}) = \frac{P(\mathbf{d}|\mathbf{p})P(\mathbf{p})}{P(\mathbf{d})} \tag{9.10}$$

The denominator is a simple normalization factor, and thus does not affect the location of the likelihood function's peak or its width. Assuming a uniform prior $P(\mathbf{p})$, one then has that $P(\mathbf{p}|\mathbf{d}) \propto P(\mathbf{d}|\mathbf{p})$, where $P(\mathbf{d}|\mathbf{p})$ is the likelihood function, which we can predict given a sufficiently well-studied theory. One can often assume a Gaussian likelihood function

$$P(\mathbf{d}|f_{\mathrm{NL}}^{\mathrm{loc}}) = \frac{1}{2\pi\sqrt{\det \mathbf{C}}} \exp\left(-\frac{1}{2}(\mathbf{d}-\boldsymbol{\mu})\mathbf{C}^{-1}(\mathbf{d}-\boldsymbol{\mu})\right) \tag{9.11}$$

where $\mathbf{d}$ is the data vector, $\boldsymbol{\mu} = \boldsymbol{\mu}(f_{\mathrm{NL}}^{\mathrm{loc}})$ is the average data vector of our model at a given f_{NL}, and $\mathbf{C}$ is the covariance matrix. The sharper the likelihood function is peaked at its maximum, the more sensitive it is to the cosmological parameters under consideration. Along these lines, the data can be thought of as coming from a theory with $f_{\mathrm{NL}}^{\mathrm{loc}} = 0$, using the average data vector over 800 simulations. One then evaluates the likelihood function for each statistic using $f_{\mathrm{NL}}^{\mathrm{loc}} = 0, 10, 50$ for $\boldsymbol{\mu}(f_{\mathrm{NL}}^{\mathrm{loc}})$, also averaging over 800 simulations. Here the covariance matrix $\mathbf{C}$ is calculated using 2000 $f_{\mathrm{NL}}^{\mathrm{loc}} = 0$ simulations. The likelihood function is then well-approximated by a Gaussian in $f_{\mathrm{NL}}^{\mathrm{loc}}$. Resulting 1σ "constraints" are shown in Table 9.1. We observe that the PDs are almost twice as sensitive as the Betti number curves, which are in turn [23] more discerning than the genus.

9.4 TDA for the String Landscape

String theory naturally lives in 10 dimensions. To make contact with our 4-dimensional universe, one must then consider strings propagating in a background geometry that features compact dimensions. The large number of choices of background geometries and additional discrete ingredients (branes, fluxes) implies that the set of string vacua, called the *string landscape*, is vast.[5] In [14] it was noted that the presence of multiple quantized fluxes leads to a corresponding *discretuum* of values for physical observables like the cosmological constant. Thus one can argue for the existence of vacua satisfying certain criteria, but finding them remains difficult. To understand the distribution of vacua with different properties, [40] proposed a statistical approach. Subsequent work used statistical arguments and searched sets of explicit constructions to make arguments about stringy naturalness, especially regarding to the scale of supersymmetry breaking, the presence of various symmetries, and the cosmological constant [2, 8, 10, 31, 34, 35, 37–39, 41–43, 50, 60, 64, 74, 96].

More recently, advances in machine learning and data science have inspired applications of these techniques to the string landscape [1, 17, 19, 32, 59, 67, 70, 88, 100], [7, 18, 28, 54, 56, 78], see [89] for a review. In this context, we can view persistent homology as a tool for pattern recognition and characterization. As such, in [30] (see also [24]) persistent homology was used to study the distribution of vacua in complex structure moduli space for Type IIB flux vacua. In this section we review some of the results of [30]. In Sect. 9.4.1 we review the features of Type IIB flux vacua relevant for study with persistent homology. In Sect. 9.4.2 we examine the structure of a simple toy model, the rigid Calabi-Yau. In this context, we find it useful to introduce a persistence diagram dressed with information regarding the intersection of destroying simplices.

9.4.1 Flux Vacua

As previously mentioned, string theory must be compactified on a background geometry in order to make contact with 4-dimensional physics. The compactification is well-controlled if some amount of supersymmetry is preserved. This can be achieved by compactifying on a Calabi-Yau manifold. One then has an effective 4-dimensional theory. In the effective theory, there are massless fields, called moduli. These arise for example from certain deformations of the internal manifold. If the moduli remain massless, they present problems for cosmology. Therefore, they must be stabilized, i.e. given a potential. A particular modulus is fixed by its potential to lie at a local minimum. More precisely, the vacuum expectation value (VEV) is fixed to lie at this minimum.

In this section, we will consider type IIB string theory on a Calabi-Yau orientifold. There are then several classes of moduli: the axiodilaton, a universal modulus that includes the string coupling g_s; the complex structure moduli (which parametrize roughly speaking the "shapes" of the internal manifold); the Kähler moduli (which control the "sizes" of the internal (sub)manifolds); and open string moduli. The

[5] For a sense of scale, it is estimated that the number of flux vacua for a typical geometry is around 10^{500} [8, 34] and can be as large as $10^{272,000}$ [98]. The number of geometries in a particular of F-theory ensemble is bounded below by $\frac{4}{3} \times 2.96 \times 10^{755}$ [55].

axiodilaton and complex structure moduli are stabilized by turning on 3-form fluxes that thread the internal manifold. We will study the resulting distribution of stabilized axiodilaton and complex structure moduli VEVs using persistent homology. There are of course other interesting quantities to study, including the VEVs of other moduli (which must be stabilized via other means), masses of various moduli, and the magnitude of the flux superpotential (to be defined shortly). As we will see, restricting to the axiodilaton and complex structure moduli VEVs already gives plenty of structure for us to explore.

We now review the relevant formulas for generating distributions of vacua. Readers who are not familiar with the terminologies of supergravity may find the following discussion somewhat technical but the basic idea is simple, in that we are looking for local minima of the potential energy. The most important equations for our purposes are (9.19), (9.20), and (9.22). We follow the conventions of [36]. Consider a Calabi-Yau threefold M with $h_{2,1}$ complex structure moduli. Take a symplectic basis $\{A^a, B_b\}$ for the $b_3 = 2h_{2,1} + 2$ three-cycles, with $a, b = 1, \ldots, h_{2,1} + 1$. We have dual cohomology elements α_a, β^b satisfying

$$\int_{A^a} \alpha_b = \delta^a_b, \quad \int_{B_b} \beta^a = -\delta^a_b, \quad \int_M \alpha_a \wedge \beta^b = \delta^b_a \tag{9.12}$$

From the unique holomorphic three-form Ω, we have the periods $z^a \equiv \int_{A_a} \Omega$, $\mathcal{G}_b \equiv \int_{B_b} \Omega$, which form the b_3-vector $\Pi(z) \equiv (\mathcal{G}_b, z^a)$. Additionally

$$\int_M \Omega \wedge \overline{\Omega} = \overline{z}^a \mathcal{G}_a - z^a \overline{\mathcal{G}}_a = -\Pi^\dagger \cdot \Sigma \cdot \Pi \tag{9.13}$$

where we have the symplectic matrix

$$\Sigma = \begin{pmatrix} 0 & 1 \\ -1 & 0 \end{pmatrix} \tag{9.14}$$

whose entries are $(h_{2,1} + 1) \times (h_{2,1} + 1)$ matrices. The NSNS and RR 3-form fluxes are quantized and may be written in the α, β basis

$$F_3 = -(2\pi)^2\alpha'(f_a\alpha_a + f_{a+h_{2,1}+1}\beta^a), \quad H_3 = -(2\pi)^2\alpha'(h_a\alpha_a + h_{a+h_{2,1}+1}\beta^a) \tag{9.15}$$

where we have defined the integer-valued b_3-vectors f and h. From now on we set $(2\pi)^2\alpha' = 1$. The Kähler potential truncated to the axiodilaton and complex structure moduli is

$$\mathcal{K} = -\log\left(i\int_M \Omega \wedge \overline{\Omega}\right) - \log\left(-i(\phi - \overline{\phi})\right) = -\log(-i\Pi^\dagger \cdot \Sigma \cdot \Pi) - \log(-i(\phi - \overline{\phi})) \tag{9.16}$$

The fluxes induce a superpotential for the complex structure moduli and axiodilaton $\phi \equiv C_0 + ie^{-\varphi}$ [51]

$$W = \int_M G_3 \wedge \Omega(z) = (f - \phi h) \cdot \Pi(z) \tag{9.17}$$

where $G_3 \equiv F_3 - \phi H_3$. The superpotential and Kähler potential are related to the scalar potential via

$$V = e^{\mathcal{K}} \left(g^{a\bar{b}} D_a W D_{\bar{b}} \overline{W} - 3\, W|^2 \right) \tag{9.18}$$

where $D_a W \equiv \partial_a W + W \partial_a \mathcal{K}$, and a runs over all complex structure moduli. (At tree-level, the Kähler moduli do not appear in the scalar potential.) We are interested in vacua with vanishing F-terms

$$D_\phi W = \frac{1}{\overline{\phi} - \phi}(f - \overline{\phi} h) \cdot \Pi(z) = 0 \tag{9.19}$$

$$D_a W = (f - \phi h) \cdot (\partial_a \Pi(z) + \Pi(z) \partial_a \mathcal{K}) = 0 \tag{9.20}$$

These so-called F-flatness conditions (9.19) and (9.20) imply that the (3,0) and (1,2) parts of the fluxes vanish, so that G_3 is imaginary self-dual, $\star_6 G_3 = i G_3$.

The fluxes also contribute to the D3-brane charge

$$N_{\text{flux}} = \int_M F_3 \wedge H_3 = f \cdot \Sigma \cdot h \tag{9.21}$$

Since the corresponding flux threads a compact space, the total D3-brane charge must vanish, a consistency condition called tadpole cancellation. For imaginary self-dual fluxes, we have that $N_{\text{flux}} > 0$. Therefore tadpole cancellation requires the presence of negative D3-brane charges. A fixed amount of negative charge is induced by orientifolding. Orientifolding is a particular discrete identification of the internal manifold giving rise to fixed points with negative D-brane charge. For cancellation, any difference $L - N_{\text{flux}}$ can be made up by mobile D3-branes spanning the four-dimensional spacetime. We will not consider an explicit orientifold, but will instead take the orientifold charge as a free parameter $L_{\max}$, so that

$$0 < N_{\text{flux}} \leq L_{\max} \tag{9.22}$$

As we will see, $L_{\max}$ sets the scale at which we observe interesting structure in the moduli space distribution. To summarize, for a given Calabi-Yau, we choose fluxes satisfying (9.22) and use the F-flatness conditions (9.19), (9.20) to solve for the VEVs of corresponding stabilized moduli. We then study the distribution of these moduli using persistent homology.

One subtlely is that we must fix a gauge, since multiple flux choices can be equivalent under modular transformations. We will fix our gauge by mapping moduli VEVs to a fundamental domain. It is sensible to perform our gauge fixing in the moduli space because this is where we will perform our persistent homology analysis. If we were to perform gauge-fixing by imposing certain conditions on the fluxes, some inequivalent vacua with the same moduli VEVs would be mapped to different parts of the moduli space. Moreover, if we were to choose to fix our gauge by restricting the moduli to a collection of regions other than a domain (e.g.. multiple disconnected regions), the resulting structure of the distribution would depend on our particular choice of regions. On the other hand, as was checked in [30], the choice of particular domain does not affect the persistent homology results.

Finally, there is physical motivation for gauge-fixing in this way. Given the large number of vacua in the landscape, one might find themselves with an attitude of "anything goes," i.e. any consistent 4-dimensional physics is realized. However, as has recently been emphasized by the swampland program [99], this does not seem to be true. In other words, requiring that a 4-dimensional theory arises from UV-complete quantum gravity imposes nontrivial restrictions. In our context, these restrictions arise as the presence of correlations in distributions of vacua and an inability to fine-tune certain quantities beyond a particular resolution. This latter point, the existence of a minimum resolution, is best captured for moduli VEVs when one fixes gauge in the way we have proposed.

9.4.2 *Rigid Calabi-Yau*

Consider a toy Calabi-Yau with no complex structure moduli, studied in [8, 34, 36]. We have $b_3 = 2$. A symplectic basis for $H_3(M)$ can be written as $\{\alpha, \beta\}$. We take the periods of the holomorphic three-form Ω to be

$$\int_\beta \Omega = 1, \quad \int_\alpha \Omega = i \tag{9.23}$$

The superpotential is

$$W = A\phi + B \tag{9.24}$$

where $A \equiv -h_1 - ih_2$, $B \equiv f_1 + if_2$. The D3-brane charge induced by the fluxes is

$$N_{\text{flux}} = f_1 h_2 - f_2 h_1 \tag{9.25}$$

and the vacuum equation is

$$D_\phi W = A\overline{\phi} + B = 0 \tag{9.26}$$

which is solved by

$$\phi = -\overline{\left(\frac{B}{A}\right)} \tag{9.27}$$

We are interested in the distribution of flux vacua on the axiodilaton moduli space at fixed $L_{\max}$. We fix a gauge by restricting the axiodilaton to the $SL(2,\mathbb{Z})$ fundamental domain:

$$\phi \in \mathcal{F}_D = \left\{ z \in \mathbb{C}_+ : -\frac{1}{2} < \mathrm{Re}(z) \leq \frac{1}{2},\ |z| \geq 1 \right\} \tag{9.28}$$

This can be accomplished by using the $SL(2,\mathbb{Z})$ transformation

$$\phi \to \frac{a\phi + b}{c\phi + d}, \quad \begin{pmatrix} f \\ h \end{pmatrix} \to \begin{pmatrix} a\ b \\ c\ d \end{pmatrix}, \quad a, b, c, d \in \mathbb{Z}, \quad ad - bc = 1 \tag{9.29}$$

It is easy to check that the scalar potential is unchanged by such a transformation. Despite the simplicity of the toy model, the distribution of ϕ exhibits rich structure, see Fig. 9.7. In particular, there are "voids" with no vacua except for at accumulation points in their centers. In fact, each point lies at the center of its own void.

This structure arises from a combination of flux quantization, tadpole cancellation, and the F-term equation. The F-term equation (9.27) tells us that when we project onto ϕ, we are introducing some degeneracy, since multiple flux configurations give the same stabilized value for ϕ. Specific values for ϕ correspond to 2-dimensional hyperplanes (intersecting with some gauge-fixing conditions) in the 4-dimensional flux space. Flux quantization forces these slopes to be rational. For example, $\phi = i\beta$ for rational β corresponds to the hyperplane $f_1 = \beta h_2,\ f_2 = -\beta h_1$. Moving around in the axiodilaton moduli space corresponds to rotating these hyperplanes along two axes. However, tadpole cancellation with finite $L_{\max}$ means that most points in the axiodilaton moduli space are not represented in the distribution, since their corresponding hyperplanes fail to hit integer points in flux space before reaching the limits in flux space imposed by tadpole cancellation. These ϕ values are not present in the resulting distribution.

Consider for example $\phi = i\beta$. If such a ϕ is present one of the corresponding vacua takes the form $\frac{if_1}{h_2}$. Without loss of generality assume that f_1 and h_2 have no common factors and that both are positive. The nearest vacuum on the imaginary axis takes the form $i\left(\frac{f_1}{h_2} - \frac{a}{b}\right)$ with $a, b > 0$. (One should use a plus sign for $f_1 = h_2$ to stay in the fundamental domain.) To find the nearest neighbor, one minimizes $\frac{a}{b}$ over the naturals subject to the constraint $N_{\text{flux}} = b^2 f_1 h_2 - abh_2^2 \leq L_{\max}$. It can be shown that

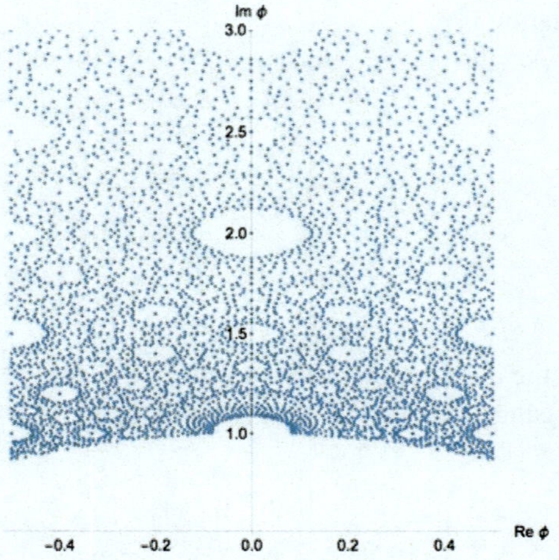

Fig. 9.7 Rigid Calabi-Yau vacua projected onto axiodilaton with $L_{\max} = 150$. The relative sizes of the voids depend on number-theoretic aspects of the complex rational value of the axiodilaton at the center

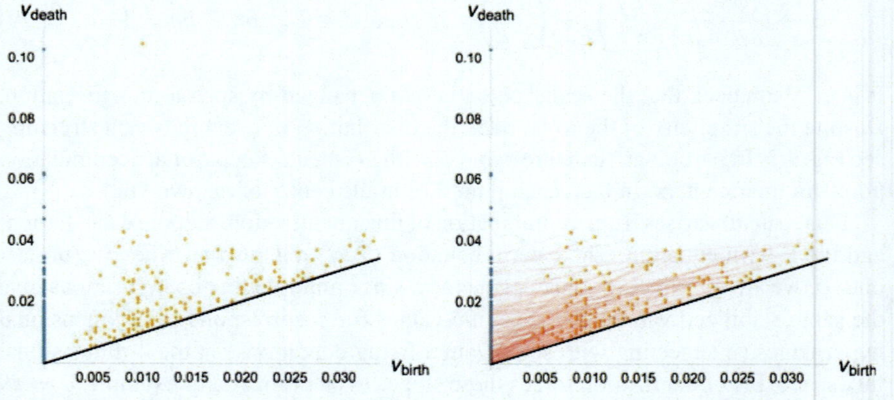

Fig. 9.8 Left: persistence diagram for rigid Calabi-Yau flux vacua projected onto axiodilaton with $L_{\max} = 150$, using a lazy witness complex with 700 landmark points. Witness complexes (and their lazy variants) are useful when one needs to subsample the data set for computational reasons, see [33] for their definition. The orange points represent 1-cycles and the blue points represent 0-cycles. We observe many long-lived 1-cycles, corresponding to voids in the distribution, which seem to correlate with long-lived 0-cycles. Right: we link with a red line 0-cycles and 1-cycles whose destroying simplices overlap

the constrained minimization is achieved for $a = 1$, $b = \lfloor \frac{h_2}{2f_1} + \sqrt{\left(\frac{h_2}{2f_1}\right)^2 + \frac{L_{\max}}{f_1 h_2}} \rfloor$. Since b scales as $\sqrt{L_{\max}}$, we recover the previously known fact that the voids shrink as $\sqrt{1/L_{\max}}$ as $L_{\max}$ is increased [34, 36].

In terms of persistent homology, we expect long-lived 1-cycles corresponding to large voids. Additionally, we expect the death of the 1-cycle corresponding to the void will be correlated with the death of a 0-cycle corresponding to the isolated vacuum in the center of the void. Persistence diagrams for a subregion of the rigid Calabi-Yau axiodilaton distribution are shown in Fig. 9.8. The longer-lived 1-cycles correspond to the larger voids in Fig. 9.7.

9.4.2.1 Dressed Persistence Diagrams

While the correlation between 1-cycle and 0-cycle deaths is suggestive, it is not enough to reconstruct the void/isolated point structure of the distribution. To recover more fine-grained information about the point cloud, [30] introduced a dressed persistence diagram. This representation visualizes information from the persistence pairing that is not shown in a standard persistence diagram. Recall that for a filtration, one generates a persistence diagram by computing a set of persistence pairs (σ, τ). A given homology class is created when σ is added to the filtration and destroyed when τ is added. We might then be able to visualize correlations between different topological features by using information contained in the persistence pairs. Note that while elements of the homology groups are of equivalence classes of p-cycles, the simplices in the corresponding persistence pair are unique. There is no need for us to choose a canonical cycle. Specifically, one may draw a line between cycles in the persistence diagram whose destroying simplices overlap, and this is a unique procedure. In the present context, this proves useful because the death of the 0-cycle corresponding to an isolated vacuum in the center of a void is caused by the addition of an edge connecting that vacuum to a point on the edge of the void. This same edge is contained in the triangles that fill in the void, causing the 1-cycle's death. This dressed persistence diagram is shown in Fig. 9.8, demonstrating further correlations not evident in a standard persistence diagram.

The rigid Calabi-Yau has a moduli space of real dimension 2, and as such can simply be examined by eye. In [30] examples of moduli spaces with 4 real dimensions were also considered, in particular the symmetric $T^6 = (T^2)^3$ and a particular hypersurface in weighted projective space.

9.5 Conclusion

Much of theoretical physics involves finding the right structures to efficiently describe and understand a system. For example, the existence and identification of conserved quantities can simplify one's description of a theory greatly. Studying physical data sets using persistent homology seems a promising avenue for discovering patterns that allow for such a simplified description. Moreover, it has recently been shown that persistent homology can efficiently detect and classify phases of matter, providing quantitative and interpretable order parameters for phase transitions [27]. This suggests that persistent homology has potentially wider applications in identifying the structures of data.

In this chapter, we reviewed two such applications. In the first, persistent homology was used to characterize the effect of primordial non-Gaussianity on the Cosmic Microwave Background. A statistical pipeline based on persistent homology demonstrated impressive increase in sensitivity compared to previous topological observables, motivating future work into making analytic predictions. In the second application, persistent homology was used to characterize distributions of string vacua. In this context, it proved helpful to introduce a dressed persistence diagram that shows the relationships between cycles of different dimension.

Developing further these applications appears promising. As mentioned briefly, applying persistent homology in a cosmological context would prove more powerful if we had a stronger grasp on analytic properties of the persistent topology of random fields. Moreover, in addition to the CMB and the LSS there is the 21-cm field, the topology of which is an interesting probe of the epoch of reionization [45]. In terms of the string landscape, it would be interesting to use persistent homology to study data sets in dimensions higher than four, and to further understand the distribution of vacua with phenomenologically interesting properties.

Disclaimer: Views and opinions expressed are those of the authors and do not necessarily represent official positions of their respective companies.

References

1. Abel, S., Rizos, J.: Genetic algorithms and the search for viable string vacua. JHEP **08**, 010 (2014). https://doi.org/10.1007/JHEP08(2014)010
2. Acharya, B.S., Denef, F., Valandro, R.: Statistics of M theory vacua. JHEP **06**, 056 (2005). https://doi.org/10.1088/1126-6708/2005/06/056
3. Ackley, D.H., Hinton, G.E., Sejnowski, T.J.: A learning algorithm for boltzmann machines. Cogn. Sci. **9**(1), 147–169 (1985)
4. Adams, H., Emerson, T., Kirby, M., Neville, R., Peterson, C., Shipman, P., Chepushtanova, S., Hanson, E., Motta, F., Ziegelmeier, L.: Persistence images: a stable vector representation of persistent homology. J. Mach. Learn. Res. **18**(1), 218–252 (2017)
5. Ade, P.A.R., et al.: Planck 2015 results XVII constraints on primordial non-gaussianity. Astron. Astrophys. **594**, A17 (2016). https://doi.org/10.1051/0004-6361/201525836
6. Albrecht, A., Steinhardt, P.J.: Cosmology for grand unified theories with radiatively induced symmetry breaking. Phys. Rev. Lett. **48**, 1220–1223 (1982). https://doi.org/10.1103/PhysRevLett.48.1220
7. Ashmore, A., He, Y.H., Ovrut, B.A.: Machine learning Calabi-Yau metrics (2019)
8. Ashok, S., Douglas, M.R.: Counting flux vacua. JHEP **01**, 060 (2004). https://doi.org/10.1088/1126-6708/2004/01/060
9. Babich, D., Creminelli, P., Zaldarriaga, M.: The shape of non-gaussianities. JCAP **0408**, 009 (2004). https://doi.org/10.1088/1475-7516/2004/08/009
10. Banks, T., Dine, M., Gorbatov, E.: Is there a string theory landscape? JHEP **08**, 058 (2004). https://doi.org/10.1088/1126-6708/2004/08/058
11. Bardeen, J.M., Steinhardt, P.J., Turner, M.S.: Spontaneous creation of almost scale-free density perturbations in an inflationary universe. Phys. Rev. D **28**(4), 679 (1983)
12. Biagetti, M., Cole, A., Shiu, G.: The persistence of large scale structures I: primordial non-gaussianity **9**, (2020)

13. Bond, J.R., Kofman, L., Pogosyan, D.: How filaments of galaxies are woven into the cosmic web. Nature **380**(6575), 603–606 (1996)
14. Bousso, R., Polchinski, J.: Quantization of four form fluxes and dynamical neutralization of the cosmological constant. JHEP **06**, 006 (2000). https://doi.org/10.1088/1126-6708/2000/06/006
15. Brüel-Gabrielsson, R., Nelson, B.J., Dwaraknath, A., Skraba, P., Guibas, L.J., Carlsson, G.: A topology layer for machine learning. arXiv preprint arXiv:1905.12200 (2019)
16. Bubenik, P.: Statistical topological data analysis using persistence landscapes. J. Mach. Learn. Res. **16**(1), 77–102 (2015)
17. Bull, K., He, Y.H., Jejjala, V., Mishra, C.: Machine learning CICY threefolds (2018)
18. Bull, K., He, Y.H., Jejjala, V., Mishra, C.: Getting CICY high. Phys. Lett. **B795**, 700–706 (2019). https://doi.org/10.1016/j.physletb.2019.06.067
19. Carifio, J., Halverson, J., Krioukov, D., Nelson, B.D.: Machine learning in the string landscape. JHEP **09**, 157 (2017). https://doi.org/10.1007/JHEP09(2017)157
20. Carriere, M., Chazal, F., Ike, Y., Lacombe, T., Royer, M., Umeda, Y.: Perslay: a neural network layer for persistence diagrams and new graph topological signatures. Stat **1050**, 17 (2019)
21. Chazal, F., Michel, B.: An introduction to topological data analysis: fundamental and practical aspects for data scientists. arXiv preprint arXiv:1710.04019 (2017)
22. Chen, X., Huang, M.x., Kachru, S., Shiu, G.: Observational signatures and non-Gaussianities of general single field inflation. JCAP **0701**, 002 (2007). https://doi.org/10.1088/1475-7516/2007/01/002
23. Chingangbam, P., Park, C., Yogendran, K.P., van de Weygaert, R.: Hot and cold spot counts as probes of non-gaussianity in the cosmic microwave background. Astrophys. J. **755**, 122 (2012). https://doi.org/10.1088/0004-637X/755/2/122
24. Cirafici, M.: Persistent homology and string vacua. JHEP **03**, 045 (2016). https://doi.org/10.1007/JHEP03(2016)045
25. Codis, S., Pogosyan, D., Pichon, C.: On the connectivity of the cosmic web: theory and implications for cosmology and galaxy formation. Mon. Not. Roy. Astron. Soc. **479**(1), 973–993 (2018). https://doi.org/10.1093/mnras/sty1643, https://doi.org/10.1093/mnras/stz3535, [Erratum: Mon. Not. Roy. Astron. Soc.491, no.4,5794(2020)]
26. Cohen-Steiner, D., Edelsbrunner, H., Harer, J.: Stability of persistence diagrams. Discrete Comput. Geom. **37**(1), 103–120 (2007)
27. Cole, A., Loges, G.J., Shiu, G.: Quantitative and interpretable order parameters for phase transitions from persistent homology 9 (2020)
28. Cole, A., Schachner, A., Shiu, G.: Searching the Landscape of Flux Vacua with Genetic Algorithms. JHEP **11**, 045 (2019). https://doi.org/10.1007/JHEP11(2019)045
29. Cole, A., Shiu, G.: Persistent homology and non-gaussianit. JCAP **1803**(03), 025 (2018). https://doi.org/10.1088/1475-7516/2018/03/025
30. Cole, A., Shiu, G.: Topological data analysis for the string landscape. JHEP **03**, 054 (2019). https://doi.org/10.1007/JHEP03(2019)054
31. Conlon, J.P., Quevedo, F.: On the explicit construction and statistics of Calabi-Yau flux vacua. JHEP **10**, 039 (2004). https://doi.org/10.1088/1126-6708/2004/10/039
32. Constantin, A., He, Y.H., Lukas, A.: Counting String Theory Standard Models (2018)
33. De Silva, V., Carlsson, G.E.: Topological estimation using witness complexes. SPBG **4**, 157–166 (2004)
34. Denef, F., Douglas, M.R.: Distributions of flux vacua. JHEP **05**, 072 (2004). https://doi.org/10.1088/1126-6708/2004/05/072
35. Denef, F., Douglas, M.R.: Distributions of nonsupersymmetric flux vacua. JHEP **03**, 061 (2005). https://doi.org/10.1088/1126-6708/2005/03/061
36. DeWolfe, O., Giryavets, A., Kachru, S., Taylor, W.: Enumerating flux vacua with enhanced symmetries. JHEP **02**, 037 (2005). https://doi.org/10.1088/1126-6708/2005/02/037
37. Dienes, K.R.: Statistics on the heterotic landscape: Gauge groups and cosmological constants of four-dimensional heterotic strings. Phys. Rev. D **73**, 106010 (2006). https://doi.org/10.1103/PhysRevD.73.106010

38. Dine, M., O'Neil, D., Sun, Z.: Branches of the landscape. JHEP **07**, 014 (2005). https://doi.org/10.1088/1126-6708/2005/07/014
39. Dine, M., Gorbatov, E., Thomas, S.D.: Low energy supersymmetry from the landscape. JHEP **08**, 098 (2008). https://doi.org/10.1088/1126-6708/2008/08/098
40. Douglas, M.R.: The Statistics of string / M theory vacua. JHEP **05**, 046 (2003). https://doi.org/10.1088/1126-6708/2003/05/046
41. Douglas, M.R.: Statistical analysis of the supersymmetry breaking scale (2004)
42. Douglas, M.R., Shiffman, B., Zelditch, S.: Critical points and supersymmetric vacua. Commun. Math. Phys. **252**, 325–358 (2004). https://doi.org/10.1007/s00220-004-1228-y
43. Douglas, M.R., Taylor, W.: The Landscape of intersecting brane models. JHEP **01**, 031 (2007). https://doi.org/10.1088/1126-6708/2007/01/031
44. Edelsbrunner, H., Harer, J.: Computational topology: an introduction. Am. Math. Soc. (2010)
45. Elbers, W., van de Weygaert, R.: Persistent topology of the reionization bubble network – i. formalism and phenomenology. Mon. Not. Roy. Astron. Soc. **486**(2), 1523–1538 (2019). https://doi.org/10.1093/mnras/stz908
46. Elsner, F., Wandelt, B.D.: Improved simulation of non-gaussian temperature and polarization cosmic microwave background maps. Astrophys. J. **184**(2), 264 (2009), http://stacks.iop.org/0067-0049/184/i=2/a=264
47. Feldbrugge, J., van Engelen, M., van de Weygaert, R., Pranav, P., Vegter, G.: Stochastic homology of gaussian vs. non-gaussian random fields: graphs towards betti numbers and persistence diagrams. JCAP **1909**(09), 052 (2019). https://doi.org/10.1088/1475-7516/2019/09/052
48. Gay, C., Pichon, C., Pogosyan, D.: Non-Gaussian statistics of critical sets in 2 and 3D: Peaks, voids, saddles, genus and skeleton. Phys. Rev. D **85**, 023011 (2012). https://doi.org/10.1103/PhysRevD.85.023011
49. Gay, C., Pichon, C., Pogosyan, D.: Non-gaussian statistics of critical sets in 2d and 3d: Peaks, voids, saddles, genus, and skeleton. Phys. Rev. D **85**(2), 023011 (2012)
50. Gmeiner, F., Blumenhagen, R., Honecker, G., Lust, D., Weigand, T.: One in a billion: MSSM-like D-brane statistics. JHEP **01**, 004 (2006). https://doi.org/10.1088/1126-6708/2006/01/004
51. Gukov, S., Vafa, C., Witten, E.: CFT's from Calabi-Yau four folds. Nucl. Phys. B **584**, 69–108 (2000). https://doi.org/10.1016/S0550-3213(01)00289-9, https://doi.org/10.1016/S0550-3213(00)00373-4, [Erratum: Nucl. Phys. B608,477(2001)]
52. Guth, A.H.: Inflationary universe: A possible solution to the horizon and flatness problems. Phys. Rev. D **23**, 347–356 (Jan 1981). 10.1103/PhysRevD.23.347, https://link.aps.org/doi/10.1103/PhysRevD.23.347
53. Guth, A.H., Pi, S.Y.: Fluctuations in the new inflationary universe. Phys. Rev. Lett. **49**(15), 1110 (1982)
54. Halverson, J., Long, C.: Statistical Predictions in String Theory and Deep Generative Models (2020)
55. Halverson, J., Long, C., Sung, B.: Algorithmic universality in F-theory compactifications. Phys. Rev. D **96**(12), 126006 (2017). https://doi.org/10.1103/PhysRevD.96.126006
56. Halverson, J., Nelson, B., Ruehle, F.: Branes with Brains: Exploring String Vacua with Deep Reinforcement Learning. JHEP **06**, 003 (2019). https://doi.org/10.1007/JHEP06(2019)003
57. Hatcher, A.: Algebraic Topology. Cambridge University Press (2002)
58. Hawking, S.: The development of irregularities in a single bubble inflationary universe. Phys. Lett. B **115**(4), 295–297 (1982)
59. He, Y.H.: Deep-Learning the Landscape (2017)
60. He, Y.H.: The Calabi-Yau Landscape: from Geometry, to Physics, to Machine-Learning (2018)
61. Hikage, C., Komatsu, E., Matsubara, T.: Primordial Non-Gaussianity and Analytical Formula for Minkowski Functionals of the Cosmic Microwave Background and Large-scale Structure. Astrophys. J. **653**, 11–26 (2006). https://doi.org/10.1086/508653
62. Hofer, C., Kwitt, R., Dixit, M., Niethammer, M.: Connectivity-optimized representation learning via persistent homology. arXiv preprint arXiv:1906.09003 (2019)

63. Hofer, C., Kwitt, R., Niethammer, M., Uhl, A.: Deep learning with topological signatures. In: Advances in Neural Information Processing Systems. pp. 1634–1644 (2017)
64. Kallosh, R., Linde, A.D.: Landscape, the scale of SUSY breaking, and inflation. JHEP **12**, 004 (2004). https://doi.org/10.1088/1126-6708/2004/12/004
65. Kim, K., Kim, J., Kim, J.S., Chazal, F., Wasserman, L.: Efficient topological layer based on persistent landscapes. arXiv preprint arXiv:2002.02778 (2020)
66. Kimura, Y., Imai, K.: Quantification of lss using the persistent homology in the sdss fields. Adv. Space Res. **60**(3), 722–736 (2017)
67. Klaewer, D., Schlechter, L.: Machine Learning Line Bundle Cohomologies of Hypersurfaces in Toric Varieties (2018)
68. Klypin, A., Shandarin, S.F.: Percolation technique for galaxy clustering. Astrophys. J. **413**, 48–58 (1993)
69. Komatsu, E., Spergel, D.N.: Acoustic signatures in the primary microwave background bispectrum. Phys. Rev. D **63**, 063002 (2001). https://doi.org/10.1103/PhysRevD.63.063002
70. Krefl, D., Seong, R.K.: Machine Learning of Calabi-Yau Volumes. Phys. Rev. D **96**(6), 066014 (2017). https://doi.org/10.1103/PhysRevD.96.066014
71. Liguori, M., Hansen, F.K., Komatsu, E., Matarrese, S., Riotto, A.: Testing primordial non-gaussianity in cmb anisotropies. Phys. Rev. D **73**, 043505 (2006). https://doi.org/10.1103/PhysRevD.73.043505
72. Linde, A.D.: A new inflationary universe scenario: a possible solution of the horizon, flatness, homogeneity, isotropy and primordial monopole problems. Phys. Lett. **108B**, 389–393 (1982). https://doi.org/10.1016/0370-2693(82)91219-9
73. Maldacena, J.M.: Non-Gaussian features of primordial fluctuations in single field inflationary models. JHEP **05**, 013 (2003). https://doi.org/10.1088/1126-6708/2003/05/013
74. Marchesano, F., Shiu, G., Wang, L.T.: Model building and phenomenology of flux-induced supersymmetry breaking on D3-branes. Nucl. Phys. B **712**, 20–58 (2005). https://doi.org/10.1016/j.nuclphysb.2005.01.046
75. Matsubara, T.: Analytic minkowski functionals of the cosmic microwave background: second-order non-gaussianity with bispectrum and trispectrum. Phys. Rev. D **81**, 083505 (2010). https://doi.org/10.1103/PhysRevD.81.083505
76. Mecke, K.R., Buchert, T., Wagner, H.: Robust morphological measures for large scale structure in the universe. Astron. Astrophys. **288**, 697–704 (1994)
77. Milnor, J.: Morse theory (AM-51), vol. 51. Princeton University Press (2016)
78. Mütter, A., Parr, E., Vaudrevange, P.K.S.: Deep learning in the heterotic orbifold landscape (2018)
79. Mukhanov, V.F.: Gravitational instability of the universe filled with a scalar field. JETP Lett. **41**(9), 493–496 (1985)
80. Mukhanov, V.F., Chibisov, G.: Quantum fluctuations and a nonsingular universe. JETP Lett. **33**(10), 532–535 (1981)
81. Mukhanov, V.F., Chibisov, G.: Vacuum energy and large-scale structure of the universe. Sov. Phys.-JETP (Engl. Transl.);(United States) **56**(2) (1982)
82. Munkres, J.R.: Elements of Algebraic Topology. CRC Press (2018)
83. Perea, J.A.: A Brief History of Persistence. ArXiv e-prints (Sep 2018)
84. Poulenard, A., Skraba, P., Ovsjanikov, M.: Topological function optimization for continuous shape matching. In: Computer Graphics Forum. vol. 37, pp. 13–25. Wiley Online Library (2018)
85. Pranav, P., Adler, R.J., Buchert, T., Edelsbrunner, H., Jones, B.J.T., Schwartzman, A., Wagner, H., van de Weygaert, R.: Unexpected topology of the temperature fluctuations in the cosmic microwave background. Astron. Astrophys. **627**, A163 (2019). https://doi.org/10.1051/0004-6361/201834916
86. Pranav, P., Edelsbrunner, H., van de Weygaert, R., Vegter, G., Kerber, M., Jones, B.J.T., Wintraecken, M.: The topology of the cosmic web in terms of persistent betti numbers. Mon. Not. Roy. Astron. Soc. **465**(4), 4281–4310 (2017). https://doi.org/10.1093/mnras/stw2862

87. Pranav, P., van de Weygaert, R., Vegter, G., Jones, B.J.T., Adler, R.J., Feldbrugge, J., Park, C., Buchert, T., Kerber, M.: Topology and geometry of gaussian random fields i: on betti numbers, euler characteristic and minkowski functionals. Mon. Not. Roy. Astron. Soc. **485**(3), 4167–4208 (2019). https://doi.org/10.1093/mnras/stz541
88. Ruehle, F.: Evolving neural networks with genetic algorithms to study the String Landscape. JHEP **08**, 038 (2017). https://doi.org/10.1007/JHEP08(2017)038
89. Ruehle, F.: Data science applications to string theory. Phys. Rep. (2019)
90. Schmalzing, J., Gorski, K.M.: Minkowski functionals used in the morphological analysis of cosmic microwave background anisotropy maps. Mod. Not. R Astron. Soc. **297**, 355–365 (1998). https://doi.org/10.1046/j.1365-8711.1998.01467.x
91. Schmalzing, J., Buchert, T.: Beyond genus statistics: a unifying approach to the morphology of cosmic structure. Astrophys. J. Lett. **482**(1), L1 (1997), http://stacks.iop.org/1538-4357/482/i=1/a=L1
92. Schneider, A., Teyssier, R., Potter, D., Stadel, J., Onions, J., Reed, D.S., Smith, R.E., Springel, V., Pearce, F.R., Scoccimarro, R.: Matter power spectrum and the challenge of percent accuracy. JCAP **1604**(04), 047 (2016). https://doi.org/10.1088/1475-7516/2016/04/047
93. Sousbie, T.: DisPerSE: robust structure identification in 2D and 3D (2013)
94. Starobinsky, A.A.: A new type of isotropic cosmological models without singularity. Phys. Lett. **91B**, 99–102 (1980). https://doi.org/10.1016/0370-2693(80)90670-X
95. Starobinsky, A.A.: Dynamics of phase transition in the new inflationary universe scenario and generation of perturbations. Phys. Lett. B **117**(3–4), 175–178 (1982)
96. Susskind, L.: Supersymmetry breaking in the anthropic landscape **1745–1749**, (2004)
97. Szegedy, C., Zaremba, W., Sutskever, I., Bruna, J., Erhan, D., Goodfellow, I., Fergus, R.: Intriguing properties of neural networks. arXiv preprint arXiv:1312.6199 (2013)
98. Taylor, W., Wang, Y.N.: The F-theory geometry with most flux vacua. JHEP **12**, 164 (2015). https://doi.org/10.1007/JHEP12(2015)164
99. Vafa, C.: The String landscape and the swampland (2005)
100. Wang, Y.N., Zhang, Z.: Learning non-Higgsable gauge groups in 4D F-theory (2018)
101. van de Weygaert, R., Pranav, P., Jones, B.J., Bos, E., Vegter, G., Edelsbrunner, H., Teillaud, M., Hellwing, W.A., Park, C., Hidding, J., et al.: Probing dark energy with alpha shapes and betti numbers. arXiv preprint arXiv:1110.5528 (2011)
102. Winitzki, S., Kosowsky, A.: Minkowski functional description of microwave background gaussianity. New Astron. **3**(2), 75–99 (1998). https://doi.org/10.1016/S1384-1m076(97)00046-8, http://www.sciencedirect.com/science/article/pii/S1384107697000468
103. Zomorodian, A.J.: Topology for computing, vol. 16. Cambridge University Press (2005)

Chapter 10
A Review of Two Decades of Correlations, Hierarchies, Networks and Clustering in Financial Markets

Gautier Marti, Frank Nielsen, Mikołaj Bińkowski, and Philippe Donnat

Abstract We review the state of the art of clustering financial time series and the study of their correlations alongside other interaction networks. The aim of the review is to gather in one place the relevant material from different fields, e.g. machine learning, information geometry, econophysics, statistical physics, econometrics, behavioral finance. We hope it will help researchers to use more effectively this alternative modeling of the financial time series. Decision makers and quantitative researchers may also be able to leverage its insights. Finally, we also hope that this review will form the basis of an open toolbox to study correlations, hierarchies, networks and clustering in financial markets.

10.1 Introduction

Since the seminal paper of Mantegna in 1999, many works have followed, and in many directions (e.g. statistical methodology, fundamental understanding of markets, risk, portfolio optimization, trading strategies, alphas), over the last two decades. We felt the need to track the developments and organize them in this present review.

G. Marti
HKML Research Limited, 5/F., Bonham Trade Centre, 50 Bonham Strand, Sheung Wan, Hong Kong
e-mail: gautier.marti@hkml-research.com

F. Nielsen (✉)
Sony Computer Science Laboratories, Inc, Tokyo, Japan
e-mail: Frank.Nielsen@acm.org

M. Bińkowski
Imperial College London, Exhibition Rd., South Kensington, London SW7 2BU, United Kingdom
e-mail: mikolaj.binkowski14@imperial.ac.uk

P. Donnat
Hellebore Capital Ltd., Michelin House, 81 Fulham Rd., London SW3 6RD, United Kingdom
e-mail: philippe.donnat@helleborecapital.com

F. Nielsen (ed.), *Progress in Information Geometry*,
Signals and Communication Technology,
https://doi.org/10.1007/978-3-030-65459-7_10

10.2 The Standard and Widely Adopted Methodology

The methodology which is widely adopted in the literature stems from Mantegna's seminal paper [89] (cited more than 1550 times as of 2019) and Chap. 13 of the book [90] (cited more than 4400 times as of 2019) published in 1999. We describe it below:

- Let N be the number of assets.
- Let $P_i(t)$ be the price at time t of asset i, $1 \leq i \leq N$.
- Let $r_i(t)$ be the log-return at time t of asset i:

$$r_i(t) = \log P_i(t) - \log P_i(t-1).$$

- For each pair i, j of assets, compute their correlation:

$$\rho_{ij} = \frac{\langle r_i r_j \rangle - \langle r_i \rangle \langle r_j \rangle}{\sqrt{\left(\langle r_i^2 \rangle - \langle r_i \rangle^2\right)\left(\langle r_j^2 \rangle - \langle r_j \rangle^2\right)}}.$$

- Convert the correlation coefficients ρ_{ij} into distances:

$$d_{ij} = \sqrt{2(1 - \rho_{ij})}.$$

- From all the distances d_{ij}, compute a minimum spanning tree (MST) using, for example, Algorithm 1:

Algorithm 1 Kruskal's algorithm

procedure BUILDMST($\{d_{ij}\}_{1 \leq i,j \leq N}$)
    ▷ Start with a fully disconnected graph $G = (V, E)$
    $E \leftarrow \emptyset$
    $V \leftarrow \{i\}_{1 \leq i \leq N}$
    ▷ Try to add edges by increasing distances
    for $(i, j) \in V^2$ ordered by increasing d_{ij} **do**
        ▷ Verify that i and j are not already connected by a path
        if not *connected*(i, j) **then**
            ▷ Add the edge (i, j) to connect i and j
            $E \leftarrow E \cup \{(i, j)\}$
    ▷ G is the resulting MST
    return $G = (V, E)$

Several other algorithms are available to build the MST [62].

The methodology described above builds a tree, i.e. a connected graph with $N - 1$ edges and no loop. This tree is unique as soon as all distances d_{ij} are different. The resulting MST also provides a unique indexed hierarchy [90] which corresponds to the one given by the dendrogram obtained using the Single Linkage Clustering Algorithm.

10.3 Methodological Concerns and Extensions

Several papers have highlighted the potential shortcomings of the original methodology. We found many references which raise concerns, and suggest alternative methods. However, to this day, it does not seem that any alternatives have gained a broad enough acceptance so that they are systematically adopted for empirical studies of market correlations.

10.3.1 Concerns About the Standard Methodology

We list below the concerns that have been raised about the standard methodology during the last 20 years:

- The clusters obtained from the MST (or equivalently, the Single Linkage Clustering Algorithm (SLCA)) are known to be **unstable** (small perturbations of the input data may cause big differences in the resulting clusters) [95].
- The clustering instability may be partly due to the algorithm (MST/Single Linkage are known for the **chaining phenomenon** [26]).
- The clustering instability may be partly due to the correlation coefficient (**Pearson linear correlation**) defining the distance which is known for being **brittle to outliers**, and, more generally, not well suited to distributions other than the Gaussian ones [37].
- **Theoretical results** providing the statistical reliability of hierarchical trees and correlation-based networks are still **not available** [148].
- One might expect that the higher the correlation associated to a link in a correlation-based network is, the higher the **reliability** of this link is. In [144], authors show that this is **not always observed empirically**.
- Changes affecting specific links (and clusters) during prominent crises are of **difficult interpretation** due to the **high level of statistical uncertainty** associated with the correlation estimation [130].
- The standard method is somewhat **arbitrary**: A change in the method (e.g. using a different clustering algorithm or a different correlation coefficient) may yield a huge change in the clustering results [81, 95]. As a consequence, it implies huge variability in portfolio formation and perceived risk [81].

Notice that Benjamin F. King in his 1966 paper [74] (the first paper, to the best of our knowledge, about clustering stocks based on their historical returns; apparently unknown to Mantegna and his colleagues who reinvented a similar method) adds a final footnote which serves both as an advice and a warning for future work and applications:

> One final comment on the method of analysis: this study has employed techniques that rely on finite variances and stationary processes when there is considerable doubt about the existence of these conditions. It is believed that a convincing argument has been made for acceptance of the hypothesis that a small number of factors, market and industry, are sufficient to explain the essential comovement of a large group of stock prices; it is possible, however, that more satisfactory results could be obtained by methods that are distribution free. Here we are thinking of a factor-analytic analogue to median regression and non-parametric analysis of variance, where the measure of distance is something other than expected squared deviation. In future research we would probably seriously consider investing some time in the exploration of distribution free methods.

It is only but recently that researchers have started to focus on these shortcomings as we will observe through the research contributions detailed in the next section.

10.3.2 Contributions for Improving the Methodology

To alleviate some of the shortcomings mentioned in the previous section, researchers have mainly proposed alternative algorithms and enhanced distances. Some refinements of the methodology as a whole, alongside efforts to tackle the concerns about statistical soundness, have been proposed.

On algorithms

Several alternative algorithms have been proposed to replace the minimum spanning tree and its corresponding clusters:

- **Average Linkage Minimum Spanning Tree (ALMST)** [144]; Authors introduce a spanning tree associated to the Average Linkage Clustering Algorithm (ALCA); It is designed to remedy the unwanted chaining phenomenon of MST/SLCA.
- **Planar Maximally Filtered Graph (PMFG)** [3, 143] which strictly contains the Minimum Spanning Tree (MST) but encodes a larger amount of information in its internal structure.
- **Directed Bubble Hierarchal Tree (DBHT)** [131, 132] which is designed to extract, without parameters, the deterministic clusters from the PMFG.
- **Triangulated Maximally Filtered Graph (TMFG)** [96]; Authors introduce another filtered graph more suitable for big datasets.
- Clustering using **Potts super-paramagnetic transitions** [77]; When anti-correlations occur, the model creates repulsion between the stocks which modify their clustering structure.
- Clustering using **maximum likelihood** [52, 53]; Authors define the likelihood of a clustering based on a simple 1-factor model, then devise parameter-free methods to find a clustering with high likelihood.

- Clustering using **Random Matrix Theory (RMT)** [118]; Eigenvalues help to determine the number of clusters, and eigenvectors their composition.
- [88] proposes network-based community detection methods whose null hypothesis is consistent with RMT results on cross-correlation matrices for financial time series data, unlike existing community detection algorithms.
- Clustering using the ***p*-median problem** [75]; With this construction, every cluster is a star, i.e. a tree with one central node.

On distances

At the heart of clustering algorithms is the fundamental notion of distance that can be defined upon a proper representation of data. It is thus an obvious direction to explore. We list below what has been proposed in the literature so far:

- Distances that try to quantify how one financial instrument provides information about another instrument:
  - Distance using **Granger causality** [14],
  - Distance using **partial correlation** [72],
  - Study of asynchronous, **lead-lag relationships by using mutual information** instead of Pearson's correlation coefficient [47, 124],
  - The correlation matrix is normalized using the affinity transformation: the correlation between each pair of stocks is normalized according to the correlations of each of the two stocks with all other stocks [71].
- Distances that aim at including non-linear relationships in the analysis:
  - Distances using mutual information, mutual information rate, and other **information-theoretic** distances [6, 9, 48, 54, 56, 124],
  - The Brownian distance [154],
  - **Copula-based** [19, 41, 94] and **tail dependence** [40, 84] distances.
- Distances that aim at taking into account multivariate dependence:
  - Each stock is represented by a bivariate time series: its returns and traded volumes [20]; a distance is then applied to an *ad hoc* transform of the two time series into a symbolic sequence,
  - Each stock is represented by a multivariate time series, for example the daily (high, low, open, close) [79]; Authors use the **Escoufier's RV coefficient** (a multivariate extension of the Pearson's correlation coefficient).
- A distance taking into account both the **correlation** between returns **and** their **distributions** [37].
- Unlike recent studies which claim that the existence of nonlinear dependence between stock returns have effects on network characteristics, [58] documents that "most of the apparent **nonlinearity is due to univariate non-Gaussianity**.

Further, strong non-stationarity in a few specific stocks may play a role. In particular, the sharp decrease of some stocks during the global financial crisis in 2008" gives rise to apparent negative tail dependence among stocks. When constructing unweighted stock networks, they suggest to use linear correlation "on marginally normalized data", that is Spearman's rank correlation. In fact, this is similar to the idea of splitting apart the dependence information from the distribution one as in [37], where Spearman's rank correlation stems from using a Euclidean distance between the uniform margins of the underlying bivariate copula. Following previous studies, and unlike in [37], the distribution information is discarded when constructing the network.

On other methodological aspects

Besides research contributions on algorithms and distances, other methodological aspects have been pushed further.

- Reliability and statistical uncertainty of the methods:
  - A **bootstrap** approach is used to estimate the statistical reliability of both hierarchical trees [92, 146] and correlation-based networks [101, 144],
  - Reference [91] suggests the use of generative adversarial networks (**GANs**) for sampling realistic yet artificial financial correlation matrices to compare and test the robustness of results, **an alternative approach to bootstrap** methods,
  - **Consistency** proof of clustering algorithms for recovering clusters defined by nested block correlation matrices; Study of empirical **convergence rates** [92],
  - **Kullback-Leibler divergence** is used to estimate the amount of filtered information between the sample correlation matrix and the filtered one [147],
  - **Cophenetic correlation** is used between the original correlation distances and the hierarchical cluster representation [115],
  - Several measures between successive (in time) clusters, dendrograms, networks are used to estimate **stability** of the methods, e.g. cophenetic correlation between dendrograms in [114], adjusted Rand index (ARI) between clusters in [95], mutual information (MI) of link co-occurrence between networks in [130].
  - In [68], authors claim that clustering still cannot compete with "fundamental" industry classifications in terms of performance due to inherent out-of-sample instabilities, and thus propose to improve such given "fundamental" industry classification via further clustering large sub-industries at the most granular level; Similarly, Lopez de Prado in [86] finds that "empirical correlation matrices are **unstable and backward-looking**, and proposes a more sophisticated approach to tackle this issue: A method which fits a correlation matrix constrained by a **hierarchical structure encoding an economic and potentially forward-looking hypothesis**. Avellaneda, in [4], builds on similar motivations to propose a hierarchical PCA yielding principal components which are con-

strained by prior (hierarchical) market partition (e.g. into sectors); these principal components are thus more stable and interpretable.

- Preprocessing of the time series:
  - **Subtract the market mode** before performing a cluster or network analysis on the returns [17],
  - Encode both **rank statistics** and a **distribution histogram** of the returns into a **representative vector** [37],
  - Fit an ARMA(p,q)-FIEGARCH(1,d,1)-cDCC process (**econometric preprocessing**) to obtain dynamic correlations instead of the common approach of rolling window Pearson correlations [128],
  - Use a clustering of successive correlation matrices to **infer a market state** [59, 100, 115, 123, 136].
- **Clustering** of the time series based on **parameters of estimated models**:
  - using coefficients of GARCH processes [110]
  - using a Wasserstein distance between probability distributions obtained from a time series change-point model [152]
- Use of **other types of networks**: threshold networks [109], influence networks [50], partial-correlation networks [70, 72], Granger causality networks [14, 151], cointegration-based networks [141], bipartite networks [149], multilayer networks [8, 34], Bayesian networks and other probabilistic graphical models [36], etc.
- Understanding of the drivers of synchronous correlations using the properties of the collective stock dynamics at **shorter time scales** [33] by using **directed networks** of lagged correlations [32, 33].

10.4 Dynamics of Correlations, Hierarchies, Networks and Clustering

Many of the empirical studies are based on the whole period available from the data. Some researchers have started to investigate the dynamics of the empirical correlations, and also the hierarchies, networks and clusters extracted from them (cf. [108] as one of the earliest work). This dynamic setting which has the potential to track changes of the market structure is more interesting for practitioners (e.g. risk managers, traders, regulatory agencies). This research is still in its infancy and we think its results are still hardly exploitable in practice. For instance, an interesting but difficult question is the following: Are **changes** in the correlation structure due to statistical **noise** and data artifacts **or** do they provide a **real signal**?

No predominant methodology has emerged for now but the naive one which consists in:

- Computing Pearson correlations on a **rolling window of arbitrary length**,
- then independently computing a network or a clustering based on the rolling empirical correlation matrix.

Some promising avenue of research may be the use of temporal networks and temporal centrality measures [156].

Besides the shortcomings of Pearson correlation detailed above, this approach is brittle due to its strong **dependence *a priori* on:**

- the **sampling frequency** (e.g., intraday, daily, weekly),
  - Concerning the sampling frequency, authors in [15] notice that at intraday frequency level some time is needed before the cluster organization emerges completely. According to the paper, "the changes observed in the structure of the MST and of the hierarchical tree suggest that the intrasector correlation decreases faster than intersector correlation between pairs of stocks" when sampling frequency increases. In [17, 95], authors observe that the clusters obtained using daily returns are similar to the ones obtained with weekly timescales, and even to some extent to the ones using monthly returns. Most of the empirical studies focus on daily returns and only a few explore intraday data: [15, 17, 33, 66, 80, 155]. Working with higher frequencies (e.g. at the transaction or quote level) brings further difficulties such as coping with asynchronous data and the Epps effect [45].
- authors study a large propthe **length** T of the rolling window,
  - What is the right length for the rolling window? No clear-cut answer has yet been proposed and, in most studies, its length is set somewhat arbitrarily. In [108], authors posit that "the choice of window width is a trade-off between too noisy and too smoothed data for small and large window widths, respectively" and that they "have explored a large scale of different values for both parameters, and the given values were found optimal". What are the proper criteria for setting the window length? The choice can be driven by the goal (e.g. time investment horizon), by regulatory rules (e.g. computing Value-at-Risk using 1-year historical data), by the stability of clusters [95], by a statistical convergence rate [92], by economic regimes or by a trade-off of the preceding criteria.
- the **number** N **of assets** studied.
  - The number of considered assets has also a significant impact on the results: the ratio T/N drives the precision of the correlation matrix estimation and ultimately the clustering [18, 21, 22, 92].

This dependence makes it difficult to fully understand and analyze results. Once these 'parameters', i.e. the sampling frequency, T, and N, are chosen, one can study

- the **dynamics of correlations**:
  - In [71], authors are using a sliding window of $T = 22$ days to measure and monitor the eigenvalue entropy of the stock correlation matrices (estimated

using daily returns, for $N = 25$ (Tel-Aviv stock market), and $N = 455$ (from S&P 500)). They also propose a 3D visualization to monitor the configuration of stocks using a 3D PCA.

- Reference [98] notices three regime shifts during the period 1989-2011 by monitoring eigenvalues and eigenvectors of the empirical correlation matrices (estimated using quarterly recorded prices from the US housing market; $T = 60$, $N = 51$, the number of US states).

- the **dynamics of the MST** and other hierarchical trees:
Using summary statistics:

  - The MST which evolves over time is monitored using summary statistics (also called topological features) [107] such as the normalized tree length [108], the mean occupation layer [108], the tree half-life [108], a survival ratio of the edges [66, 106, 128], node degree, strength [128], eigenvector, betweenness [138], closeness centrality [128], the agglomerative coefficient [97].
  - Using these statistics, [108] notices that:

    the MST strongly shrinks during a stock market crisis,
    the optimal Markowitz portfolio lies practically at all times on the outskirts of the tree,
    the normalized tree length and the investment diversification potential are very strongly correlated.

  - And [128] notices that in the Asia-Pacific stock market:

    the DST (a dynamic MST built from dynamic correlations) shrinks over time,
    Hong Kong is found to be the key financial market,
    the DST has a significantly increased stability in the last few years,
    the removal of the key player has two effects: there is no clear key market any longer and the stability of the DST significantly decreases.

  - In [67], authors observe that for the Japanese and Korean stock markets, there is a decrease of grouping by industry categories.

  Using distances or similarity measures between successive dendrograms:

  - Cophenetic correlation coefficient. In [97], authors propose a cophenetic analysis of public debt dendrograms in the European Union ($N = 29$ countries) computed using Pearson correlation of quarterly debt-to-GDP ratios between 2000 Q1 and 2014 Q1 ($T = 57$) with a sliding window of size $w = 15$.

- the **dynamics of clusters**:

  - The paper [75] finds that the cluster structures are more stable during crises (using the p-median problem, an alternative clustering methodology).

- Authors in [66] notice that there is an "ecology of clusters": They "can survive for finite periods of time during which time they may evolve in some identifiable way before eventually dissipating or dying".
- In [115], the authors track the merging, splitting, birth, death, contraction, and growth of the clusters in time.

10.5 Clusters, Hierarchies, and Networks Based on Alternative Data

In this review, we cited a number of publications computing and studying clusters, hierarchies and networks based on time series of returns and their correlations, i.e. readily available and cheap data. Another avenue of research consists of using alternative data to estimate lead-lag or contemporaneous relations between companies.

10.5.1 Based on Textual Data

- In [55], authors suggest that media network based investors' attention is a powerful predictor of market premium: In brief, the more frequent the stocks are co-mentioned by media news, the more non-shareholders attention is triggered, and the higher probability of overvaluation for connected stocks. They create an attention index, by aggregating across all the stocks in the market on monthly basis, and find that their index can forecast the market premium with a significantly negative coefficient, a 5.97 and a 5.80% monthly in-sample and out-of-sample R^2 respectively. In addition, they show that the findings hold when controlling for alternative attention proxies, news-based predictors, fundamental information predictors and other standard factors. Their indicator consistently provides negative return forecasts for both time-series and cross-sectional portfolios. As a final test, they also provide evidence that their indicator captures investor attention by sorting cross-sectional portfolios on news co-occurrence frequencies and by checking the performance of average correlation of Google search and Bloomberg search frequencies.
- Authors in [49] propose to derive a company risk indicator from news-sentiment networks. For their research, they gathered from Seeking Alpha a set of articles, which contain a body of text and the author's self-reported sentiment regarding the targeted entities in the articles. The sentiment in the articles, can be seen as a reflection of the author's future expectations. They build co-occurrence networks from the parsed news articles on a quarterly basis, starting from Q1 2011 until the end of Q2 2016. These co-occurrences can cover partnerships, joint ventures, competitors, and suppliers, but can also cover other type of relationships, such as for example an author listing his or her opinion on current best stock picks or

current worst stock picks. They show that the highest quarterly risk value outputted by their risk model is correlated to a higher chance of stock price decline up to 70 days after a quarterly risk measurement.

- Authors of [125, 126] suggest to study the co-occurrence network of European banks mentioned in financial discussions to better understand systemic risk in the financial system. Unlike conventional approaches that estimate interdependence based either on non-publicly disclosed information (such as interbank asset and liability exposures) or publicly available co-movements in market data which only measure their dependence indirectly (co-movements may be driven by other factors) and may not be forward-looking (co-movements are estimated on past market data), their text-to-network process can be applied to publicly available data (such as discussions in financial forums) which may be forward-looking. However, authors document some of their methodology limits: "the loose definition of co-mention context, as an entire post, lessens reliability of relations extracted, while yielding more relations. Provided enough source data, the context could be narrowed down, to increase the likelihood that a relation pair is actually meaningfully related." They find that centrality in the network can be explained by a set of standard variables including size (measures of total assets and total deposits).
- In [133], authors aim at predicting the probability that an edge is inserted or removed in the (correlation-based) financial network estimated on past returns. To do so, they use both the past past structure of this financial network and also a social network estimated from the correlations of Twitter sentiment time series provided by PsychSignal.com on the same set of stocks. They find that considering both networks considerably improves the predictability of future financial networks over a naive benchmark of persistent structure. They also find that the so-called social networks are harder to predict and much less persistent than the financial ones.
- Following [46, 61] builds similarity networks of US listed firms from text-based analysis of the product descriptions included in the Securities Exchange Commission (SEC) filings (Form 10-K). Authors use topological features of these networks as input to machine learning models to predict the future success or failure of the firms. When these features are combined with typical financial data such as the book-to-market ratio, measures of leverage, profitability, default risk of the firms, momentum and liquidity of their stocks, accuracy of the predictions are significantly higher.

In the next three subsections, we list a couple of papers exploring interactions of different kinds: supply chain, payments, business partnerships, financial contracts, and mutual ownership.

10.5.2 Based on Supply Chain Data

- Reference [153] explores "the interactions from the perspective of an economy-wide supply-chain network, and propose a list of network centrality measures to capture the relative importance of each company within this network." Using FactSet Supply Chain Relationships database, the author estimates dynamic supplier networks and compute centrality measures for each company. From there, the author constructs a supplier central portfolio of stocks based on the top ten companies with the highest centrality in the network. The author finds that the stock performance of supplier central portfolios tends to predict the movements of the overall stock market.
- In [24], authors study the network structure implied by firms' cash flows on the cross-section of expected returns. They find that firms which are more central have lower P/D ratios and higher expected returns.
- Reference [1] find that there is significant credit risk propagation along supply chains by analyzing credit default swap spreads.

10.5.3 Based on Transaction Data

- In [83], authors study a large proprietary dataset which consists of a network built from transactional data of the payment platform of a major European bank. These transactions link around 2.4 million Italian firms whose credit risk rating is known for a large fraction of them. Authors find that this network, like many financial networks, is sparse but made of a single component, scale free and verifies the small world property. The main contribution of the authors is to document significant correlations between local topological properties of a node (firm) and its risk. They also employ machine learning techniques to build classifiers for predicting the risk rating using as inputs network properties (e.g. degree, community) alone, i.e. no balance sheet information or any other features. Their classification method outperforms significantly its benchmark, i.e. random assignments. Another contribution is to show the existence of an homophily of risk, i.e. the tendency of firms with similar risk profile to be statistically more connected among themselves through payments. Risk is therefore not spread uniformly on the network, but rather concentrated in specific areas. This implies that an idiosyncratic shock on a single firm can propagate more or less quickly depending on the local network structure and the community the node belongs to.

10.5.4 Based on Key People

- In [111], authors build a co-occurrence network of people mentioned in 21,578 Reuters news articles mostly focused on economics published in 1987: People are represented as vertices and two persons are connected if they co-occur in the same article. They observed that this network has small-world features with power-law degree distribution. The network is disconnected and the component size distribution has power-law characteristics. They compare the importance of these people back in 1987 in the news co-occurrence network to the importance they have as of 2007 proxied by their Wikipedia article. They find medium level Spearman's rank correlations between these two measures.

10.5.5 Based on Investors

- Traders and investors are socially connected and have access to comparable sources of information. Reference [29] investigates social connections and information linkages, and demonstrates how they can predict patterns of trade correlation: trades generated by "neighbor" traders are positively correlated and trades generated by "distant" traders are negatively correlated.
- Using a dataset of all trades on the Istanbul Stock Exchange in 2005, [112] identifies traders with similar trading behavior as linked in an empirical investor network. They find that central investors earn higher returns and trade earlier than peripheral investors with respect to information events (e.g. earnings), consistent with the view that information diffusion among the investor population influences trading behavior and returns.
- In [7], authors study investor clusters in the first two years after the IPO filing in the Helsinki Stock Exchange by using a statistically validated network method to infer investor links based on the co-occurrences of investors' trade timing for 69 IPO stocks. They find that the clusters of investors based on their trading are stable between new and mature stocks. Authors identify a highly persistent institutional investor cluster, which they consider as evidence about institutional herding.
- Reference [8] investigate trading decisions for Finnish investors using multilayer networks. They find that households in the capital have high centrality in investor networks, which, under the theory of information channels in investor networks, suggests that they are well-informed investors.
- Using a Kinetic Ising model, [25] reconstructs a lead-lag network of traders client of a major dealer in the Foreign Exchange market. They identify leading players in this market who are typically leading the order flow on the 5 minutes timescale. Based on this lead-lag network, authors also define a herding measure based on the observe and inferred traders' opinions. They find a causal link between herding and the liquidity in the inter-dealer market used by dealers to rebalance their inventories.

10.5.6 Based on Several Corporate Networks

- Reference [34] combines different types of networks (ownership links, social ties through board members, research collaborations, and stock correlations) into a single multiplex structure. Authors employs several methods to show the significance of this multilayer network. Their initial results indicate a relation between company performance and multiplex centrality.

10.6 Financial Applications

Though many of the academic studies focus on the MST or the clusters *per se*, some papers try to extend their use beyond the filtering of empirical correlation matrices. It has been proposed to leverage them for making financial policies, optimizing portfolios, computing alternative Value-at-Risk measures, residualizing expected returns, grouping and selecting quantitative trading alphas, etc.

10.6.1 Portfolio Design

- Reference [108] finds that the Markowitz portfolio layer in the MST is higher than the mean layer at all times.
- As the stocks of the minimum risk portfolio are found on the outskirts of the tree [108, 119], authors expect larger trees to have greater diversification potential.
- In [114, 139], authors compare the Markowitz portfolios from the filtered empirical correlation matrices using the clustering approach, the RMT approach and the shrinkage approach.
- References [116, 122] propose to invest in different part of the MST depending on the estimated market conditions.
- Authors show that there is no inner-mathematical relationship between the minimum variance portfolio from Markowitz theory and the portfolios designed from the minimum spanning tree [64]. Empirical evidence of such relations found by previous studies is essentially a stylized fact of financial returns correlations and time series, not a general property of correlation matrices.
- It appears that a large number of stocks are unnecessary for building an index of market change [74].
- The paper [38] describes methods for index tracking and enhanced index tracking based on clusters of financial time series.
- Reference [40] introduces a procedure to design portfolios which are diversified in their tail behavior by selecting only a single asset in each cluster.

- Reference [5] investigates several network and hierarchy based active portfolio optimizations, and find their out-of-sample performance competitive with respect to conventional ones.
- Reference [82] presents the performance of seven portfolios created using clustering analysis techniques to sort out assets into categories and then applying classical optimization inside every cluster to select best assets inside each asset category.
- Reference [51] applies a hierarchical clustering to a set of government bond factors (investment styles such as carry, momentum, slope, convexity, reversal, real rate vs. growth), and finds that most factors benefit from a positive carry.
- References [35, 84, 85, 117, 120, 121] leverage hierarchical clustering to build diversified portfolios that outperform out-of-sample by refining the risk parity and the equal risk contribution methods to take into account the hierarchical correlations of assets.
- Reference [65] compares the Hierarchical Risk Parity (HRP) from [85] to more traditional risk-based portfolios such as inverse volatility weighted, minimum variance and maximum diversification portfolios. The main takeaway from the study is that **results strongly depend on the estimation of the covariance matrix**. If estimates are crude, then inverse volatility weighted portfolios outperform the ones obtained from the other methods as less sensitive to the covariance misspecification. HRP stands in between this simple yet robust method and the optimization-based ones. Rebalancing frequency of the portfolios also impacts the relative performance of the different methods, HRP being superior at minimizing out-of-sample portfolio variance with longer rebalancing horizons. Note that the study only considers the HRP [85] to represent the class of machine learning-based portfolios whereas other methods are available in the literature such as the Hierarchical Clustering based Asset Allocation (HCAA) [120] and the Hierarchical Equal Risk Contribution (HERC) [121]. Note also that the empirical results described in the paper were obtained for portfolios of 10 assets only. HRP, and related methods, would benefit from larger portfolios, since not inverting the covariance matrix.

10.6.2 *Trading Strategies*

- In [115], they suggest that tracking the merging, splitting, birth, and death of the clusters in time could be the basis for pairs-like reversal trading strategies but with pairs corresponding to clusters.
- One can build a simple mean-reversion statistical arbitrage strategy whereby one assumes that stocks in a given industry move together, cross-sectionally demeans stock returns within said industry, shorts stocks with positive residual returns and goes long stocks with negative residual returns [68].
- Earnings per share forecasts prepared on the basis of statistically grouped data (clusters) outperform forecasts made on data grouped on traditional industrial criteria as well as forecasts prepared by mechanical extrapolation techniques [44].

- Reference [127] suggests that one may design a new set of Ricci curvature network based-strategies in statistical arbitrage (e.g. for mean-reverting portfolios).
- Reference [103] finds the existence of significant relations between past changes in the market correlation structure and future changes in the market volatility.
- In [68], authors suggest the use of clustering to build statistical classification of quantitative trading alphas for which there is no analog of "fundamental" industry classifications such as the GICS, BICS, ICB, NAICS, SIC, etc.
- References [27, 76] describe a behavioral bias: investors overly rely on the standard industry classifications (e.g., SIC, NAICS). Corporate managers can exploit this behavioral bias to steer their company towards a more favorable industry and therefore benefit from a lower cost of capital. The article [27] does not discuss this behavioral bias from the investor point of view, but it can pay to have a different (statistical) view of the standard industry classification to avoid such opportunistic window dressing. In [76], authors claim that long-short strategies exploiting mispricing due to the industry categorization bias generate statistically significant and economically sizable risk-adjusted excess returns.
- In [61], authors show that considering alternative industry classifications (for example, text-based ones) can enhance the returns of well-known quantitative strategies such as the industry momentum one which is driven, according to the paper, by inattention to less visible horizon peers.
- The same authors find in [60] that their text-based classification of peers significantly outperform traditional industry peers in explaining firm valuations and peer comovement in the stock market: Firms with peers whose products more closely match theirs have higher stock-market comovement; Firms that have more unique products relative to their peers have higher stock market valuations.
- In [153], the author investigates the supply chain network using several centrality measures. He finds that "the stock performance of supplier central portfolios tends to predict the movements of the overall stock market." More generally, the author suggests that "shocks to the supply chain can generate ripple effects, which can show up potentially as lead-lag predictive relations in security returns, earning surprises, default probabilities, and/or distinctive term structure patterns in realized and option implied volatilities and credit default swaps."
- Authors of [87] suggests using hierarchical clustering to estimate two parameters required by their "False Strategy" theorem, namely the number K of effectively independent tests, and the variance of the Sharpe ratios across the K effectively independent tests. This number K corresponds to the number of clusters of strategies found considering a distance based on the correlation of their returns. Using these estimates, they can conclude how likely a strategy is spurious.
- In [31], authors monitor the correlation network statistics of the constituents of a stock index, and find that an abrupt change in the network is predictive of significant falls in the price of this index. Reference [134] aims at predicting such abrupt network changes.

10.6.3 Risk Management

How much money a given portfolio can lose? in normal market conditions? in stressed market conditions? in the presence of systemic risk?

To answer these questions, the use of clusters and networks can help. As presented previously, the clustering hierarchy can be used to filter a correlation [139, 142] or a tail dependence [40] matrix, which helps to measure the risk in normal and stressed market conditions respectively. The systemic risk as defined by the Bank for International Settlements is the risk that a failure of a participant to meet its contractual obligations may in turn cause other participants to default, with the chain reaction leading to broader financial difficulties. Networks seem thus a particularly relevant tool to study this kind of risk.

- Study of **systemic risk**:
  - In [57], authors assert that the diminution of regulation has removed barriers between sectors and regions allowing bank to diversify their risk, but it also increased the economic risk through increased interdependencies.
  - The paper [78] is focused on energy derivative markets, and their market integration which can be seen as a necessary condition for the propagation of price shocks. The MST is used to "identify the most probable and the shortest path for the transmission of price shocks".
  - Authors in [12, 13] identify the set of indices possessing influence over the Asian region, namely the Hong Kong, Singapore and South Korea ones. Authors claim that the findings of their study can be utilized in effective systemic risk management and for the selection of an optimally-diversified portfolio, resilient to system-level shocks.
  - Authors in [98] focus on the US housing market. According to the paper, "dramatic increases in the systemic risk are usually accompanied by regime shifts, which provide a means of early detection of housing bubbles." They find a sharp increase in housing market correlations over the past decade, indicating that systemic market risk has also greatly increased; They observe that prices diffuse in complex ways that do not require geographical clusters unlike worldwide stock markets which exhibit clear geographical clustering [130].
  - The paper [154] is focused on the shipping market. Authors explore the connections between the shipping market and the financial market: The shipping market can provide efficient warning before market downturn. Alike many economic systems which have been exhibiting an increase in the correlation between different market sectors, a factor that exacerbates the level of systemic risk, the three major world shipping markets, (i) the new ship market, (ii) the second-hand ship market, and (iii) the freight market, have experienced such an increase. Authors show it using the MST, Granger causality analysis, and Brownian distance on the prices of the real shipping market, and the stock prices of publicly-listed shipping companies.

- Reference [14] investigates the monthly returns of hedge funds, banks, broker/dealers, and insurance companies. They find that all four sectors have become highly interrelated over the past decade, likely increasing the level of systemic risk.
- Reference [127] shows that Ricci curvature may serve as an indicator of fragility in the context of financial networks.
- Reference [115] detects distinct correlation regimes between 1998 and 2013. These correlation regimes have been significantly different since the financial crisis of 2008 than they had been previously. Cluster tracking shows that asset classes are now less separated. Correlation networks help the authors to identify "risk-on" and "risk-off" assets.
- For authors in [100], the identification of market states based on correlation matrices can be helpful to build an "early warning system" for financial markets. This system could be implemented by comparing the current state to previous similar states or monitoring rapid changes in the correlation structure.
- In [102], authors study the clusters' composition evolution, and their persistence. They observe that the clustering structure is quite stable in the early 2000s becoming gradually less persistent before the unfolding of the 2007-2008 crisis. The correlation structure eventually recovers persistence in the aftermath of the crisis, settling up a new phase which is distinct from the pre-crisis structure one, where the market structure is less related to industrial sector activity.
- Reference [63] finds that financial institutions which have, in the correlation networks, greater node strength, larger node betweenness centrality, larger node closeness centrality and larger node clustering coefficient tend to be associated with larger systemic risk contributions.
- References [11, 135] discuss the detection of early-warning signals of the 2008 crisis via the analysis of the properties of interbank networks.
- Authors in [2] define the structural entropy of a network, which is simply the entropy of the probability vector encoding the proportional size of the clusters in the network. They propose to use structural entropy to monitor the structure of correlation-based networks over time. They suggest this quantity could be an early warning indicator of financial crises. They observe "a remarkably high linear correlation between the new measure and the volatility of the assets' prices over time".
- In [28], authors highlight that the underlying financial network required to study systemic risk is only partially observable in general. They propose a method to reconstruct such a network, i.e. to build a set of (directed and weighted) dependencies among the constituents of a complex system.

- Risk management **methods**:

  - In [42], authors design clusters that tend to be comonotonic in their extreme low values: To avoid contagion in the portfolio during risky scenarios, an investor should diversify over these clusters.

- As far as diversification is concerned, portfolio managers should probably focus on the most stable parts of the graph [78].
- In [99], authors postulate the existence of a hierarchical structure of risks which can be deemed responsible for both stock multivariate dependency structure and univariate multifractal behaviour, and then propose a model that reproduces the empirical observations (entanglement of univariate multi-scaling and multivariate cross-correlation properties of financial time series). The interplay between multi-scaling and average cross-correlation is confirmed in [23].
- Industries (e.g. clusters as statistical industry classification) can be used as risk factors in multifactor risk models [68].
- Clusters (statistical industry classification) can be an alternative to sometimes unavailable "fundamental" industry classifications (e.g. in emerging or small markets) [68].
- In [10], authors apply the TMFG for building sparse forecasting models and for financial applications such as stress-testing and risk allocation.
- Reference [83] predicts credit risk based on local properties of the network of payments between firms.

We found that the risk literature using correlation networks and clusters consists essentially in descriptive studies. For now, there are only too few propositions in the academic literature to build effective network-based or cluster-based risk systems.

10.6.4 Financial Policy Making

Clusters and networks can help designing financial policies. Several papers propose to leverage them to detect risky market environments, develop indicators that can predict forthcoming crisis or economic recovery [155], improve economic nowcasting [43], or find key markets and assets that drive a whole region, and on which stimulus can be applied effectively.

- Authors of [57] claim that "separation prevents failure propagation and connections increase risks of global crises" whereas the prevailing view in favor of deregulation is that banks, by investing in diverse sectors, would have greater stability. To support their argument, using financial networks, they study the aftermath of the Glass-Steagall Act (1933) repeal by Clinton administration in 1999. They find that erosion of the Glass–Steagall Act, and cross sector investments eliminated "firewalls" that could have prevented the housing sector decline from triggering a wider financial and economic crisis:

  > Our analysis implies that the investment across economic sectors itself creates increased cross-linking of otherwise much more weakly coupled parts of the economy, causing dependencies that increase, rather than decrease, risk.

- According to [14], bank and insurance capital requirements and risk management practices based on VaR, which are intended to ensure the soundness of individual financial institutions, may amplify aggregate fluctuations if they are widely adopted:

> For example, if the riskiness of assets held by one bank increases due to heightened market volatility, to meet its VaR requirements the bank will have to sell some of these risky assets. This liquidation may restore the bank's financial soundness, but if all banks engage in such liquidations at the same time, a devastating positive feedback loop may be generated unintentionally. These endogenous feedback effects can have significant implications for the returns of financial institutions, including autocorrelation, increased correlation, changes in volatility, Granger causality, and, ultimately, increased systemic risk, as our empirical results seem to imply.

- In [78], authors find that the move towards integration started some time ago and there is probably no way to stop or refrain it. However, regulation authorities may act in order to prevent prices shocks from occurring, especially in places where their impact may be important.
- In [36], the author shows how Bayesian networks (and probabilistic graphical models in general) can be used to model complex networks of debt dependencies between firms. These particular types of networks allow for causal and counterfactual reasoning: What would happen if a set of institutions defaults? What would happen if this institution is bailed out?

10.7 Practical Fruits of Clusters, Networks, and Hierarchies

10.7.1 Stylized Facts

Stylized facts can be described as follows [30][1]:

> A set of [statistical] properties, common across many instruments, markets and time periods, [which] has been observed by independent studies.

From the papers we reviewed, we can list the following stylized facts:

- Firms belonging to some economic sectors are strongly connected within themselves, whereas others are much less connected.
- The Energy and Financial sectors are examples of strong connections whereas elements belonging to the Conglomerates, Consumer cyclical, Transportation, and Capital Goods sectors are weakly connected.
- General Electric is at the center of US stocks networks (for several centrality criteria) [15, 20, 89, 108].
- The Energy, Technology, and Basic Materials sectors are sectors of elements significantly connected among them but weakly interacting with stocks belonging to different economic sectors.

[1] reference to the book *Practical Fruits of Econophysics* [137]

- The Financial sector is strongly connected within, but also to others.
- The assets of the classic Markowitz portfolio are always located on the outer leaves of the tree [108, 116, 119].
- The maximum eigenvalue of the correlation matrix, which carries most of the correlations, is very large during market crashes [39] (increased value of the mean correlation).
- The MST shrinks during market crashes [108] and contains a low number of clusters [97].
- The MST provides a taxonomy which is well compatible with the sector classification provided by an outside institution [90, 108].
- Scale free structure of the MST (i.e. the degree of vertices is power law distributed $f(n) \sim n^{-\alpha}$) [16, 73, 108, 150], but the scaling exponent depends on market period and window width [106].
- The MST obtained with the one-factor model is very different from the one obtained using real data [16]. This invalidates the Capital Asset Pricing Model which is based on the one-factor model $r_i(t) = \alpha_i + \beta_i r_M(t) + \epsilon_i(t)$.
- Stocks compose a hierarchical system progressively structuring as the sampling time horizon increases [17, 145].
- The correlation among market indices presents both a fast and a slow dynamics. The slow dynamics is a gradual growth associated with the development and consolidation of globalization. The fast dynamics is associated with events that originate in a specific part of the world and rapidly (in less than 3 months) affect the global system [98, 130].
- Removing the dynamics of the center of mass decreases the level of correlations, but also makes the cluster structure more evident [17].
- Scale invariance of correlation structure (by subtraction of the market mode) might have important implications for risk management, because it suggests that correlations on short time scales might be used as a proxy for correlations on longer time-horizons [17].
- The MST is star-like in low-volatility segments, and chain-like in high-volatility segments [155].
- Volatility shocks always start at the fringe and propagate inwards [155].
- The "post-subprime" regime correlation matrix shows markedly higher absolute correlations than the others [115].
- In [115], authors find far less asset class separation in the post-subprime period.
- One can distinguish three types of topological configurations for the companies: (i) important nodes, (ii) links and (iii) dangling ends [150].
- A node keeps the majority of its neighbours. The non-randomness of the stock market topology is thus a robust property [150].
- The largest eigenvector of the correlation matrix is strongly non-Gaussian, tending to uniform - suggesting that all companies participate. Authors find indeed that all components participate approximately equally to the largest eigenvector. This implies that every company is connected with every other company. In the stock market problem, this eigenvector conveys the fact that the whole market "moves"

together and indicates the presence of correlations that pervade the entire system [118].

- The measure of the average length of shortest path in the PMFG shows a small world effect present in the networks at any time horizon [145].
- Among the 100 largest market capitalization stocks in the NYSE, the auto and lagged intraday correlations play a much more prominent role in 2011-2013 than in 2001-2003 [33].
- Authors in [33] find striking periodicities in the validated lagged correlations, characterized by surges in network connectivity at the end of the trading day.
- At short time scales, measured synchronous correlations among stock returns tend to be lower in magnitude [45], but lagged correlations among assets may become non-negligible [32, 140].
- Banks may be of more concern than hedge funds from the perspective of connectedness [14].
- A lack of distinct sector identity in emerging markets [105, 113]; Few largest eigenvalues deviate from the bulk of the spectrum predicted by RMT (far fewer than for the NYSE) [104, 113].
- Emergence of an internal structure comprising multiple groups of strongly coupled components is a signature of market development [113].

10.7.2 Moot Points and Controversies

Though most of the conclusions of empirical studies do agree, we find some claims that seem to be contradictory:

- Reference [69] finds that eccentricity-based risk budgeting portfolios have improved return to risk ratios, hence better invest in centrality (of the minimum spanning tree). On the contrary, [108, 116, 119] conclude that it is better to invest in the peripheries (of the minimum spanning tree).
- Volatility shocks always start at the fringe and propagate inwards [155], but in [129], authors assert that the credit crisis spreads among affected stocks from more centralized to more outer ones, as spread the news about the extent of damage to the global economy.
- One might expect that the higher the correlation associated to a link in a correlation-based network is, the higher the reliability of the link is. The paper [144] shows that it is not always observed empirically. However, the Cramér–Rao lower bound (CRLB) for correlation [93] points out that the higher the correlation, the easier its estimation, i.e. less statistical uncertainty for high correlations.
- For filtering the correlation matrix, SLCA is more stable than ALCA according to [148], but ALCA is more stable and appropriate than SLCA according to [115, 139].

- During a crisis period, is there an increase or decrease of clusters stability? Most papers find a decrease (e.g. [78, 102]), but at least one [75] (using an alternative clustering methodology, the p-median problem) advocates for an increase.
- In many papers, researchers observe that Minimum Variance portfolios tend to select the same assets as Network-based portfolios [108, 116, 119]; In [64], authors show that in general there are no relations between the two methods, suggesting any empirical relations observed come from special properties of the financial correlations. Which ones? This question is left open to these days.

10.8 Conclusion

Correlation networks (and complex networks in general) provide a useful set of quantitative tools for several tasks in finance, e.g. monitoring systemic risk and acting preemptively on a few institutions to prevent cascading effects; building diversified portfolios of assets or strategies; designing statistical arbitrage strategies; residualizing returns by removing the main hierarchical risk factors; finding less crowded cross-sectional risk premia.

However, in our opinion, researchers in this field need to overcome several challenges so that results and techniques exposed in this review can become part of the standard toolbox of the practitioner. We can observe that there is i) a lack of reproducibility for a couple of studies (hence some contradictory claims); ii) difficulty to compare methods due to re-implementation biases and lack of open source code; and iii) no widely accepted common tasks or benchmarks; iv) data is not shared and most of the time confidential.

To tackle these issues, we recommend researchers to provide code and data (or at least synthetic datasets). When the underlying data is confidential, the use of Generative Adversarial Networks (GANs) or other generative models could be a solution to i) share anonymized synthetic data having properties similar to the original dataset; ii) define a common task and iii) set benchmark results enticing other researchers to challenge. This approach has hugely benefited the progress of machine learning as a research field. We think it should help researchers in correlation networks (and complex networks in general) having more impact as well.

Acknowledgements Many thanks to the researchers which have contributed to improve this review (in chronological order): David Matesanz, Tiziana Di Matteo, Diego Garlaschelli, Damiano Brigo, Fabrizio Lillo, Quang Nguyen, Thomas Guhr.

Disclaimer: Views and opinions expressed are those of the authors and do not necessarily represent official positions of their respective companies.

References

1. Agca, S., Babich, V., Birge, J.R., Wu, J.: Credit risk propagation along supply chains: Evidence from the cds market. Georgetown McDonough School Bus. Res. Paper (3078752) (2017)
2. Almog, A., Shmueli, E.: Structural entropy: Monitoring correlation-based networks over time with application to financial markets. Sci. Rep. **9**(1), 1–13 (2019)
3. Aste, T., Di Matteo, T., Hyde, S.T.: Complex networks on hyperbolic surfaces. Phys. A: Stat. Mech. Appl. **346**(1), 20–26 (2005)
4. Avellaneda, M.: Hierarchical PCA and applications to portfolio management. Revista mexicana de economía y finanzas **15**(1), 1–16 (2020)
5. Baitinger, E., Papenbrock, J.: Interconnectedness Risk and Active Portfolio Management (20160
6. Baitinger, E., Papenbrock, J.: Interconnectedness risk and active portfolio management: The information-theoretic perspective (2017)
7. Baltakys, K., Kanniainen, J., Pedreschi, D., Lillo, F., et al.: Clusters of investors around initial public offering. Tech. Rep. (2019) arXiv. org
8. Baltakys, K., Kanniainen, J., Emmert-Streib, F.: Multilayer aggregation with statistical validation: Application to investor networks. Sci. Rep. **8**(1), 8198 (2018)
9. Barbi, A.Q., Prataviera, G.A.: Nonlinear dependencies on brazilian equity network from mutual information minimum spanning trees (2017). arXiv:1711.06185
10. Barfuss, W., Massara, G.P., Di Matteo, T., Aste, T.: Parsimonious modeling with information filtering networks. Phys. Rev. E **94**(6), 062306 (2016)
11. Battiston, S., Farmer, J.D., Flache, A., Garlaschelli, D., Haldane, A.G., Heesterbeek, H., Hommes, C., Jaeger, C., May, R., Scheffer, M.: Complexity theory and financial regulation. Science **351**(6275), 818–819 (2016)
12. Bhattacharjee, B., Shafi, M., Acharjee, A.: Investigating the evolution of linkage dynamics among equity markets using network models and measures: The case of asian equity market integration. Data **2**(4), 41 (2017)
13. Bhattacharjee, B., Shafi, M., Acharjee, A.: Network mining based elucidation of the dynamics of cross-market clustering and connectedness in asian region: An mst and hierarchical clustering approach. J. King Saud Univ.-Comput. Inf, Sci (2017)
14. Billio, M., Getmansky, M., Lo, A.W., Pelizzon, L.: Econometric measures of connectedness and systemic risk in the finance and insurance sectors. J. Financial Econ. **104**(3), 535–559 (2012)
15. Bonanno, G., Lillo, F., Mantegna, R.N., et al.: High-frequency cross-correlation in a set of stocks. Quantitative Finance **1**(1), 96–104 (2001)
16. Bonanno, G., Caldarelli, G., Lillo, F., Mantegna, R.N.: Topology of correlation-based minimal spanning trees in real and model markets. Phys. Rev. E **68**(4), 046130 (2003)
17. Borghesi, C., Marsili, M., Miccichè, S.: Emergence of time-horizon invariant correlation structure in financial returns by subtraction of the market mode. Phys. Rev. E **76**(2), 026104 (2007)
18. Borysov, P., Hannig, J., Marron, J.S.: Asymptotics of hierarchical clustering for growing dimension. J. Mult. Anal. **124**, 465–479 (2014)
19. Brechmann, E.C. et al.: *Hierarchical Kendall copulas and the modeling of systemic and operational risk*. Ph.D. thesis, Universitätsbibliothek der TU München (2013)
20. Brida, J.G., Risso, W.A.: Multidimensional minimal spanning tree: The dow jones case. Phys. A: Stat. Mech. Appl. **387**(21), 5205–5210 (2008)
21. Bun, J., Allez, R., Bouchaud, J.-P., Potters, M.: Rotational invariant estimator for general noisy matrices. IEEE Trans. Inf. Theory **62**(12), 7475–7490 (2016)
22. Bun, J., Bouchaud, J.-P., Potters, M.: Cleaning large correlation matrices: tools from random matrix theory. Phys. Rep. **666**, 1–109 (2017)
23. Buonocore, R.J., Mantegna, R.N., Di Matteo, T.: On the interplay between multiscaling and average cross-correlation (2018). arXiv:1802.01113

24. Buraschi, A., Porchia, P.: Dynamic networks and asset pricing. In: AFA 2013 San Diego Meetings Paper (2012)
25. Campajola, C., Lillo, F., Tantari, D.: Unveiling the relation between herding and liquidity with trader lead-lag networks (2019). arXiv:1909.10807, 2019
26. Carlsson, G., MÃŠmoli, F.: Characterization, stability and convergence of hierarchical clustering methods. J. Mach. Learn. Res. **11**, 1425–1470 (2010)
27. Chen, H., Cohen, L., Lou, D.: Industry window dressing. Rev. Financial Stud. **29**(12), 3354–3393 (2016)
28. Cimini, G., Squartini, T., Garlaschelli, D., Gabrielli, A.: Systemic risk analysis on reconstructed economic and financial networks. Sci. Rep. **5**, (2015)
29. Colla, P., Mele, A.: Information linkages and correlated trading. Rev. Financial Stud. **23**(1), 203–246 (2009)
30. Rama Cont. Empirical properties of asset returns: stylized facts and statistical issues (2001)
31. Cordoba, A., Castillejo, C., García-Machado, J.J., Lara, A.M.: Anticipating abrupt changes in complex networks: Significant falls in the price of a stock index. In: Nonlinear Systems, vol. 1, pp. 317–338. Springer, Berlin (2018)
32. Curme, C., Tumminello, M., Mantegna, R.N., Stanley, H.E., Kenett, D.Y.: Emergence of statistically validated financial intraday lead-lag relationships. Quantitative Finance **15**(8), 1375–1386 (2015)
33. Curme, C., Tumminello, M., Mantegna, R.N., Stanley, H.E., Kenett, D.Y.: How lead-lag correlations affect the intraday pattern of collective stock dynamics. Office Financial Res. Working Paper, 15 (2015)
34. van Lidth de Jeude, J., Aste, T., Caldarelli, G.: The multilayer structure of corporate networks (2019). arXiv:1901.07411
35. de Prado, M.L.: Advances in Financial Machine Learning. Wiley, New York (2018)
36. Denev, A.: Probabilistic Graphical Models: A New Way of Thinking in Financial Modelling. Risk Books (2015)
37. Donnat, P., Marti, G., Very, P.: Toward a generic representation of random variables for machine learning. Pattern Recognit. Lett. **70**, 24–31 (2016)
38. Dose, C., Cincotti, S.: Clustering of financial time series with application to index and enhanced index tracking portfolio. Phys. A: Stat. Mech. Appl. **355**(1), 145–151 (2005)
39. Drożdż, S., Grümmer, F., Górski, A.Z., Ruf, F., Speth, J.: Dynamics of competition between collectivity and noise in the stock market. Phys. A: Stat. Mech. Appl. **287**(3), 440–449 (2000)
40. Durante, F., Foscolo, E., Pappadà, R., Wang, H.: A Portfolio Diversification Strategy Via Tail Dependence Measures (2015)
41. Durante, F., Pappada, R.: Cluster analysis of time series via kendall distribution. In: Strengthening Links Between Data Analysis and Soft Computing, pp. 209–216. Springer, Berlin (2015)
42. Durante, F., Pappadà, R., Torelli, N.: Clustering of financial time series in risky scenarios. Adv. Data Anal. Classif. **8**(4), 359–376 (2014)
43. Elshendy, M., Colladon, A.F.: Big data analysis of economic news: Hints to forecast macroeconomic indicators. Int. J. Eng. Bus. Manag. **9**, 1847979017720040 (2017)
44. Elton, E.J., Gruber, M.J.: Improved forecasting through the design of homogeneous groups. J. Bus. **44**(4), 432–450 (1971)
45. Epps, T.W.: Comovements in stock prices in the very short run. J. Am. Stat. Assoc. **74**(366a), 291–298 (1979)
46. Fan, J., Cohen, K., Shekhtman, L.M., Liu, S., Meng, J., Louzoun, Y., Havlin, S.: Topology of products similarity network for market forecasting. Appl. Netw. Sci. **4**(1), 1–15 (2019)
47. Fiedor, P.: Information-theoretic approach to lead-lag effect on financial markets. Eur. Phys. J. B **87**(8), 1–9 (2014)
48. Fiedor, P.: Networks in financial markets based on the mutual information rate. Phys. Rev E **89**(5), 052801 (2014)
49. Forss, T., Sarlin, P.: News-Sentiment Networks as a Risk Indicator (2017). arXiv:1706.05812, 2017

50. Gao, Y.-C., Zeng, Y., Cai, S.-M.: Influence network in the Chinese stock market. J. Stat. Mech.: Theory Experiment **2015**(3), P03017 (2015)
51. Gava, J., Lefebvre, W., Turc, J.: Beyond carry and momentum in government bonds. Available at SSRN **3446653**, (2019)
52. Giada, L., Marsili, M.: Data clustering and noise undressing of correlation matrices. Phys. Rev. E **63**(6), 061101 (2001)
53. Giada, L., Marsili, M.: Algorithms of maximum likelihood data clustering with applications. Phys. A: Stat. Mech. Appl. **315**(3), 650–664 (2002)
54. Goh, Y.K., Hasim, H.M., Antonopoulos, C.G.: Inference of financial networks using the normalised mutual information rate. PloS One **13**(2), e0192160 (2018)
55. Guo, L., Peng, L., Tao, Y., Tu, J.: Media Network Based Investors' Attention: A Powerful Predictor of Market Premium (2018)
56. Guo, X., Hu, Z., Tianhai, T.: Development of stock correlation networks using mutual information and financial big data. PloS One **13**(4), e0195941 (2018)
57. Harmon, D., Stacey, B., Bar-Yam, Y., Bar-Yam, Y.: Networks of Economic Market Interdependence and Systemic Risk (2010). arXiv:1011.3707
58. Hartman, D., Hlinka, J.: Nonlinearity in Stock Networks (2018). arXiv:1804.10264
59. Heckens, A.J., Krause, S.M., Guhr, T.: Uncovering the Dynamics of Correlation Structures Relative to the Collective Market Motion (2020). arXiv:2004.12336
60. Hoberg, G., Phillips, G.: The stock market, product uniqueness, and comovement of peer firms. SSRN eLibrary (2012)
61. Hoberg, G., Phillips, G.: Text-Based Industry Momentum (2017)
62. Huang, F., Gao, P., Wang, Y.: Comparison of Prim and Kruskal on Shanghai and Shenzhen 300 Index hierarchical structure tree. In: International Conference on Web Information Systems and Mining. WISM 2009, pp. 237–241. IEEE (2009)
63. Huang, W.-Q., Zhuang, X.-T., Yao, S., Uryasev, S.: A financial network perspective of financial institutions' systemic risk contributions. Phys. A: Stat. Mech. Appl. **456**, 183–196 (2016)
64. Hüttner, A., Mai, J.-F. Mineo, S.: Portfolio selection based on graphs: Does it align with markowitz-optimal portfolios? Dependence Modeling (2018)
65. Jain, P., Jain, S.: Can machine learning-based portfolios outperform traditional risk-based portfolios? the need to account for covariance misspecification. Risks **7**(3), 74 (2019)
66. Johnson, N.F., McDonald, M., Suleman, O., Williams, S., Howison, S.: What shakes the FX tree? understanding currency dominance, dependence, and dynamics (keynote address). In: SPIE Third International Symposium on Fluctuations and Noise, pp. 86–99. International Society for Optics and Photonics (2005)
67. Jung, W.-S., Kwon, O., Wang, F., Kaizoji, T., Moon, H.-T., Stanley, H.E.: Group dynamics of the Japanese market. Phys. A: Stat. Mec. Appl. **387**(2), 537–542 (2008)
68. Kakushadze, Z., Yu, W.: Statistical Industry Classification (2016)
69. Kaya, H.: Eccentricity in asset management. J. Netw. Theory Finance **1**(1), 1–32 (2014)
70. Kenett, D.Y., Preis, T., Gur-Gershgoren, G., Ben-Jacob, E.: Dependency network and node influence: application to the study of financial markets. Int. J. Bifurcat. Chaos **22**(07), 1250181 (2012)
71. Kenett, D.Y., Shapira, Y., Madi, A., Bransburg-Zabary, S., Gur-Gershgoren, G., Ben-Jacob, E.: Dynamics of stock market correlations. AUCO Czech Econ. Rev. **4**(3), 330–341 (2010)
72. Kenett, D.Y., Tumminello, M., Madi, A., Gur-Gershgoren, G., Mantegna, R.N., Ben-Jacob, E.: Dominating clasp of the financial sector revealed by partial correlation analysis of the stock market. PloS One **5**(12), e15032 (2010)
73. Kim, H.J., Kim, I.M., Lee, Y., Kahng, B.: Scale-free network in stock markets. J.-Korean Phys. Soc. **40**, 1105–1108 (2002)
74. King, B.F.: Market and industry factors in stock price behavior. J. Bus. **39**(1), 139–190 (1966)
75. Kocheturov, A., Batsyn, M., Pardalos, P.M.: Dynamics of cluster structures in a financial market network. Phys. A: Stat. Mech. Appl. **413**, 523–533 (2014)
76. Krüger, P., Landier, A., Thesmar, D.: Categorization bias in the stock market. Available SSRN **2034204**, (2012)

77. Kullmann, L., Kertesz, J., Mantegna, R.N.: Identification of clusters of companies in stock indices via potts super-paramagnetic transitions. Phys. A: Stat. Mech. Appl. **287**(3), 412–419 (2000)
78. Lautier, D., Raynaud, F.: Systemic risk in energy derivative markets: a graph-theory analysis. Available at SSRN **1579629**, (2011)
79. Lee, G.S., Djauhari, M.A.: Multidimensional stock network analysis: An Escoufier's RV coefficient approach. AIP Conference Proceedings **1**, 550–555 (2013)
80. Lee, J., Youn, J., Chang, W.: Intraday volatility and network topological properties in the Korean stock market. Phys. A: Stat. Mech. Appl. **391**(4), 1354–1360 (2012)
81. Lemieux, V., Rahmdel, P.S., Walker, R., Wong, B.L., Flood, M.: Clustering techniques and their effect on portfolio formation and risk analysis. In: Proceedings of the International Workshop on Data Science for Macro-Modeling, pp. 1–6. ACM (2014)
82. León, D., Aragón, A., Sandoval, J., Hernández, G., Arévalo, A., Niño, J.: Clustering algorithms for risk-adjusted portfolio construction. Procedia Comput. Sci. **108**, 1334–1343 (2017)
83. Letizia, E., Lillo, F.: Corporate Payments Networks and Credit Risk Rating (2018)
84. Lohre, H., Rother, C., Schäfer, K.A.: Hierarchical risk parity: Accounting for tail dependencies in multi-asset multi-factor allocations. Mach. Learn, Asset Manag (2020)
85. de Prado, M.L.: Building Diversified Portfolios that Outperform Out-Of-Sample (2016)
86. de Prado, M.L.: Estimation of theory-implied correlation matrices. Available at SSRN (2019)
87. de Prado, M.L., Lewis, M.J.: Detection of False Investment Strategies Using Unsupervised Learning Methods (2018)
88. MacMahon, M., Garlaschelli, D.: Community detection for correlation matrices. Phys. Rev. X **5**, 021006 (2015)
89. Mantegna, R.N.: Hierarchical structure in financial markets. Eur. Phys. J. B-Condensed Matter Complex Syst. **11**(1), 193–197 (1999)
90. Mantegna, R.N., Stanley, H.E.: Introduction to Econophysics: Correlations and Complexity in Finance. Cambridge University Press, Cambridge (1999)
91. Marti, G.: Corrgan: Sampling Realistic Financial Correlation Matrices Using Generative Adversarial Networks (2019)
92. Marti, G., Andler, S., Nielsen, F., Donnat, P.: Clustering financial time series: How long is enough? In: Proceedings of the Twenty-Fifth International Joint Conference on Artificial Intelligence, IJCAI 2016, pp. 2583–2589. New York, NY, USA (2016)
93. Marti, G., Andler, S., Nielsen, F., Donnat, P.: Optimal transport vs. fisher-rao distance between copulas for clustering multivariate time series. IEEE Statistical Signal Processing Workshop. SSP 2016, pp. 1–5. Palma de Mallorca, Spain (2016)
94. Marti, G., Nielsen, F., Donnat, P.: Optimal copula transport for clustering multivariate time series. In: 2016 IEEE International Conference on Acoustics, Speech and Signal Processing (ICASSP), pp. 2379–2383. IEEE (2016)
95. Marti, G., Very, P., Donnat, P., Nielsen, F.: A proposal of a methodological framework with experimental guidelines to investigate clustering stability on financial time series. 14th IEEE International Conference on Machine Learning and Applications. ICMLA 2015, pp. 32–37. FL, USA, Miami (2015)
96. Massara, G.P., Di Matteo, T., Aste, T.: Network filtering for big data: triangulated maximally filtered graph. J. Complex Netw. **5**(2), 161–178 (2016)
97. Matesanz, D., Ortega, G.J.: Sovereign public debt crisis in europe. a network analysis. Phys. A: Stat. Mech. Appl. **436**, 756–766 (2015)
98. Meng, H., Xie, W.-J., Jiang, Z.-Q., Podobnik, B., Zhou, W.-X., Stanley, H.E.: Systemic risk and spatiotemporal dynamics of the US housing market. Sci. Rep. **4**, 3655 (2014)
99. Morales, R., Di Matteo, T., Aste, T.: Dependency structure and scaling properties of financial time series are related. Sci. Rep. **4**(4589) (2014)
100. Münnix, M.C., Shimada, T., Schäfer, R., Leyvraz, F., Seligman, T.H., Guhr, T., Stanley, H.E.: Identifying states of a financial market. Sci. Rep. **2**, 644 (2012)
101. Musciotto, F., Marotta, L., Miccichè, S., Mantegna, R.N.: Bootstrap validation of links of a minimum spanning tree (2018). arXiv:1802.03395

102. Musmeci, N., Aste, T., Di Matteo, T.: Risk diversification: a study of persistence with a filtered correlation-network approach. Netw. Theory Finance **1**(1), 77–98 (2015)
103. Musmeci, N., Aste, T., Di Matteo, T.: Interplay between past market correlation structure changes and future volatility outbursts. Sci. Rep. **6**, (2016)
104. Nguyen, H.T., Tran, P.N., Nguyen, Q.: An analysis of eigenvectors of a stock market cross-correlation matrix. In: International Econometric Conference of Vietnam, pp. 504–513. Springer, Berlin (2018)
105. Nguyen, Q.: One-factor model for the cross-correlation matrix in the vietnamese stock market. Phys. A: Stat. Mech. Appl. **392**(13), 2915–2923 (2013)
106. Onnela, J.-P., Chakraborti, A., Kaski, K., Kertiész, J.: Dynamic asset trees and portfolio analysis. Eur. Phys. J. B-Condensed Matter Complex Syst. **30**(3), 285–288 (2002)
107. Onnela, J.-P., Chakraborti, A., Kaski, K., Kertesz, J.: Dynamic asset trees and black monday. Phys. A: Stat. Mech. Appl. **324**(1), 247–252 (2003)
108. Onnela, J.-P., Chakraborti, A., Kaski, K., Kertesz, J., Kanto, A.: Dynamics of market correlations: Taxonomy and portfolio analysis. Phys. Rev. E **68**(5), 056110 (2003)
109. Onnela, J.-P., Kaski, K., Kertész, J.: Clustering and information in correlation based financial networks. Eur. Phys. J. B-Condensed Matter Complex Syst. **38**(2), 353–362 (2004)
110. Otranto, E.: Clustering heteroskedastic time series by model-based procedures. Comput. Stat. Data Anal. **52**(10), 4685–4698 (2008)
111. Özgür, A., Cetin, B., Bingol, H.: Co-occurrence network of reuters news. Int. J. Modern Phys. C **19**(05), 689–702 (2008)
112. Ozsoylev, H.N., Walden, J., Yavuz, M.D., Bildik, R.: Investor networks in the stock market. Rev. Financial Stud. **27**(5), 1323–1366 (2013)
113. Pan, R.K., Sinha, S.: Collective behavior of stock price movements in an emerging market. Phys. Rev. E **76**(4), 046116 (2007)
114. Panton, D.B., Lessig, V.P., Joy, O.M.: Comovement of international equity markets: a taxonomic approach. J. Financial Quantitative Anal. **11**(03), 415–432 (1976)
115. Papenbrock, J., Schwendner, P.: Handling risk-on/risk-off dynamics with correlation regimes and correlation networks. Financial Markets Portfolio Manag. **29**(2), 125–147 (2015)
116. Peralta, G., Zareei, A.: A network approach to portfolio selection. J. Emp. Finance (2016)
117. Pfitzinger, J., Katzke, N., et al.: A constrained hierarchical risk parity algorithm with cluster-based capital allocation. Techn, Rep (2019)
118. Plerou, V., Gopikrishnan, P., Rosenow, B., Amaral, L.N., Stanley, H.E.: A random matrix theory approach to financial cross-correlations. Phys. A: Stat. Mech. Appl. **287**(3), 374–382 (2000)
119. Pozzi, F., Di Matteo, T., Aste, T.: Spread of risk across financial markets: better to invest in the peripheries. Sci. Rep. **3**, (2013)
120. Raffinot, T.: Hierarchical clustering based asset allocation. Available at SSRN **2840729**, (2016)
121. Raffinot, T.: The hierarchical equal risk contribution portfolio. Available at SSRN **3237540**, (2018)
122. Ren, F., Lu, Y.-N., Li, S.-P. Jiang, X.-F. Zhong, L.-X., Qiu, T.: Dynamic portfolio strategy using clustering approach (2016). arXiv:1608.03058
123. Rinn, P., Stepanov, Y., Peinke, J., Guhr, T., Schäfer, R.: Dynamics of quasi-stationary systems: Finance as an example. EPL (Europhysics Letters) **110**(6), 68003 (2015)
124. Rocchi, J., Tsui, E.Y.L., Saad, D.: Emerging interdependence between stock values during financial crashes (2016). arXiv:1611.02549
125. Rönnqvist, S., Sarlin, P.: From text to bank interrelation maps. In: 2104 IEEE Conference onComputational Intelligence for Financial Engineering and Economics (CIFEr), pp. 48–54. IEEE (2014)
126. Rönnqvist, S., Sarlin, P.: Bank networks from text: interrelations, centrality and determinants. Quantitative Finance **15**(10), 1619–1635 (2015)
127. Sandhu, R., Georgiou, T., Tannenbaum, A.: Market fragility, systemic risk, and Ricci curvature (2015). arXiv:1505.05182

128. Sensoy, A., Tabak, B.M.: Dynamic spanning trees in stock market networks: The case of Asia-Pacific. Phys. A: Stat. Mech. Appl. **414**, 387–402 (2014)
129. Smith, R.: The spread of the credit crisis: view from a stock correlation network. J. Korean Phys. Soc. **54**(6), 2460–2463 (2009)
130. Song, D.-M., Tumminello, M., Zhou, W.-X., Mantegna, R.N.: Evolution of worldwide stock markets, correlation structure, and correlation-based graphs. Phys. Rev. E **84**(2), 026108 (2011)
131. Song, W.-M., Di Matteo, T., Aste, T.: Hierarchical information clustering by means of topologically embedded graphs. PLoS One **7**(3), e31929 (2012)
132. Song, W.-M., Di Matteo, T., Aste, T.: Nested hierarchies in planar graphs. Discret. Appl. Math. **159**(17), 2135–2146 (2011)
133. Souza, T.T.P., Aste, T.: Predicting future stock market structure by combining social and financial network information (2018). arXiv:1812.01103
134. Spelta, A.: Financial market predictability with tensor decomposition and links forecast. Appl. Netw. Sci. **2**(1), 7 (2017)
135. Squartini, T., Van Lelyveld, I., Garlaschelli, D.: Early-warning signals of topological collapse in interbank networks. Sci. Rep. **3**, (2013)
136. Stepanov, Y., Rinn, P., Guhr, T., Peinke, J., Schäfer, R.: Stability and hierarchy of quasi-stationary states: financial markets as an example. J. Stat. Mech.: Theory Exp. **2015**(8), P08011 (2015)
137. Takayasu, H.: Practical Fruits of Econophysics. Springer, Berlin (2006)
138. Tang, Y., Xiong, J.J., Jia, Z.-Y., Zhang, Y.-C.: Complexities in financial network topological dynamics: Modeling of emerging and developed stock markets. Complexity (2018)
139. Tola, V., Lillo, F., Gallegati, M., Mantegna, R.N.: Cluster analysis for portfolio optimization. J. Econ. Dyn. Control **32**(1), 235–258 (2008)
140. Tóth, B., Kertész, J.: Accurate estimator of correlations between asynchronous signals. Phys. A Stat. Mech. Appl. **388**(8), 1696–1705 (2009)
141. Chengyi, T.: Cointegration-based financial networks study in chinese stock market. Phys. A: Stat. Mech. Appl. **402**, 245–254 (2014)
142. Tumminello, M., Mantegna, R.N., Lillo, F.: Shrinkage and spectral filtering of correlation matrices: A comparison via the Kullback-Leibler distance. Acta Phys. Polonica. Ser. B **39**(1), 4079–4088 (2008)
143. Tumminello, M., Aste, T., Di Matteo, T., Mantegna, R.N.: A tool for filtering information in complex systems. Proc. Natl. Acad. Sci. U. S. A. **102**(30), 10421–10426 (2005)
144. Tumminello, M., Coronnello, C., Lillo, F., Micciche, S., Mantegna, R.N.: Spanning trees and bootstrap reliability estimation in correlation-based networks. Int. J. Bifurcat. Chaos **17**(07), 2319–2329 (2007)
145. Tumminello, M., Di Matteo, T., Aste, T., Mantegna, R.N.: Correlation based networks of equity returns sampled at different time horizons. Eur. Phys. J. B **55**(2), 209–217 (2007)
146. Tumminello, M., Lillo, F., Mantegna, R.N.: Hierarchically nested factor model from multivariate data. EPL (Europhysics Letters) **78**(3), 30006 (2007)
147. Tumminello, M., Lillo, F., Mantegna, R.N.: Kullback-leibler distance as a measure of the information filtered from multivariate data. Phys. Rev. E **76**(3), 031123 (2007)
148. Tumminello, M., Lillo, F., Mantegna, R.N.: Correlation, hierarchies, and networks in financial markets. J. Econ. Beh. Organ. **75**(1), 40–58 (2010)
149. Tumminello, M., Miccichè, S., Lillo, F., Piilo, J., Mantegna, R.N.: Statistically validated networks in bipartite complex systems. PloS One **6**(3), e17994 (2011)
150. Vandewalle, N., Brisbois, F., Tordoir, X., et al.: Non-random topology of stock markets. Quantitative Finance **1**(3), 372–374 (2001)
151. Výrost, T., Lyócsa, V., Baumöhl, E.: Granger causality stock market networks: Temporal proximity and preferential attachment. Phys. A: Stat. Mech. Appl. **427**, 262–276 (2015)
152. Whiteley, N.: Dynamic time series clustering via volatility change-points (2019). arXiv:1906.10372
153. Wu, L.: Centrality of the Supply Chain Network (2015)

154. Zhang, X., Podobnik, B., Kenett, D.Y., Stanley, H.E.: Systemic risk and causality dynamics of the world international shipping market. Phys. A: Stat. Mech. Appl. **415**, 43–53 (2014)
155. Zhang, Y., Lee, G.H.T., Wong, J.C., Kok, J.L., Prusty, M., Cheong, S.A.: Will the us economy recover in 2010? a minimal spanning tree study. Phys. A: Stat. Mech. Appl. **390**(11), 2020–2050 (2011)
156. Zhao, L., Wang, G.-J., Wang, M., Bao, W., Li, W., Stanley, H.E.: Stock market as temporal network (2017). arXiv:1712.04863